Cancer Prevention with Molecular Target Therapies

Cancer Prevention with Molecular Target Therapies

Editor

Laura Paleari

MDPI • Basel • Beijing • Wuhan • Barcelona • Belgrade • Manchester • Tokyo • Cluj • Tianjin

Editor
Laura Paleari
Research, Innovation and HTA unit
Liguria Health Authority A.Li.Sa.
Genova
Italy

Editorial Office
MDPI
St. Alban-Anlage 66
4052 Basel, Switzerland

This is a reprint of articles from the Special Issue published online in the open access journal *International Journal of Molecular Sciences* (ISSN 1422-0067) (available at: www.mdpi.com/journal/ijms/special_issues/Cancer_Prevention).

For citation purposes, cite each article independently as indicated on the article page online and as indicated below:

LastName, A.A.; LastName, B.B.; LastName, C.C. Article Title. *Journal Name* **Year**, *Volume Number*, Page Range.

ISBN 978-3-0365-7927-6 (Hbk)
ISBN 978-3-0365-7926-9 (PDF)

Contents

About the Editor vii

Preface to "Cancer Prevention with Molecular Target Therapies" ix

Laura Paleari, Mariangela Rutigliani, Giacomo Siri, Nicoletta Provinciali, Nicoletta Colombo and Andrea Decensi
Aromatase Inhibitors as Adjuvant Treatment for ER/PgR Positive Stage I Endometrial Carcinoma: A Retrospective Cohort Study
Reprinted from: *Int. J. Mol. Sci.* **2020**, *21*, 2227, doi:10.3390/ijms21062227 1

Elizabeth A. Fry, Gloria E. Niehans, Robert A. Kratzke, Fumitake Kai and Kazushi Inoue
Survival of Lung Cancer Patients Dependent on the LOH Status for *DMP1*, *ARF*, and *p53*
Reprinted from: *Int. J. Mol. Sci.* **2020**, *21*, 7971, doi:10.3390/ijms21217971 9

Manlio Tolomeo and Antonio Cascio
The Multifaced Role of STAT3 in Cancer and Its Implication for Anticancer Therapy
Reprinted from: *Int. J. Mol. Sci.* **2021**, *22*, 603, doi:10.3390/ijms22020603 25

Yi-Hsun Chen, Wei-Kai Wu and Ming-Shiang Wu
Microbiota-Associated Therapy for Non-Alcoholic Steatohepatitis-Induced Liver Cancer: A Review
Reprinted from: *Int. J. Mol. Sci.* **2020**, *21*, 5999, doi:10.3390/ijms21175999 51

Alex Friedlaender, Alfredo Addeo, Alessandro Russo, Vanesa Gregorc, Diego Cortinovis and Christian D. Rolfo
Targeted Therapies in Early Stage NSCLC: Hype or Hope?
Reprinted from: *Int. J. Mol. Sci.* **2020**, *21*, 6329, doi:10.3390/ijms21176329 71

Maria Krchniakova, Jan Skoda, Jakub Neradil, Petr Chlapek and Renata Veselska
Repurposing Tyrosine Kinase Inhibitors to Overcome Multidrug Resistance in Cancer: A Focus on Transporters and Lysosomal Sequestration
Reprinted from: *Int. J. Mol. Sci.* **2020**, *21*, 3157, doi:10.3390/ijms21093157 83

Martina Minoli, Mirjam Kiener, George N. Thalmann, Marianna Kruithof-de Julio and Roland Seiler
Evolution of Urothelial Bladder Cancer in the Context of Molecular Classifications
Reprinted from: *Int. J. Mol. Sci.* **2020**, *21*, 5670, doi:10.3390/ijms21165670 103

Gaetano Marverti, Gaia Gozzi, Eleonora Maretti, Angela Lauriola, Leda Severi and Francesca Sacchetti et al.
A Peptidic Thymidylate-Synthase Inhibitor Loaded on Pegylated Liposomes Enhances the Antitumour Effect of Chemotherapy Drugs in Human Ovarian Cancer Cells
Reprinted from: *Int. J. Mol. Sci.* **2020**, *21*, 4452, doi:10.3390/ijms21124452 127

Frank Waldbillig, Katja Nitschke, Abdallah Abdelhadi, Jost von Hardenberg, Philipp Nuhn and Malin Nientiedt et al.
Phosphodiesterase *SMPDL3B* Gene Expression as Independent Outcome Prediction Marker in Localized Prostate Cancer
Reprinted from: *Int. J. Mol. Sci.* **2020**, *21*, 4373, doi:10.3390/ijms21124373 149

Kensuke Harada, Ryuya Ohashi, Kyoko Naito and Keita Kanki
Hedgehog Signal Inhibitor GANT61 Inhibits the Malignant Behavior of Undifferentiated Hepatocellular Carcinoma Cells by Targeting Non-Canonical GLI Signaling
Reprinted from: *Int. J. Mol. Sci.* **2020**, *21*, 3126, doi:10.3390/ijms21093126 159

Yuko Oya, Hiroaki Kuroda, Takeo Nakada, Yusuke Takahashi, Noriaki Sakakura and Toyoaki Hida
Efficacy of Immune Checkpoint Inhibitor Monotherapy for Advanced Non-Small-Cell Lung Cancer with *ALK* Rearrangement
Reprinted from: *Int. J. Mol. Sci.* **2020**, *21*, 2623, doi:10.3390/ijms21072623 . . . . **173**

Mohammed O. Altonsy, Anutosh Ganguly, Matthias Amrein, Philip Surmanowicz, Shu Shun Li and Gilles J. Lauzon et al.
Beta3-Tubulin Is Critical for Microtubule Dynamics, Cell Cycle Regulation, and Spontaneous Release of Microvesicles in Human Malignant Melanoma Cells (A375)
Reprinted from: *Int. J. Mol. Sci.* **2020**, *21*, 1656, doi:10.3390/ijms21051656 . . . . **187**

About the Editor

Laura Paleari

I have more than twenty years of experience in translational, pre-clinical, and clinical cancer research and scientific/medical writing, and have robust skills in molecular/cellular biology and in vivo testing (including the generation ok knockout, conditional knockout, and pdx murine models). I am a member of the regional HTA Network and a consultant for the National Health Agency (Agenas). I have gained skills in pharmaco-economics and in the management of health systems.

I have served as a member of peer review panels for the following major international and national research funding calls: the CCM (National Center for Prevention and Disease Control); the translational co-fund call of ERACoSysMed, ERANET on System Medicine under Horizon2020; and the MINDED program (7th Framework program, Marie Curie Actions). I am the author of 70+ scientific papers, being the first/last/corresponding author of the majority.

Preface to "Cancer Prevention with Molecular Target Therapies"

Personalized medicine plays an important role in cancer prevention. To date, it is clear that many cancers are molecularly distinct subtypes, and different therapeutic approaches would be required for each. Indeed, the identification of cancer susceptibility genes permits identifying patients "at risk" of developing neoplasia, and supports modifying individual risk behaviors or the choice of preventive therapy. Additionally, the efficacy of various targeted therapies in different cancer subtypes suggests that treatment choices in the near future will be more and more centered on molecular signatures. Data from preclinical, clinical, and observational studies have revealed the ability to prevent cancer development for compounds with different indications than cancer. The concept of drug repurposing permits combinations that can target several critical pathways of a specific disease, decreasing the risk of resistance observed when using single-agent targeted therapy.

This open-access Special Issue brings together original research and review articles on molecular oncology with attention to the early detection and prevention of cancer. It highlights new findings, methods, and technical advances in molecular cancer research. The main feature of this Special Issue is to provide an open-source sharing of significant works in the field of molecular oncology that can increase our understanding of cancer development, which may lead to the discovery of new molecular diagnostic technologies and targeted therapeutics.

Laura Paleari
Editor

International Journal of
Molecular Sciences

MDPI

Article

Aromatase Inhibitors as Adjuvant Treatment for ER/PgR Positive Stage I Endometrial Carcinoma: A Retrospective Cohort Study

Laura Paleari [1,*], Mariangela Rutigliani [2], Giacomo Siri [3], Nicoletta Provinciali [4], Nicoletta Colombo [5,6] and Andrea Decensi [4,7]

1 A.Li.Sa., Liguria Region Health Authority, 16121 Genoa, Italy
2 Pathology Unit, Galliera Hospital, 16128 Genoa, Italy; mariangela.rutigliani@galliera.it
3 Office of the Scientific Director, Galliera Hospital, 16128 Genoa, Italy; giacomo.siri@galliera.it
4 Medical Oncology Unit, Galliera Hospital, 16128 Genoa, Italy; nicoletta.provinciali@galliera.it (N.P.); andrea.decensi@galliera.it (A.D.)
5 Gynecology Program, European Institute of Oncology, 20141 Milan, Italy; nicoletta.colombo@ieo.it
6 School of Medicine and Surgery, University Milan Bicocca, 20126 Milan, Italy
7 Barts School of Medicine, Queen Mary University of London, London E1 4NS, UK
* Correspondence: laura.paleari@regione.liguria.it; Tel.: +39-010-548-4243

Received: 14 February 2020; Accepted: 18 March 2020; Published: 23 March 2020

Abstract: Objective: Although endometrial cancer (EC) is a hormone dependent neoplasm, there are no recommendations for the determination of steroid hormone receptors in the tumor tissue and no hormone therapy has ever been assessed in the adjuvant setting. The purpose of this study was to explore the effect of adjuvant aromatase inhibitors (AIs) on progression-free survival (PFS) and overall survival (OS) in patients with early stage and steroid receptors-positive EC. Methods: We retrospectively analyzed clinical and pathological factors in 73 patients with high-risk (49.3%) or low-risk (50.7%) stage I (n = 71) or II (n = 2) endometrial cancer who received by their preference after counseling either no treatment (reference group) or AI. Prognostic factors were well balanced between groups. Expression of estrogen receptor (ER), progesterone receptor (PgR), and Ki-67 index was correlated with clinical outcomes. Results: Univariate and multivariate Cox proportional regression analyses, adjusted for age, grade, stage, depth of myometrial invasion, lymphovascular space invasion, BMI, ER, PgR and Ki-67 labeling index levels, showed that PFS and OS had a trend to be longer in patients receiving AI than in the reference group HR= 0.23 (95% CI; 0.04–1.27) for PFS and HR= 0.11 (95% CI; 0.01–1.36) for OS. Conclusion: Compared with no treatment, AI exhibited a trend toward a benefit on PFS and OS in patients with early stage hormone receptor-positive EC. Given the exploratory nature of our study, randomized clinical trials for ER/PgR positive EC patients are warranted to assess the clinical benefit of AI and the potential predictive role of steroid receptors and Ki-67.

Keywords: endometrial cancer; steroid receptors; aromatase inhibitors; progression free survival; overall survival

1. Introduction

Endometrial cancer (EC) is the most common cancer of the female reproductive organs affecting mainly postmenopausal women with an average age at diagnosis of 60 years. The American Cancer Society estimates about 58,000 new cases and more than 10,000 deaths in 2018 in the US [1]. EC represents over 90% of uterine cancer and develops along two different pathways with distinct molecular alterations, histologic and clinical types [2]. The majority (~80%) of ECs have endometrioid differentiation, are associated with an excess of estrogens related to obesity and insulin resistance (type I) and are usually detected at an early stage [3]. Conversely, the remaining ~20% ECs of unknown etiology (type II) are diagnosed at a more advanced stage, tend to be more aggressive and their risk factors are less well identified but may include women with inherited microsatellite instability [2]. Estrogen (ER) and progesterone (PgR) receptor expression is linked to the grade of histology differentiation in EC with 70%, 55% and 41% of grade 1, 2 and 3, respectively [4]. The recent analysis of the Cancer Genome Atlas Project (TCGA) confirmed an increased expression of ER and PgR in association with the grade of histological differentiation in EC [4]. The incidence of EC has been found to directly correlate with the increase in body mass index (BMI) which is an independent risk factor for this disease [2]. The relative risk of cancer-specific mortality for obese women (BMI = 30–34.9 kg/m^2) is more than doubled compared with women with a normal range BMI (RR = 2.53 (95% CI; 2.02–3.18)) and is much greater for women with BMI > 40.0 kg/m^2 (RR = 6.25 (95% CI; 3.75–10.42)) [2]. In addition, uterine endometrial cell proliferation is under the control of both estrogen and progesterone [5] and previous clinical, biological and epidemiological studies have demonstrated that the excess of exogenous/endogenous estrogens represents one of the main risk factors for EC [6]. Moreover, prospective cohort studies have shown that increased serum estrogen levels double the risk of EC incidence particularly for type I [7,8]. Aromatase, the enzyme responsible for a key step in the biosynthesis of estrogens, produces the majority of circulating estrogen in postmenopausal women [9]. Studies have demonstrated how high levels of aromatase are expressed in EC with respect to normal endometrium leading to the hypothesis that the enzyme functions through a paracrine mechanism [10,11]. Selective estrogen receptor modulators (SERMs) or aromatase inhibitors (AIs) are being used in the palliative treatment of advanced EC, but their effect as adjuvant treatment is unknown [12]. Recently, the National Comprehensive Cancer Network (NCCN) has updated the guidelines for EC management including the use of hormone therapy for advanced low-grade endometrioid histology, preferably in patients with small tumor volume or an indolent growth pace, although recommendations are category 2A because of the lack of definitive trials [13]. This retrospective cohort study aimed to probe the effect of AI in the adjuvant setting of EC and to explore the prognostic/predictive significance of ER/PgR expression and Ki-67 levels.

2. Results

2.1. Study Population

In total, 73 patients met our inclusion criteria. The overall median age was 74.7 years (SD + 10.4). The median expression levels of steroid receptors were 80% (range, 20–40) for ER and 70% (range, 30–90) for PgR and 40% (range, 30–65) for Ki-67. A depth ≥ 50% of myometrial invasion was present in 45.2% of patients while lymphovascular space invasion was present in 5.5%, both prognostic factors for EC were well-balanced in the AIs and no-treatment group. Overall, 50.7% were low risk and 49.3% were at high risk according the ESMO guidelines [14,15]. The 38.3% of the cohort population had healthy weight, 28.3% overweight and 33.3% were obese. Adjuvant treatment included AIs (exemestane or letrozole, 71% and 29%, respectively). The choice of the AIs was based on patient preference after medical counseling on the toxicity profile of each compound, and was offered to all the patients [16,17]. The main patient characteristics and prognostic factors were evenly distributed between groups and are summarized in Table 1.

Table 1. Patients' characteristics.

Variable		Overall N = 73 (100%)	Hormone Therapy N = 31 (42.5%)	No Treatment N = 42 (57.5%)	*p*-Value
Age, mean (SD)		74.6 (10.4)	73.4 (9.7)	75.5 (10.9)	0.396 [2]
Age, N (%)	<70 yrs	25 (34.3)	11 (35.5)	14 (33.3)	1.000 [1]
	>70 yrs	48 (65.8)	20 (64.5)	28 (66.7)	
BMI, N (%)	healthy weight	22 (30.1%)	9 (29.0%)	13 (30.9%)	0.912 [1]
	over weight	31 (42.5%)	14 (45.2%)	17 (40.5%)	
	obesity	20 (27.4%)	8 (25.8%)	12 (28.6%)	
Adjuvant treatment:					
Exemestane		22 (71%)	22 (71%)	0 (0%)	-
Letrozole		9 (29%)	9 (29%)	0 (0%)	
Depth of myometrial invasion, N (%)	<50%	40 (54.8)	18 (58.1)	22 (52.4)	0.644 [1]
	>50%	33 (45.2)	13 (41.9)	20 (47.6)	
Stage, N (%)	I	56 (82.4)	26 (86.7)	30 (79.0)	0.720 [1]
	II	7 (10.3)	2 (6.7)	5 (13.2)	
	III	5 (7.4)	2 (6.7)	3 (7.9)	
Grade, N (%)	<G2	36 (50.7)	14 (46.7)	22 (53.7)	0.617 [1]
	>G3	35 (49.3)	16 (53.3)	19 (46.3)	
Lymphovascular space invasion, N (%)	No	69 (94.5)	29 (93.5)	40 (95.2)	1.000 [1]
	Yes	4 (5.5)	2 (6.5)	2 (4.8)	
ER, median (IQR)		80 (20–90)	80 (40–90)	80 (20–90)	0.499 [3]
PgR, median (IQR)		70 (30–90)	80 (50–90)	70 (10–90)	0.358 [3]
Ki-67, median (IQR)		40 (30–65)	40 (20–60)	50 (30–70)	0.308 [3]

(1) Fisher's Exact for frequencies; (2) *t*-test for means; (3) Mann-Whitney test for equality of medians.

2.2. *Efficacy*

At the time of the present analysis, 62 out of 73 (86.3%) patients were alive. Kaplan–Meier survival curves for median progression-free survival (PFS) and overall survival (OS) are shown in Figure 1. The choice for hormone therapy was offered to all the patients and the type of treatment was decided by patient preference but it is interesting to note that the treatment arms with AIs and no-therapy were well balanced in terms of prognostic factors. Multivariate analysis, adjusted for age, BMI, grade, depth of myometrial invasion, lymphovascular space invasion, ER and PgR expression levels and Ki-67 labeling index, showed a ~80% relative reduction in the risk of progression in patients treated with AIs over the untreated group in terms of PFS (HR = 0.23 (95% CI; 0.04–1.27); $p = 0.089$) (Figure 1a). Likewise, women in the AIs group exhibited an 89% relative reduction in the risk of death compared with the untreated group (HR = 0.11 (95% CI; 0.01–1.36); $p = 0.047$) (Figure 1b).

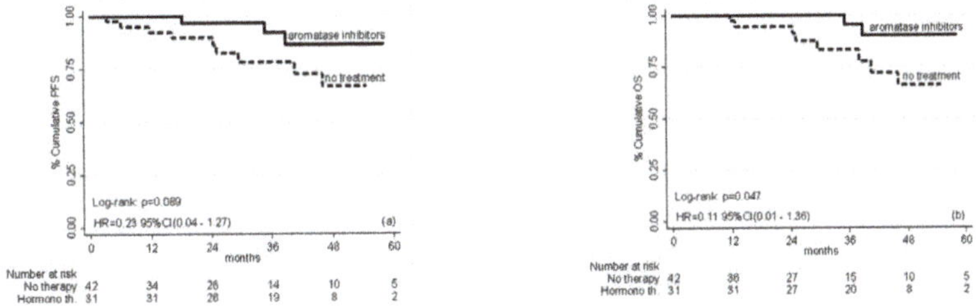

(a) (b)

Figure 1. (**a**) Progression-free survival (left panel) and (**b**) overall survival (right panel) in patients treated with adjuvant hormone therapy or no treatment; HR adjusted for age, BMI, grade, depth of myometrial invasion, grading, lymphovascular space invasion, ER and PgR expression levels and Ki-67 labeling index.

In addition, patients with age > 70 years and high Ki-67 levels were associated with a higher risk of shorter PFS and OS (Tables 2 and 3). Intriguingly, while ER expression was associated with longer PFS and OS, PgR expression had a direct association with shorter PFS and OS (Tables 2 and 3). No severe adverse events have been reported and the most common toxicity was a mild arthralgia which has never led to treatment disruption.

Table 2. Association between variables and progression-free survival (PFS) in multivariate Cox model.

PFS	HR	95% CI	*p*-Value
AI vs. no therapy	0.23	0.04–1.27	0.092
Age > 70	9.89	0.93–105.53	0.058
BMI 25–29 vs. BMI < 25	0.28	0.05–1.55	0.144
BMI > 30 vs. BMI < 25	0.44	0.06–3.12	0.409
Myometrium invasion (yes vs. no)	0.87	0.19–3.94	0.854
Grade > G2	1.02	0.23–4.53	0.974
Lymphovascular space invasion (yes vs. no)	3.33	0.32–34.49	0.314
ER	0.97	0.93–1.01	0.178
PgR	1.03	0.98–1.09	0.209
Ki67	1.04	1.01–1.09	0.026

Table 3. Association between variables and overall survival (OS) in multivariate Cox model.

OS	HR	95% CI	*p*-Value
AI vs. no therapy	0.11	0.01–1.36	0.085
Age > 70	Na *	Na *	Na *
BMI 25–29 vs. BMI < 25	0.18	0.03–0.97	0.046
BMI > 30 vs. BMI < 25	0.21	0.01–3.03	0.253
Myometrium invasion (yes vs. no)	1.04	0.22–4.92	0.962
Grade > G2	0.80	0.14–4.71	0.803
Lymphovascular space invasion (yes vs. no)	2.66	0.21–33.14	0.446
ER	0.96	0.90–1.02	0.199
PgR	1.05	0.97–1.13	0.204
Ki67	1.06	1.01–1.11	0.027

(*) not estimable due to collinearity with the outcome.

3. Discussion

Although EC is a hormone-dependent neoplasm, to date the standard treatment of early stage EC includes mainly surgery and/or radiotherapy and/or chemotherapy. Hormone therapy is recommended by the NCCN only in advanced EC patients, with low grade, low volume and slow growth rate tumors. It is also considered for early stage patients who are not suitable for primary surgery, in selected cases. [13]. Hormone therapy could represent a tolerable and effective therapeutic option in the adjuvant setting but the lack of RCTs for evaluating its efficacy does not allow clarifying the clinical benefit in terms of PFS and OS. It is well known that estrogen is the most significant risk factor for type I EC and the results of genome analysis confirmed the association of high ER/PgR expression levels with endometrioid histology suggesting a prognostic role of steroid receptors. Thus, we performed a retrospective study to explore the effect of AIs in terms of PFS and OS versus no therapy in ER/PgR positive early stage EC patients and to determine the prognostic and predictive role of ER/PgR to adjuvant endocrine therapy. Treatment with letrozole or exemestane was offered to all the patients based on clinical characteristics and patients opted for either AIs or no treatment based on their preference after accurate counseling. The two groups were well balanced for prognostic factors and our prior experience with a similar patient-decision approach in women with breast ductal carcinoma in situ [18] and in breast intraepithelial neoplasia using low dose tamoxifen indicates that the results are in line with those obtained in a subsequent RCT [19]. In this context, our results suggest a potential clinical benefit in AIs arm compared to untreated patients with regard to PFS (HR = 0.23 (95% CI; 0.04–1.27)) and OS (HR = 0.11 (95% CI; 0.01–1.36)) and are worth exploring in a randomized trial.

Interestingly, no clinical trial of AIs in an adjuvant setting of early stage EC has ever been completed or is underway [20]. The reason for this gap in this unmet medical area is unclear, but may in part be due to the lack of commercial interest of AIs, which are all out of patent, thus making a non-profit clinical trial difficult to conduct for financial constraints. Tamoxifen could also be an alternative to AIs given its favorable safety profile, especially at a lower dose where efficacy is comparable to the standard dose in early breast cancer [19]. Importantly, a progestin such as medroxyprogesterone acetate was assessed in a randomized trial in the early 1990s [21]. The results were negative in terms of PFS. Our findings seem to suggest that PgR expression is associated with worse prognosis and this might explain why giving a progestin may not be effective. Our findings also advise that Ki-67 labeling index may be a prognostic factor in EC, in line with previous data [14,22–24].

Study limitations are the non-randomized nature of the study and the small sample size, which only provides exploratory evidence. However, despite the retrospective nature of the study, there are no significant differences in terms of prognostic factors and/or possible confounders, i.e., the two treatment arms are comparable for each other factor different from the treatment ($p > 0.3$). The Multiple Cox model, adjusted for possible confounders, confirms the findings of the univariate analysis presented in Table 1 and in the survival curves (Figure 1). In conclusion, our results suggest that AI is a useful treatment modality to prolong PFS and possibly OS in the adjuvant treatment of patients with steroid receptors-positive EC. Randomized trials assessing the efficacy of AIs are warranted to confirm our findings.

4. Materials and Methods

4.1. Patients

This is a retrospective survival study conducted between January 2011 and December 2017 on 73 women operated for EC at Galliera Hospital. The study was approved by the Regional Ethical Committee (code 214-2018, 25 March 2019). Demographic, clinical, pathologic, and follow-up data were obtained from patients' medical records. The study only included patients with ER/PgR positive tumors, criterion for prescription of hormone therapy and the median follow-up was 39 months. The choice for hormone therapy was offered to all the patients and the decision to undergo an aromatase inhibitor or no treatment in women with stage I EC was based on patient preference after careful medical counseling by a single medical oncologist (ADC). Patients were treated with Exemestane 25 mg once daily or Letrozole 2.5 mg once daily for 2 years. The inclusion criteria were: (1) histologically diagnosed stage I or II EC and positive hormone receptor expression; (2) patients operated for EC in the period 2011–2017 at the Galliera Hospital; (3) age > 18 years; (4) no contraindication to AIs use, including any prior cancer, prior cardiovascular disease, osteoporosis, grade 2 or higher biochemical alterations, prior use of selective estrogen receptor modulators or aromatase inhibitors, or mental disorders. Patient response to therapy and survival rates were correlated with demographic and clinic-pathologic variables to evaluate factors associated with the response to AIs, including a high risk of relapse according to the ESMO guideline which comprises FIGO stage and lymphovascular space invasion while the kind of surgery, the nodal status and the number of positive nodes are not included for the determination of the risk of recurrence [15,25].

4.2. *Tissue Sampling, Histopathological Analysis and Immunohistochemistry (IHC)*

Tumor sampling was performed immediately after hysterectomy. For histopathologic examination, 2 µm-thick FFPE sections were stained with the conventional hematoxylin and eosin stain. Two pathologists confirmed the diagnosis of endometrial carcinoma. IHC was performed on 2 µm sections with an automated IHC staining system (Ventana BenchMark ULTRA, Ventana Medical Systems, Italy).

All cases were stained IHC with Estrogen (clone SP1, 1/100, Ventana), Progesterone (clone 1E2, 1/100, Ventana) and Ki67/MIB1 (clone 30-9, 1/100, Vantana). Glands and endometrial stroma constituted the internal positive control for IHC procedure. Two independent observers scored the slides.

4.3. *Statistical Analysis*

PFS refers to the period from the start of treatment administration to disease progression or death by any cause; OS is defined as the period from the start of treatment administration to death by any cause, or the last follow up. The final data cutoff date was 31 December 2018. The sample was explored with descriptive statistics in order to highlight the difference between arms of treatment. Common indices like mean (SD) and median (IQR) were adopted to analyze continuous variables, *t*-Test or Kruskal–Wallis ANOVA were used to test the equality of mean or median between the two groups of treatment. Categorical variables were described in terms of frequencies and the comparison between the two groups was done with the Fisher's exact test. PFS and OS between the two comparisons of interest, i.e., AIs versus no treatment, were assessed with Kaplan–Meier estimator and at an explorative level the difference between two survival curves was tested by using the log-rank test. An adjusted Cox PH model was adopted to quantify and test the difference of hazard between groups for both PFS and OS. Threshold of significance was set alpha = 5%, two-tailed. All the analyses were done with software STATA (Stata Corp. 2015. Release 14.2. College Station, TX: Stata Corp LP).

4.4. *Assessment of Therapeutic Option and Patients Counselling*

Patients were stratified into risk classes in accordance with the recent ESMO guidelines [15,25]; on the basis of biological characteristics of tumors, patients have been therefore assigned to three classes of risk (low, intermediate and high) of disease recurrence. Therapies were generally decided by clinicians according to practical clinical guidelines, which just take into account the risk of recurrence.

However, it has not always been possible to administer the therapies indicated according to guidelines because of the general health status, age and preferences of patients. So, in many cases, hormonal therapy was proposed as an alternative opportunity for those patients who should have undergone chemotherapy or radiotherapy but who lacked these treatments for the reasons explained above.

Author Contributions: L.P.: Conceptualization, Methodology, Validation, Investigation, Writing—Original Draft, Writing—Review & Editing, Data Curation, Project administration, Supervision, Funding acquisition. M.R.: IHC analysis, Writing—Review & Editing. G.S.: Statistical analysis, Writing—Review & Editing; N.P.: Writing—Review & Editing; N.C.: Writing—Review & Editing; A.D.: Conceptualization, Writing—Review & Editing, Supervision, Funding acquisition. All authors have read and agreed to the published version of the manuscript.

Funding: This work was supported by the Italian Association for Cancer Research (AIRC), grant IG 2018, ID 21534.

Conflicts of Interest: The authors report no conflicts of interest.

References

1. American Cancer Society, Cancer Statistics for USA. 2019. Available online: https://www.cancer.org/cancer/endometrial-cancer/about/key-statistics.html (accessed on 18 March 2020).
2. Lortet-Tiulent, J.; Ferlay, J.; Bray, F.; Jemal, A. International pattern and trends in endometrial cancer incidence, 1978–2013. *J. Natl. Cancer Inst.* **2018**, *110*, 354–361. [CrossRef] [PubMed]

3. Onstand, M.A.; Schmandt, R.E.; Lu, K.H. Addressing the role of obesity in endometrial cancer risk, prevention, and treatment. *J. Clin. Oncol.* **2016**, *34*, 4225–4230. [CrossRef] [PubMed]
4. Kandoth, N.C.; Schultz, N.; Cherniack, A.D.; Akbani, R.; Liu, Y.; Shen, H.; Robertson, A.G.; Pashtan, I.; Shen, R.; Cancer Genome Atlas Research; et al. Integrated genomic characterization of endometrial carcinoma. *Nature* **2013**, *497*, 67–73.
5. Ito, K.; Utsunomiya, H.; Yaegashi, N.; Sasano, H. Biological roles of estrogen and progesterone in human endometrial carcinoma—New development in potential endocrine therapy for endometrial cancer. *Endocr. J.* **2007**, *54*, 667–679. [CrossRef] [PubMed]
6. Grady, D.; Gebretsadik, T.; Kerlikowske, K.; Ernster, V.; Petitti, D. Hormone replacement therapy and endometrial cancer risk: A meta-analysis. *Obstet. Gynecol.* **1995**, *85*, 304–313. [CrossRef]
7. Brinton, L.A.; Trabert, B.; Anderson, G.L.; Falk, R.T.; Felix, A.S.; Fuhrman, B.J.; Gass, M.L.; Kuller, L.H.; Pfeiffer, R.M.; Rohan, T.E.; et al. Serum estrogens and estrogen metabolites and endometrial cancer risk among postmenopausal women. *Cancer Epidemiol. Biomark. Prev.* **2016**, *25*, 1081–1099. [CrossRef]
8. Allen, N.E.; Key, T.J.; Dossus, L.; Rinaldi, S.; Cust, A.; Lukanova, A.; Peeters, P.H.; Onland-Moret, N.C.; Lahmann, P.H.; Berrino, F.; et al. Endogenous sex hormones and endometrial cancer risk in women in the European Prospective Investigation into Cancer and Nutrition (EPIC). *Endocr. Relat. Cancer* **2008**, *15*, 485–497. [CrossRef]
9. Gao, C.; Wang, Y.; Tian, W.; Zhu, Y.; Xue, F. The therapeutic significance of aromatase inhibitors in endometrial carcinoma. *Gynecol. Oncol.* **2014**, *134*, 190–195. [CrossRef]
10. Watanabe, K.; Sasano, H.; Harada, N.; Ozaki, M.; Niikura, H.; Sato, S.; Yajima, A. Aromatase in human endometrial carcinoma and hyperplasia: Immunohisto chemical, in situ hybridization, and biochemical studies. *Am. J. Pathol.* **1995**, *146*, 491–500.
11. Zhao, H.; Zhou, L.; Shangguan, A.J.; Bulun, S.E. Most of the estrogen in premenopausal women is synthesized by the ovaries, while extraovarian subcutaneous adipose tissue is the predominant tissue source of estrogen after menopause. *J. Mol. Endocrinol.* **2016**, *57*, 19–33. [CrossRef]
12. Nieves-Neira, W.; Kim, J.J.; Matei, D. Hormonal strategies in gynecologic cancer: Bridging biology and therapy. *Gynecol. Oncol.* **2018**, *150*, 207–210. [CrossRef] [PubMed]
13. Available online: https://www.nccn.org/professionals/physician_gls/pdf/uterine.pdf (accessed on 19 March 2020).
14. Yerushalmi, R.; Woods, R.; Ravdin, P.M.; Hayes, M.M.; Gelmon, K.A. Ki67 in breast cancer: Prognostic and predictive potential. *Lancet Oncol.* **2010**, *11*, 174–183. [CrossRef]
15. Colombo, N.; Creutzberg, C.; Amant, F.; Bosse, T.; González-Martín, A.; Ledermann, J.; Marth, C.; Nout, R.; Querleu, D.; Mirza, M.R.; et al. ESMO-ESGO-ESTRO Consensus conference on endometrial cancer: Diagnosis, treatment and follow-up. *Ann. Oncol.* **2016**, *27*, 16–41. [CrossRef] [PubMed]
16. Smith, I.; Yardley, D.; Burris, H.; De Boer, R.; Amadori, D.; McIntyre, K.; Ejlertsen, B.; Gnant, M.; Jonat, W.; Pritchard, K.I.; et al. Comparative efficacy and safety of adjuvant letrozole versus anastrozole in postmenopausal patients with hormone receptor-positive, node-positive early breast cancer: Final results of the randomized phase III Femara Versus Anastrozole Clinical Evaluation (FACE) trial. *J. Clin. Oncol.* **2017**, *35*, 1041–1048. [PubMed]
17. Goss, P.E.; Ingle, J.N.; Pritchard, K.I.; Ellis, M.J.; Sledge, G.W.; Budd, G.T.; Rabaglio, M.; Ansari, R.H.; Johnson, D.B.; Tozer, R.; et al. Exemestane versus anastrozole in postmenopausal women with early breast cancer: NCIC CTG MA.27—A randomized controlled phase III trial. *J. Clin. Oncol.* **2013**, *31*, 1398–1404. [CrossRef]
18. Guerrieri-Gonzaga, A.; Sestak, I.; Lazzeroni, M.; Serrano, D.; Rotmensz, N.; Cazzaniga, M.; Varricchio, C.; Pruneri, G.; Leonardi, M.C.; Orecchia, R.; et al. Benefit of low-dose tamoxifen in a large observational cohort of high risk ER positive breast DCIS. *Int. J. Cancer* **2016**, *139*, 2127–2134. [CrossRef]
19. DeCensi, A.; Puntoni, M.; Guerrieri-Gonzaga, A.; Caviglia, S.; Avino, F.; Cortesi, L.; Taverniti, C.; Pacquola, M.G.; Falcini, F.; Gulisano, M.; et al. Randomized Placebo Controlled Trial of Low-Dose Tamoxifen to Prevent Local and Contralateral Recurrence in Breast Intraepithelial Neoplasia. *J. Clin. Oncol.* **2019**, *37*, 1629–1637. [CrossRef]
20. Available online: https://clinicaltrials.gov/ct2/results?cond=Endometrial+Cancer+Stage+I&term=hormone+therapy&cntry=&state=&city=&dist= (accessed on 21 December 2018).

21. De Palo, G.; Mangioni, C.; Periti, P.; del Vecchio, M.; Marubini, E. Treatment of FIGO (1971) stage I endometrial carcinoma with intensive surgery, radiotherapy and hormonotherapy according to pathological prognostic groups. Long-term results of a randomized multicentre study. *Eur. J. Cancer* **1993**, *29*, 1133–1140. [CrossRef]
22. Dowsett, M.; Nielsen, T.O.; A'Hern, R.; Bartlett, J.; Coombes, R.C.; Cuzick, J.; Ellis, M.; Henry, N.L.; Hugh, J.C.; Lively, T.; et al. Assessment of Ki67 in breast cancer: Recommendations from the international Ki67 in Breast Cancer working group. *J. Natl. Cancer Inst.* **2011**, *103*, 1656–1664. [CrossRef]
23. de Azambuja, E.; Cardoso, F.; de Castro, G., Jr.; Colozza, M.; Mano, M.S.; Durbecq, V.; Sotiriou, C.; Larsimont, D.; Piccart-Gebhart, M.J.; Paesmans, M. Ki-67 as prognostic marker in early breast cancer: A meta-analysis of published studies involving 12,155 patients. *Br. J. Cancer* **2007**, *96*, 1504–1513. [CrossRef]
24. Kitson, S.; Sivalingam, V.N.; JBolton, J.; McVey, R.; Nickkho-Amiry, M.; Powell, M.E.; Leary, A.; Nijman, H.W.; Nout, R.A.; Bosse, T. Ki-67 in endometrial cancer: Scoring optimization and prognostic relevance for window studies. *Mod. Pathol.* **2017**, *30*, 459–468. [CrossRef] [PubMed]
25. Available online: https://www.esmo.org/Guidelines/Gynaecological-Cancers/ESMO-ESGO-ESTRO-Consensus-Conference-on-Endometrial-Cancer/eUpdate-Algorithms (accessed on 30 September 2019).

International Journal of
Molecular Sciences

MDPI

Article

Survival of Lung Cancer Patients Dependent on the LOH Status for *DMP1*, *ARF*, and *p53*

Elizabeth A. Fry [1], Gloria E. Niehans [2], Robert A. Kratzke [3], Fumitake Kai [1] and Kazushi Inoue [1,*]

1 Dept. of Pathology, Wake Forest University School of Medicine, Medical Center Blvd., Winston-Salem, NC 27157, USA; eafry10@hotmail.com (E.A.F.); fkai@wakehealth.edu (F.K.)
2 Minneapolis VA Medical Center, Minneapolis, MN 55417, USA; nieha002@umn.edu
3 Dept. of Medicine, University of Minnesota Medical Center, Masonic Cancer Institute, Minneapolis, MN 55455, USA; kratz003@umn.edu
* Correspondence: kinoue2@triad.rr.com; Tel.: +1-336-407-1642; Fax: +1-336-765-2486

Received: 7 September 2020; Accepted: 19 October 2020; Published: 27 October 2020

Abstract: Lung cancer is the leading cause of cancer deaths in the world, and accounts for more solid tumor deaths than any other carcinomas. The prognostic values of *DMP1*, *ARF*, and p53-loss are unknown in lung cancer. We have conducted survival analyses of non-small cell lung cancer (NSCLC) patients from the University of Minnesota VA hospital and those from the Wake Forest University Hospital. Loss of Heterozygosity (LOH) for h*DMP1* was found in 26 of 70 cases (37.1%), that of the *ARF/INK4a* locus was found in 33 of 70 (47.1%), and that of the *p53* locus in 43 cases (61.4%) in the University of Minnesota samples. LOH for h*DMP1* was associated with favorable prognosis while that of *p53* predicted worse prognosis. The survival was much shorter for *ARF*-loss than *INK4a*-loss, emphasizing the importance of *ARF* in human NSCLC. The adverse effect of *p53* LOH on NSCLC patients' survival was neutralized by simultaneous loss of the h*DMP1* locus in NSCLC and breast cancer, suggesting the possible therapy of epithelial cancers with metastatic ability.

Keywords: DMTF1; *ARF*; INK4a; p53; Cyclin D1; YY1

1. Introduction

Lung cancer is the leading cause of cancer deaths in the world, and accounts for more solid tumor deaths than any other carcinomas. More than 22,800 new cases are diagnosed each year in the United States alone, of which 135,700 will die 2020, representing 25% of all cancer deaths [1]. Lung cancer can be categorized into two major histopathological groups: Non small-cell lung cancer (NSCLC) and small-cell lung cancer (SCLC) [2], the latter of which shows neuroendocrine features. A total of 80%–85% of lung cancers are NSCLC [2], and they are subcategorized into adenocarcinomas (AC), squamous cell carcinomas (SCC), adenosquamous carcinomas, and large-cell carcinomas [3]. SCLC and NSCLC show major differences in histopathologic characteristics that can be explained by the distinct patterns of genetic alterations found in both tumor classes [4].

The *ARF* gene was cloned as the alternate reading frame gene for $p16^{Ink4a}$ from the mouse *ARF/Ink4a* locus [5–10]. Accumulating studies showed that the *ARF* tumor suppressor is a sensor for hyperproliferative oncogenic stimuli stemming from mutant Ras, c-Myc, E2F-1, and HER2 proteins [11–14], virtually induced by all the oncogenes stimulating p53 to induce cell cycle arrest, apoptosis, and autophagy [15]. $p19^{ARF}$ ($p14^{ARF}$ in humans) and $p16^{Ink4a}$ mRNAs are generated from separate and first exons 1β and 1α (19.4 kilo base pairs (kbp) apart in humans; 12.4 kbp apart in mice) which splice into two common exons 2 and 3 [5]. These two genes are different since $p19^{ARF}$ uses only exons 1 and 2 (also $p14^{ARF}$) while $p16^{INK4a}$ uses all of the exons 1–3 for production of the protein [5,16]. This *ARF-INK4a* (*CDKN2a*) locus is located 11.5 kbp downstream of the genomic locus for *CDKN2b* that encodes for $p15^{INK4}$ [9]. All of $p15^{Ink4b}$, $p19^{ARF}$, and $p16^{Ink4a}$ are tumor suppressor genes as proved by analyses of gene knockout mice [17–20]. Since RB is regulated by $p16^{INK4a}$ and p53 is regulated by $p14^{ARF}$, the *ARF/INK4a* locus is very frequently inactivated in human cancers, second only to p53 [21,22]. *ARF* is a highly basic, insoluble protein (pI 11; [23]). Although human and mouse *ARF* differ in size (mouse 19 kDa, human 14 kDa) and show only 49% identity in amino acid sequences, the functions of the *ARF* proteins are well-conserved between the two species [6]. *ARF* is regulated at both transcriptional and protein levels by ULF, MKRN1, and SIVA1 [10].

Ectopic *ARF* arrests immortal rodent cell lines, such as NIH 3T3, as well as human cells from cancer [5,16,24]. The ability of *ARF* to inhibit cell cycle progression in a number of cell types suggested that *ARF* has powerful growth-inhibitory functions in cells, which stimulated researchers to study the in vivo activity of *ARF* to prevent tumors. *ARF* sequesters MDM2 in the nucleolus, thus preventing p53 degradation [25]. In addition, it inhibits the transcription factor E2F activity [26]. These activities lead to cell cycle arrest at G1 and G2 [16]. Itahana et al. studied the role of *ARF* in apoptosis and found that the mitochondrial protein p32/C1QBP bound to the *ARF* C-terminus, where p32 is required for *ARF* to localize to mitochondria to induce apoptosis, demonstrating the essential role of *ARF* in tumor suppression and programmed cell death [27]. Recent studies indicate that nuclear factor E2-related factor 2 is a major target of *ARF* in p53-independent tumor suppression [28].

The gene for the Dmp1 (cyclin D binding myb-like protein 1; Dmtf1) transcription factor was isolated in yeast two-hybrid screen of CTLL2 cell library with cyclin D2 bait [29]. Inoue and Sherr reported that gene expression and cell cycle arrest mediated by Dmp1 (Dmp1α) was antagonized by D-type cyclins through a Cdk-independent mechanism [30]. Importantly, Dmp1 directly binds to the Ets consensus sequence 5′ CCCGGATGC-3′ of the *ARF* promoter to activate its gene expression, thereby inducing p53-dependent cell cycle arrest ([29–47] for primary articles; [48–51] for reviews). The human *DMP1* locus encodes *DMP1α*, and its splice variants *DMP1β* and *γ* [46]. DMP1α regulates the human *ARF* promoter, the activity of which is antagonized its splice variant DMP1β [46]. Similar mechanisms must be present in human *ARF* promoter since both E2F and DMP1-consensus sequences are found in the human version as well. *Dmp1*-deficient mice were prone to spontaneous tumor development, which was accelerated when the animals were neonatally treated with ionizing radiation or dimethylbenzanthracene [34,35]. Although *Dmp1*-deficient mice develop a broad spectrum of epithelial and non-epithelial tumors, lung tumors were the most frequently encountered neoplasms in both *Dmp1*-null and *Dmp1*-hetrozygous mice [34,35]. The wild type *Dmp1* allele was retained and expressed in tumors arising from $Dmp1^{+/-}$ mice, demonstrating a haplo-insufficiency of Dmp1 in tumor suppression [35,51]. Tumors from $Dmp1^{-/-}$ or $Dmp1^{+/-}$ (*Eμ-Myc*, $K\text{-}Ras^{LA}$, *HER2* mutant) mice rarely showed mutations, deletions, or silencing of $p19^{ARF}$ or *p53*, suggesting that Dmp1 is a critical regulator of the *ARF*-p53 tumor suppressor pathway in vivo [14,34,35,39,48–51].

Activation of the *Dmp1* promoter by oncogenic Ras or mutant HER2 have been reported [11,14,36], indicating that it is a critical mediator in RAS or HER2 induced *ARF*, p53 cell cycle arrest to prevent incipient cancer cells. The *Dmp1* promoter was also activated by an inflammatory cytokine TNFα mediated by NF-κB [38] as well as dsDNA breaks [41], indicating that Dmp1 is a mediator of a variety of stress signaling. We conducted GeneChip microarray using $Dmp1^{+/+}$ and $Dmp1^{-/-}$ lungs and found that other transcriptional targets for Dmp1α include *Areg*, *Thbs1*, *JunB*, and *Egr1* [40], suggesting that it is involved in signal transduction pathways involving cell proliferation, angiogenesis, and invasion/metastasis.

Dmp1 shows it tumor suppressive activity not only transactivating the *ARF* promoter in response to oncoprotein overexpression, but also through physical interaction with p53 in response to DNA damage response [41]. Our data indicate that acceleration of DNA-binding of p53 by Dmp1 is a critical process for Dmp1 to increase the p53 function in *ARF*-deficient cells [47].

Whether h*DMP1* is involved in the pathogenesis in human cancer is a critical issue for research. We found that loss of heterozygosity (LOH) of h*DMP1* was present in ~40% of non-small cell lung carcinomas (NSCLC), especially those that retain wild type *INK4a/ARF* and/or *p53* [39]. In this study, we received specimen from the University Minnesota VA Hospital (UM) to study the survival of LOH for h*DMP1*, *ARF/INK4a*, *p53*; the impacts for overexpression of cyclin D1 and YY1 have also been studied on NSCLC survival, both progression-free survival (PFS) and total survival (TS).

The goals of this study are to clarify prognostic impact form the h*DMP1* LOH, *ARF/INK4a* LOH, p53 LOH and immunohistochemistry (IHC), Cyclin D1 (IHC), and YY1 (genomic DNA amplification).

2. Results

2.1. Impacts of LOH for hDMP1, ARF/INK4a, and p53 on NSCLC Survival

Previous publication from WFU with samples of 51 NSCLC [39] showed that LOH of the h*DMP1* gene was found in 33.3% of samples with 5′ primer, 36.1% with 3′ primer (average 35%), but signs of biallelic involvement (promoter hypermethylation or complete loss) was extremely rare (the former 2.2% and 0% for the latter) in human lung cancer indicating that h*DMP1* was haploinsufficient for tumor suppression [35,51]. Now, we have studied different set of samples ($n = 70$) with survival data (PFS and TS) from the University of Minnesota VA hospital (UM) [48,49].

With the 5′ set of h*DMP1* primers (#92465), 10 of 70 cases (14.3%) were positive for LOH; with the 3′ set of h*DMP1* primers (#198004, #176671), 24 of 70 cases were positive (34.3%) (average 24.3%). 26 of 70 cases (37.1%) were positive for either of these (Table 1). The number was close to those published in WFU samples (35%; 39). Eight of 70 cases (11.4%) were positive for LOH with both of these h*DMP1* primer sets (Table 1), suggesting that gene deletion extends the entire *DMP1* locus in these samples. LOH for h*DMP1* was associated with longer survival in both PFS ($p = 0.0029$, $\chi^2 = 8.8993$) and total survival ($p = 0.0040$, $\chi^2 = 8.3026$) in the UM samples (Figure 1A,B). Consistently, LOH for h*DMP1* was associated with favorable survival in TS in our previous WFU samples analyzed ($p = 0.0324$, $\chi^2 = 4.5230$, $n = 42$, Supplementary Figure S1A) [39]. The tendency was more prominent is squamous cell carcinoma than adenocarcinoma (Supplementary Figure S2).

Then the same NSCLC samples pairs were studied for LOH of *ARF/INK4a* and *p53*. With *ARF/INK4a* primers, 23 of 70 cases (32.9%) were positive for LOH 5′ primers close to the *ARF* locus, 19 of 70 cases (27.1%) were positive for LOH 3′ primers close to the *INK4a* locus (average 30%), and 33 of 70 showed LOH for either one of these (47.1%). Nine of 70 cases (12.9%) showed LOH with both sets of the *ARF/INK4a* primers (Table 1). Importantly, LOH for the *ARF/INK4a* locus was not associated with survival PFS ($p = 0.1718$, $\chi^2 = 1.8670$) or TFS ($p = 0.1721$, $\chi^2 = 1.8649$) although there was a trend that it was associated with worse prognosis (Figure 1C,D). The trend was the same when AC and SCC were separately analyzed (Supplementary Figure S3). The LOH for the *ARF/INK4a* locus was not associated with TS in previous WFU specimens analyzed ($p = 0.7707$, $\chi^2 = 0.08495$, $n = 43$; Supplementary Figure S1B). Forty-seven of 53 cases (88.7%) showed mutually exclusive loss of the h*DMP1* and the *ARF-INK4a* loci ($p = 0.0023$; $\chi^2 = 9.330$; 95% confidence interval, 80.1%–97.2%; Table 1). We found that LOH for *ARF* was much worse prognostic factor than that of *INK4a* (Figure 1E,F) in both PFS and TS; as a matter of fact none of the NSCLC relapsed and died within 1800 (PFS) and 3000 days (TS) of observation for *INK4a* LOH, whereas NSCLC with *ARF* deletion relapsed and died in our survival analyses (Figure 1E,F; $p = 0.0430$ for PFS and $p = 0.0448$ for TS), suggesting that *ARF* LOH has much stronger impact on NSCLC survival than that of *INK4a*. When the survival for LOH for both of *ARF* and *INK4a* was studied, it was shorter in patients with the locus involvement than those without although data were not statistically significant (Supplementary Figure S4).

Table 1. NSCLC samples of UM VA hospital. The LOH values of each genomic locus is shown. When both loci shows a single peak, real-time PCR was conducted to determine gene deletion. NME: Non-mutually exclusive, ME: Mutually exclusive. IHC: Immunohistochemistry. Not only chi square tests, but also 95% confidence interval assays were conducted to evaluate mutual exclusiveness of LOH for genomic loci. Mutually exclusive cases for LOH and immunohistochemistry (IHC), the results of confidence interval assays are shaded.

Patient ID	*hDMP1*		*INK4a/ARF*	*INK4a/ARF*	*I/A* LOH ME	*p53*		*p53* LOH ME	p53 IHC	p53 IHC ME
	#92465	#198004	#33647	#27251		#182 SE	#377 SE			
Group 1	5′	3′	5′	3′		5′	3′			
307	0.68	0.26	0.15	0.98	NME	2.01	2.04	NME	ND	
308	0.70	1.04	3.38	0.41	ME	Deletion	27.20	ME	ND	
309	1.21	2.51	1.42	1.11	ME	1.06	0.54	ME	ND	
310	0.06	4.31	0.68	0.93	ME	1.06	0.92	ME	ND	
311	1.05	1.08	1.31	1.26	None	1.50	0.84	None	ND	
313	0.94	0.47	1.46	1.32	ME	1.23	0.96	ME	ND	
314	0.46	6.09	2.41	0.63	NME	0.43	0.30	NME	ND	
315	1.09	1.87	1.12	0.98	None	9.97	0.28	ME	ND	
317	0.61	2.26	1.56	0.91	ME	Deletion	1.40	NME	ND	
318	1.23	0.90	5.48	2.21	ME	0.42	0.38	ME	ND	
319	1.86	2.29	1.30	0.82	ME	2.16	3.57	NME	ND	
321	0.77	0.29	0.69	1.03	ME	1.26	1.74	None	ND	
322	1.12	2.01	0.97	0.93	ME	1.16	0.07	NME	ND	
323	0.86	0.52	0.75	1.06	None	1.08	1.56	None	ND	
324	0.64	2.96	0.77	0.16	NME	2.31	0.31	NME	ND	
325	1.10	1.51	2.13	0.45	ME	0.34	0.29	ME	ND	
326	1.00	0.59	0.28	0.60	ME	2.37	0.62	ME	ND	
327	1.32	>10.0	1.04	1.03	ME	1.80	No deletion	ME	ND	
328	2.33	4.97	1.53	1.34	ME	No deletion	No deletion	ME	2	NME
329	0.74	0.80	1.02	2.69	ME	0.25	0.9	ME	2	ME
330	0.87	0.72	0.96	0.64	None	0.88	13.97	ME	1	None
331	1.11	1.08	0.10	4.66	ME	1.34	0.48	ME	0	None
332	0.82	0.52	0.68	0.74	None	0.97	0.40	ME	2	ME
333	1.15	0.48	0.89	1.06	ME	1.89	1.09	ME	1	ME
277	0.41	0.41	1.19	1.02	ME	0.61	No deletion	ME	1	ME

Table 1. *Cont.*

Patient ID	*hDMP1*		*INK4a/ARF*	*INK4a/ARF*	*I/A* LOH ME	*p53*		*p53* LOH ME	p53 IHC	p53 IHC ME
	#92465	#198004	#33647	#27251		#182 SE	#377 SE			
295	0.74	0.13	1.24	1.01	ME	0.99	1.02	ME	1	ME
299	0.32	2.11	1.18	0.79	ME	0.87	0.51	ME	1	ME
338	1.73	1.27	1.54	0.04	ME	13.1	4.06	ME	0	None
349	2.95	20.00	0.93	1.21	ME	1.73	1.62	ME	1	ME
351	1.69	0.41	No deletion	No deletion	ME	0.83	1.08	ME	1	ME
379	0.98	0.18	0.76	0.64	ME	0.92	0.94	ME	2	NME
380	0.33	0.33	0.67	0.80	ME	0.55	1.82	ME	ND	
271	0.66	0.59	0.49	0.91	ME	0.65	3.39	ME	2	ME
272	0.87	0.92	0.78	1.93	None	0.92	0.82	None	0	None
273	0.57	0.41	0.32	0.83	NME	0.23	Deletion	NME	2	NME
274	1.02	0.87	0.70	0.95	None	0.69	0.17	ME	1	None
275	0.55	0.90	0.18	1.04	ME	0.35	Deletion	ME	2	ME
278	0.95	No deletion	0.76	3.11	ME	0.56	14.00	ME	2	ME
282	1.12	1.02	1.03	0.95	None	0.93	4.01	ME	2	ME
284	1.02	1.02	0.35	0.92	ME	2.27	10.40	ME	0	None
290	1.03	0.92	2.48	1.03	ME	0.96	0.92	None	2	ME
293	0.94	0.99	0.60	0.73	None	1.86	0.47	ME	0	None
Group 2										
269	0.74	0.74	0.76	0.15	ME	0.71	0.33	ME	2	ME
270	1.08	1.16	1.10	0.99	None	0.73	1.06	None	2	ME
276	0.85	1.07	4.00	1.41	ME	0.15	3.08	ME	2	ME
300	1.49	0.96	1.30	0.66	None	2.23	0.57	ME	2	ME
334	0.80	0.61	0.35	0.94	ME	1.05	4.73	ME	2	ME
337	1.06	0.81	1.09	0.88	None	1.66	2.07	ME	2	ME
339	0.66	1.75	1.27	0.07	ME	0.93	0.95	None	2	ME
340	3.90	0.91	1.30	1.25	ME	1.71	No deletion	ME	1	ME
353	4.28	0.37	1.20	0.59	ME	3.62	No deletion	NME	1	ME
356	1.68	1.03	1.65	0.59	None	No deletion	No deletion	None	0	None
357	1.17	1.26	1.88	0.18	ME	2.59	2.55	ME	1	None
358	1.07	1.23	2.91	1.45	ME	2.01	0.65	ME	1	None
360	0.95	1.02	0.25	0.41	ME	0.45	3.11	ME	2	ME

Table 1. *Cont.*

Patient ID	h*DMP1*		*INK4a/ARF*	*INK4a/ARF*	*I/A* LOH ME	*p53*		*p53* LOH ME	p53 IHC	p53 IHC ME
	#92465	#198004	#33647	#27251		#182 SE	#377 SE			
362	1.64	No deletion	1.33	1.03	None	0.88	0.22	ME	2	ME
363	0.84	0.88	0.27	1.22	ME	0.35	0.18	ME	1	None
364	1.04	1.07	0.11	0.06	ME	0.75	0.16	ME	0	None
365	No deletion	0.93	1.29	0.95	None	0.75	0.93	None	1	None
366	0.87	1.26	0.44	0.99	ME	1.04	0.98	None	2	ME
367	0.92	1.04	0.78	2.15	ME	0.70	1.20	None	0	None
368	0.97	0.48	0.35	0.27	NME	1.52	0.91	ME	1	ME
369	1.49	1.06	1.40	1.27	None	3.42	No deletion	ME	0	None
370	0.55	1.11	1.76	0.26	ME	0.47	No deletion	ME	2	ME
371	1.49	0.56	8.11	2.46	ME	0.71	0.25	ME	0	None
372	2.50	1.72	0.51	0.57	ME	0.27	No deletion	NME	2	NME
381	1.30	0.99	1.02	1.19	None	0.55	0.49	ME	0	None
384	1.63	0.80	2.53	0.48	ME	1.54	1.38	None	2	ME
387	1.23	0.26	0.39	1.99	NME	2.04	1.80	NME	0	ME
389	1.14	1.23	1.47	0.17	ME	1.16	0.42	ME	1	None
Percentage		26/70 = 37.1%		33/70 = 47.1%	47/53 = 88.7%		43/70 = 61.4%	48/58 = 82.8%		29/33 = 87.9%
					p = 0.0023 X2 = 9.330			*p* = 0.0128 X2 = 6.195		
				95% CI	80.1–97.2%		95% CI	75.6–93.9%		76.7–99.0%

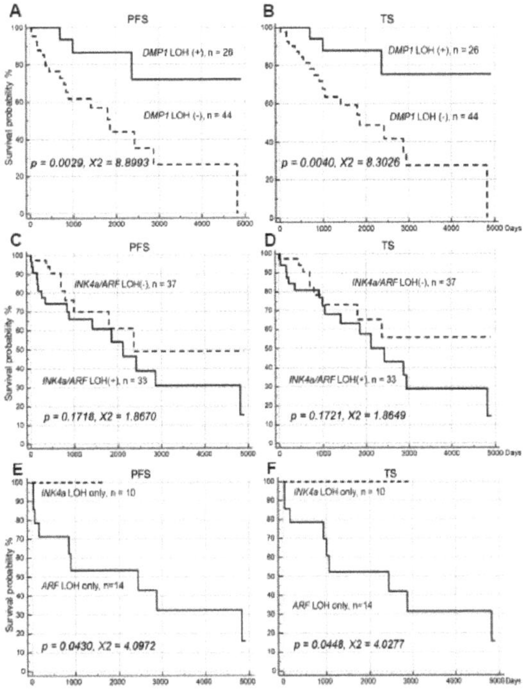

Figure 1. Progression-free Survival (PFS) and total survival (TS) of 70 cases of human non-small cell lung cancer obtained from the University of Minnesota VA hospital (UM) dependent on LOH for h*DMP1* (**A**,**B**) and *ARF/INK4a loci* (**C–F**). Kaplan-Meier analyses have been conducted to study the impact for the impact of loss of each locus on non-small cell lung cancer (NSCLC) patients' disease-free survival up to 5000 days. The Med Calc software (Mariakerke, Belgium) was used to analyze the specimens. LOH for h*DMP1* (**A**,**B**) has significantly positive impact on patient's relapse-free survival. On the other hand, LOH for *ARF/INK4a* did not have significant impact on NSCLC survival although there was a trend for association of worse prognosis (the survival indicate either of the two primers; **C**,**D**). When the *ARF/INK4a* locus was studied separately, the prognosis of patients with *ARF* LOH was much worse than that of *INK4a* only, indicating much stronger prognostic power for the former than the latter (**E**,**F**).

LOH of *p53* (incl. biallelic gene deletion) was found in 37.1% (26/70) with 5′ primers, 47.1% (33/70) with 3′ primers (average 42.1%), and 61.4% (43/70) with either primers in UM NSCLC samples (Table 1). LOH of *p53* was associated with worse prognosis in both PFS ($p = 0.0472$, $\chi^2 = 3.9372$) and TS ($p = 0.0302$, $\chi^2 = 4.6963$) in the UM samples (Supplementary Figure S5A,B). Consistently, LOH for *p53* had the trend for worse prognosis in WFU NSCLC samples ($p = 0.1079$, $\chi^2 = 2.5849$, $n = 42$; Supplementary Figure S1C). Forty-eight of 58 cases (82.8%) showed mutually exclusive loss of the h*DMP1* and the *p53* loci ($p = 0.0128$; $\chi^2 = 6.195$; 95% confidence interval, 75.6–93.9%; Table 1). The *p* values did not become smaller when AC and SCC were analyzed separately since N became smaller (Supplementary Figure S6).

We also conducted IHC analysis for p53 to study the impact of p53 expression on patients' survival (Supplementary Figure S5C,D). High expression of p53 protein (level 2, intense staining) was associated with shorter survival in PFS ($p = 0.0340$, $\chi^2 = 4.4944$, $n = 62$) and TS ($p = 0.0537$, $\chi^2 = 3.7270$) suggesting that it is an indicator worse prognosis (Supplementary Figure S5C,D). The *p* vales were relative large because there were significant number of samples where p53 IHC was not performed, and because there were samples that showed weak p53 IHC (grade 1; these cases were excluded from the study). Like the relation of LOH for h*DMP1* and *p53* loci, LOH of h*DMP1* and overexpression of the p53 protein did not to overlap each other (29/33 = 87.9% mutually exclusive; 95% confidence interval: 76.7%–99.0%; Table 1).

2.2. *The Impact of hDMP1 LOH on NSCLC Survival with p53 LOH*

The fact that LOH for h*DMP1* was mutually exclusive for that of p53 in 82.8% in NSCLC samples (48/58, Table 1) means that LOH for h*DMP1* was overlapping with that of *p53* in 17.2% of cases (10/58, 12 cases showed no LOH for *DMP1* and *p53*). We, therefore, studied the impact of h*DMP1* LOH in *p53* LOH(+) UM samples (Figure 2). We found that h*DMP1* LOH dramatically improved the survival (both PFS and TS) of *p53* LOH(+) in NSCLC patients' survival in NSCLC (Figure 2A,B). As a matter of fact, the survival of *p53* LOH(+); h*DMP1* LOH(+) was comparable to or even better than that of *p53* LOH(-) in UM samples ($p = 0.0013$, $\chi^2 = 13.2841$ in PFS; $p = 0.0012$, $\chi^2 = 13.5258$ in TS, triple survival assay), indicating that one locus h*DMP1* deletion neutralized the negative effect of *p53* LOH in NSCLC (3000 day survival of 12.0% became 87.5% in PFS (7.3 fold improvement), Figure 2A; 3000 day survival of 19.8% became 87.5% in TS (4.2 fold), Figure 2B). The same phenomenon was also observed in previous WFU NSCLC specimens as analyzed for the total survival (Supplementary Figure S7A, $p = 0.0714$, $\chi^2 = 5.2800$, $n = 42$). Indeed, no SCLC patient with double LOH for *p53* and h*DMP1* relapsed in the observation period of 1900 days. The *p* value was relatively large because two Wake Forest University Health Sciences (WFUHS) NSCLC samples with double h*DMP1*; *p53* LOH (1990-10, 2005-308 in ref. [39]) did not have any survival data. The same trend was also found in breast cancer samples in WFUHS [42] since none of the 10 patients that showed dual LOH for *p53* and h*DMP1* relapsed during the 2100 days of observation period ($p = 0.0189$, $\chi^2 = 7.9382$, $n = 105$) (Figure 2C). Conversely, the LOH for the *ARF/INK4a* locus did not improve the negative effect of *p53* LOH in NSCLC survival (Supplementary Figure S7B,C). These results show that the adverse prognostic impact of *p53* LOH in epithelial tumors (i.e., NSCLC and breast carcinoma) is greatly improved with simultaneous loss of the h*DMP1*, the result of which are consistent with favorable impact of h*DMP1* LOH in NSCLC (this study) and breast cancer [42].

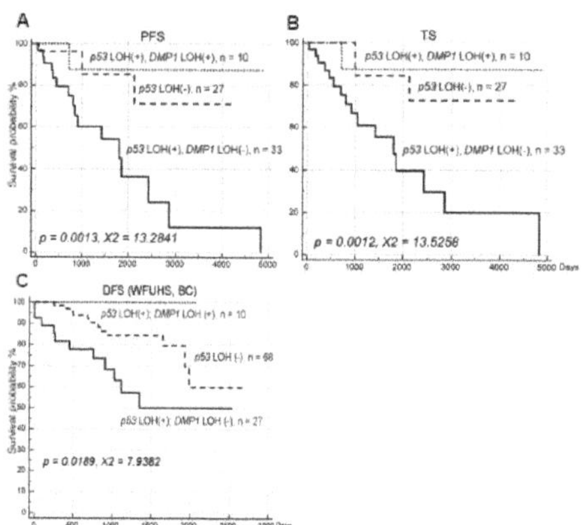

Figure 2. The triple analysis of cancer survival for p53 in University of Minnesota (UM) (NSCLC), and Wake Forest University (WFU) (breast cancer) samples. The impact of *DMP1* LOH on *p53* LOH was analyzed by Medcalc software for UM (NSCLC) and WFU (breast cancer) samples. Loss of *DMP1* neutralized the negative effect of *p53* LOH in NSCLC by moving the survival curves from *p53* LOH (+) to *p53* LOH(-) levels in both PFS and TS in UM samples. The same trend was observed in statistically significant fashion in WFU NSCLC samples ($n = 44$) or WFU breast cancer samples ($n = 105$) suggesting that the improvement of cancer survival of *p53* LOH samples with loss of the *DMP1* locus is a generalized phenomenon. (**A**) PFS of lung cancer (UM); (**B**) TS of lung cancer (UM), (**C**) DFS of breast cancer (BC, WFUHS).

2.3. Cyclin D1 Overexpression and YY1 Amplification and NSCLC Survival

Finally, we examined the impact of Cyclin D1 [52,53] overexpression at protein levels and *YY1* [54,55] expression at genomic DNA level. Overexpression of the Cyclin D1 protein (grade 2 in IHC) was found in 26 of 52 cases while 26 samples had no expression of the protein in NSCLC samples. Samples with Cyclin D1 overexpression had the trend to be associated with favorable prognosis in PFS and TS, but the data were not statistically significant ($p = 0.0875$, $\chi^2 = 2.9191$ in PFS; $p = 0.0754$, $\chi^2 = 3.1603$ in TS) (Figure 3A,B). Although both h*DMP1* LOH and Cyclin D1 overexpression were associated with favorable prognosis, the Cyclin D1 protein overexpression was not related to the LOH of h*DMP1* ($p > 0.20$) suggesting that these two events happened independently. When we conducted quadruple analysis of NSCLC for Cyclin D1 and h*DMP1* LOH, we found significantly shorter survival for the group without Cyclin D1 expression without LOH for h*DMP1* ($p = 0.0031$, $\chi^2 = 13.8559$ for PFS; $p = 0.0020$, $\chi^2 = 14.8269$ for TS; Figure 3C,D). As a matter of fact, none of the patients lived more than 1400 days in this group where clinicians are alerted for early relapse.

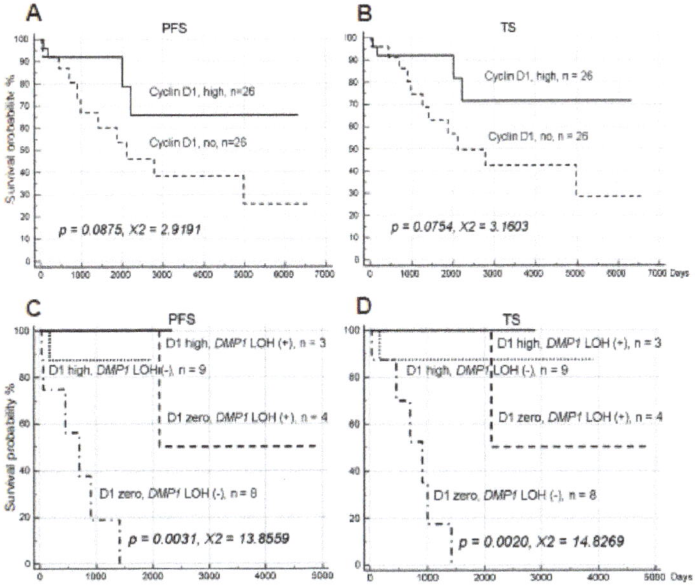

Figure 3. Progression-free survival (PFS) and total survival (TS) analysis of human NSCLC obtained from UM dependent on Cyclin D1. (**A**,**B**) PFS and TS analysis of human NSCLC obtained from UM dependent on Cyclin D1 in human NSCLC ($n = 62$). High expression of Cyclin D1 protein tended to be associated with longer survival. (**C**,**D**) The quadruple analysis of NSCLC samples on Cyclin D1 expression and LOH for *DMP1*. Quadruple analysis was conducted in UM NSCLC samples for Cyclin D1 high (2+); *DMP1* LOH (+) and (-), and Cyclin D1 no expression; *DMP1* LOH (+) and (-). Both Cyclin D1 expression and LOH for *DMP1* are associated with favorable prognosis; we could identify NSCLC patients with worst prognosis in the group of Cyclin D1, no expression; *DMP1* LOH (-) group.

The MDM2 stimulator [54] and epigenetic modifier [55]. *YY1* gene amplification (>3 folds) was found in 16 of 68 NSCLC samples (23.5%) examined. The gene amplification seemed to predict favorable outcome of patients, but neither of them (PFS: $p = 0.2833$, $\chi^2 = 1.1590$; TS: $p = 0.0973$, $\chi^2 = 2.7490$) was statistically significant (Figure 4A,B). Amplification of *YY1* genomic locus was independent of LOH for h*DMP1* ($p > 0.50$). Then we conducted quadruple analysis of NSCLC for *YY1* amplification and h*DMP1* LOH. We found shorter survival for the group without *YY1* amplification expression without LOH for h*DMP1* ($p = 0.0109$, $\chi^2 = 11.1565$ for PFS, $p = 0.0024$, $\chi^2 = 14.4367$ for TS; Figure 4C,D), indicating that *YY1* amplification (-); h*DMP1* LOH (-) genotype is an ominous sign of NSCLC.

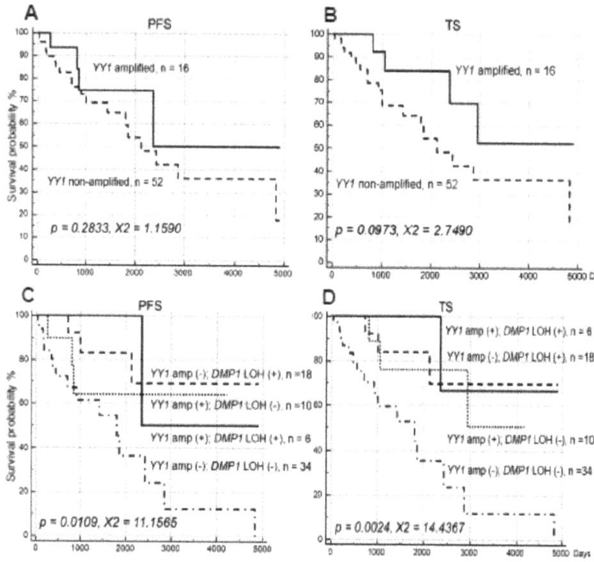

Figure 4. PFS and TS analysis of human NSCLC obtained from UM dependent on *YY1*. (**A**,**B**) PFS and TS analysis of human NSCLC obtained from UM dependent on *YY1* in human NSCLC. (**C**,**D**) The quadruple analysis of NSCLC samples on *YY1* amplification and LOH for *DMP1*. Quadruple analysis was conducted in UM NSCLC samples for *YY1* amplification (>3 folds); *DMP1* LOH (+) and (-); *YY1* no-amplification; *DMP1* LOH (+) and (-); *YY1* amplification. Both *YY1* amplification and LOH for *DMP1* are associated with favorable prognosis; we could identify NSCLC patients with worst prognosis in the group of *YY1* no-amplification; *DMP1* LOH (-) group.

In summary, both Cyclin D1 protein overexpression and *YY1* genomic amplification seem to predict favorable outcome of NSCLC patients; these do not overlap with LOH for h*DMP1* which is another favorable prognostic factor. When these are absent without LOH for h*DMP1*, NSCLC patients will not live long.

3. Discussion

We have analyzed LOH values for h*DMP1*, *ARF/INK4a*, and *p53 loci* from NSCLC specimen obtained from a different institution. The percentage is a bit different—37.1% of h*DMP1* locus, 47.1% for *ARF/INK4a*, and 61.4% for *p53* in the UM VA hospital samples collected in 1991–2000 while it was 40.8%, 36.0%, and 47.9% from the WFU hospital, 1999–2006, because of the difference of demographic distribution. Importantly, the h*DMP1* locus had positive impact on NSCLC survival in both institutions ($p = 0.0029$ in PFS, $p = 0.0040$ in TS in UM samples; $p = 0.03551$ in WFU sample) [39]. LOH for the *ARF/INK4a* locus did not have prognostic values in both institutions ($p = 0.1718$ in PFS, $p = 0.1721$ in TS in UM specimens), although there was a trend that LOH for *ARF/INK4a* locus with either of the two primers was associated with worse prognosis in UM samples. Current data show that LOH for the *ARF/INK4a* locus is mutually exclusive for that for the h*DMP1* (88.7%) in NSCLC. Published studies show that Dmp1 activates both *ARF*-p53 and Ink4a-Rb pathways in mice for tumor suppression [43]. These human and mouse data are very consistent indicating that DMP1 is in the upstream of these pathways, both having binding sites for transactivation [43].

Since *ARF/INK4a* locus 5′ probe #33647 was close to *ARF* exon 1β and #27251 was close to *INK4a* exon 1α, LOH was evaluated separately in this study. Of note, we noticed that #33647 LOH (representing the *ARF* locus; LOH, 23/70, 32.9%) had much worse impact on NSCLC survival than that of #27251 (representing the *INK4a* locus; 19/70, LOH 27.1%). This is quite unexpected because INK4a has been considered to be much more important tumor suppressor than *ARF* in human cancer [21,22]. This is possibly because the *ARF* gene is inactivated in human cancer by gene deletion, splicing alteration than *INK4a*, which is inactivated mainly by promoter methylation or coding exon point mutations [5,16]; for splicing errors for *ARF*, see ref. [2]). Frameshift caused by nucleotide insertion or deletion affects both of these genes at equal frequency [16]. Since the frequency of LOH is similar for *ARF* and *INK4a*, it is highly possible that *ARF* plays an important role as *INK4a* in suppressing tumor development in NSCLC. Importantly, LOH for *ARF/INK4a* was mutually exclusive of that for the h*DMP1* locus in UM samples (this study) as well as WFU samples [39,42] confirming that DMP1 is in the same pathway as that of *ARF*-p53 signaling.

In good contrast to LOH for h*DMP1*, LOH for *p53* had negative impact on patients' survival (Supplementary Figure S5) in both PFS and TS in UM NSCLC samples. Likewise, IHC study shows that p53 protein overexpression is associated with worse prognosis in lung cancer. Again IHC staining for p53 was mutually exclusive for that of h*DMP1* in UM specimen for NSCLC (29/33). The data are consistent our findings that Dmp1 binds directly to the p53 protein, esp. when cells receive DNA damage [41,47]. IHC for p53 (grade 2) tend not to overlap LOH for h*DMP1* in our study (87.9% exclusive), indicating that p53 overexpression is a sign of p53 mutation(s) that happen(s) with LOH for *p53*.

We also conducted survival analysis of h*DMP1*; *p53* double LOH cases ($n = 10$ in UM samples) although the analysis was difficult due to mutual exclusiveness of LOH for h*DMP1* and *p53*. We expected that when inactivation of two tumor suppressors overlap, the prognosis of patients will be even worse than *p53* LOH alone; however, the negative effect of *p53* loss was strikingly improved by simultaneous loss of h*DMP1* (Figure 2). The trend was the same regardless of the origin of NSCLC samples (UM or WFU) or the cancer types (lung cancer or breast cancer). The improvement of the negative effects of *p53* LOH in NSCLC was specific to h*DMP1* LOH because it was not found in LOH for *ARF/INK4a* (Supplementary Figure S7B,C). The quenching of poor prognosis of *p53* LOH by simultaneous loss of h*DMP1* is consistent with relatively good prognosis of NSCLC with LOH for h*DMP1* in both institutions. Although the molecular mechanism(s) underlying this phenomenon is not clear at this moment, we speculate that it is a generalized tendency of carcinoma. Our data show that Dmp1α stimulates the p53 pathway through direct physical interaction with p53 in response to dsDNA breaks [41,47]. The human *DMP1* locus encodes *DMP1α*, and its splice variants *DMP1β* and γ [46]. Our preliminary study suggests that Dmp1α binds to both wt and mutant p53 although the affinity is much higher for the latter. Thus, DMP1α is either tumor suppressive (wt p53) or oncogeic (mutant p53) dependent on the p53 status of cells. DMP1β/γ do not interact with p53, is always oncogenic [45]. DMP1α regulates the human *ARF* promoter, the activity of which is antagonized its splice variant DMP1β [46]. Whatever the mechanism is, small molecule inhibitor screening should be performed to disrupt DMP1α and mutant p53 interaction for future cancer therapy.

The present study shows that Cyclin D1 overexpression as detected by IHC is associated with longer survival/better prognosis of NSCLC in both PFS and TS. The p values were more than 0.05 because we cannot include low level of Cyclin D1 expression (grade 1) and because not all samples were stained for Cyclin D1 in the current study. The current study shows that both h*DMP1* LOH and Cyclin D1 overexpression are favorable prognostic factors associated with longer survival (Figures 1 and 4). Thus we conducted quadruple analysis of NSCLC patients and found that samples that show Cyclin D1 IHC zero and h*DMP1* LOH (-) have much worse prognosis than other three groups (Figure 4C,D). It is expected that tumor cells without LOH for h*DMP1* and Cyclin D1, low express high hDMP1. Whatever the situation is, hDMP1 overexpression combined with no Cyclin D1 expression in IHC is associated with worse prognosis of NSCLC as shown in Figure 3C,D, delineating special group of NSCLC patients who will relapse early. Dmp1α behaves as a tumor suppressor when the p53 is wild type since it binds and stabilizes p53, but it will behave like as an oncogene when p53 is mutant.

Whatever the situation is, molecular studies should be performed in the near future to explain the oncogenic role of hDMP1 in human cancer. Of note, mouse genome produces only Dmp1α, but not β or γ isoforms, explaining the difference between mice and humans.

We also studied the effects of *YY1* genomic amplification in NSCLC. Its amplification again had positive effects in NSCLC survival although the data were not statistically significant. It has been reported that YY1 augments HDM2-mediated p53 polyubiquitination, and thus be oncogenic [54]. Our study shows that Dmp1α antagonizes p53's ubiquitination by MDM2 both in vitro and in cell, and restores p53's nuclear localization that had been lost with MDM2 expression [41]; theoretically Dmp1α will antagonize the action of YY1 to show its tumor-suppressive activity. Then the LOH for h*DMP1* will nullify the effects of YY1 in NSCLC survival. As a matter of fact, the survival of *YY1*-amplified lung cancer almost the same between *YY1* amplified cases and *YY1* amplified; h*DMP1* LOH cases in PFS and TS of NSCLC. We identified a group of bad prognosis in NSCLC, i.e., NSCLC without *YY1* amplification without LOH for h*DMP1* had the worse prognosis than the other three groups.

We will analyze mRNA levels in the future.

4. Materials and Methods

4.1. Human Lung Cancer Samples

Seventy pairs of frozen human lung cancer tissues (36 cases of adenocarcinoma, 25 cases of squamous cell carcinoma, 5 cases of large cell carcinoma, and 4 cases of adenosquamous carcinoma) and their normal counterparts were obtained from the Tissue Procurement Core Facility at the Minnesota Veteran's hospital (UM VA hospital [56,57]). Fifty-one pairs of frozen human lung cancer tissues (33 adenocarcinoma, 16 squamous cell carcinoma, and 2 adenosquamous carcinoma) and their normal counterparts were obtained from the Tissue Procurement Core Facility at the Wake Forest University Comprehensive Cancer Center (WFU [39]). The samples had already been resected from patients with informed consent and had been stored in liquid nitrogen in both cases. The samples do not contain any subject identifiers. The human protocol had been approved by the Institutional Review Board.

4.2. Loss of Heterozygosity (LOH) and Sequencing Analyses of Human Lung Cancer Specimen

LOH assays have been conducted as described previously by PCR [39,42]. PCR products were visualized on a 1.2% agarose gel. Genotypes were identified by peak analysis of the fluorescent signal detected on an ABI 3700 DNA analyzer (Applied Biosystems). LOH was assessed if the qLOH value was found to be >2.0 or <0.5 [39,42]. The third set of primers (#176671; Table 1) was used to examine the status of hDMP1 gene 3′ when LOH study of h*DMP1* locus was single with those for #198004. Real-time PCR was also conducted to study the genomic status (shown as deletion, no deletion in the Table 1).

4.3. Statistical Analyses for Mutual Exclusiveness of LOH

Chi square analyses and confidence interval assays were performed as previously described [39]. IHC studies were also performed for Cyclin D1 [43,52,53] and gene amplification study was performed for exon 4 of human *YY1* [54,55]. The cut-off level for *YY1* was 3.0 folds over neighbor tissues in survival analyses.

The progression-free survival (PFS) and total survival (TS) of UM NSCLC specimen was analyzed by using MedCalc software (Ostend, Belgium; 42). The TS of WFU specimens were also analyzed by the MedCalc software.

5. Conclusions

We have done survival analyses of NSCLC patients in UM and WFU samples, the data of which were obtained by LOH and IHC analyses. LOH for h*DMP1* was associated with favorable prognosis while that of *p53* with worse prognosis. LOH for *ARF* had much negative effects than *INK4a* loss in NSCLC survival, showing the role of *ARF* tumor suppressor by gene deletion. Our data also show that the adverse effect of *p53* LOH was neutralized by simultaneous loss of the h*DMP1* locus. We are currently unaware of the mechanism(s), which will allow more bench works employing DMP1 and p53.

Supplementary Materials: Supplementary Materials can be found at http://www.mdpi.com/1422-0067/21/21/7971/s1. Supplementary Figure S1. Total survival (TS) of patients of human non-small cell lung cancer obtained from the Wake Forest Baptist hospital dependent on LOH for h*DMP1* (A), *ARF/INK4a* (B), and *p53 loci* (C). Kaplan-Meier analyses have been conducted to study the impact for of loss of each locus on non-small cell lung cancer (NSCLC) patients' total survival up to 3500 days. The MedCalc software (Mariakerke, Belgium) was used to analyze the specimens. LOH for h*DMP1* (A, $n = 42$) has significantly positive impact on patient's relapse-free survival. On the other hand, LOH for *ARF/INK4a* did not have significant impact on NSCLC survival (the TS indicate either of the two primers; B, $n = 43$). *p53* LOH had statistically significant negative impact on NSCLC patients' survival (C, $n = 42$); Supplementary Figure S2. PFS and TS of patients of human AC and SCC of the lung cancer obtained from the Minnesota VA Hospital on LOH for h*DMP1*. The quadruple analyses; Supplementary Figure S3. PFS and TS of patients of human AC and SCC of the lung cancer obtained from the Minnesota VA Hospital on LOH for *ARF/INK4a* (either). The quadruple analyses; Supplementary Figure S4. PFS and TS of patients of human non-small cell lung cancer obtained from the Minnesota VA Hospital on LOH for *ARF/INK4a* (both). PFS and TS were determined by the MedCalc software. Two groups comparison was made between that of *ARF/INK4a* LOH (-) and *ARF/INK4a* LOH, both (+) (A, PFS; B, TS); Supplementary Figure S5. PFS and TS of human NSCLC obtained from UM dependent on LOH and IHC for *p53*. Kaplan-Meier analyses have been conducted to study the impact for the impact of loss of the *p53* locus on NSCLC patients' disease-free survival up to 5000 days. The same software was used to calculate the p and chi-square values. *p53* LOH had statistically significant negative impact on NSCLC patients' survival (A,B). Similar analyses were conducted with IHC data for the *p53* protein which often overlap with LOH for *p53* (C,D). The *p* values were relatively larger because not all samples were stained for *p53*, and because grade 1 staining was excluded from the study; Supplementary Figure S6. PFS and TS of patients of human AC and SCC of the lung cancer obtained from the Minnesota VA Hospital on LOH for *p53* (either). The quadruple analyses; Supplementary Figure S7. The triple analysis of cancer survival for *p53* in WFU (NSCLC, A), UM (NSCLC; B,C) samples. The impact of *DMP1* LOH on *p53* LOH was analyzed by Medcalc software for WFU and UM NSCLC samples. Loss of *DMP1* neutralized the negative effect of *p53* LOH in NSCLC by moving the survival curves from *p53* LOH (+) to *p53* LOH (-) levels or better in WFU samples. On the other hand, both PFS and TS of *p53* LOH became worse by simultaneous LOH for *ARF/INK4a* (B,C).

Author Contributions: Conception and design: G.E.N., R.A.K. and K.I. Development of methodology: E.A.F., G.E.N., R.A.K. and F.K. Acquisition of data: E.A.F., G.E.N., R.A.K. and F.K. Analysis and interpretation of data: G.E.N., R.A.K. and K.I. Writing, review and revision of the manuscript: E.A.F., G.E.N., R.A.K. and K.I., Administrative, technical, or material support: G.E.N., R.A.K. and K.I. Acquisition for grants: K.I. All authors have read and agreed to the published version of the manuscript.

Funding: K.I. was supported by NIH/NCI 2R01CA106314, ACS RSG-07-207-01-MGO, and KG080179.

Acknowledgments: We thank Gregory Kucera for total survival data for NSCLC at WFUHS hospital. We also thank Deb Sumitra for providing mutant p53 cDNAs, all other members of Inoue's lab for sharing unpublished research data.

Conflicts of Interest: The authors declare no conflict of interest.

References

1. Siegel, R.L.; Miller, K.D.; Jemal, A. Cancer statistics, 2020. *CA Cancer J. Clin.* **2020**, *70*, 7–30. [CrossRef] [PubMed]
2. Oser, M.G.; Niederst, M.J.; Sequist, L.V.; Engelman, J.A. Transformation from non-small-cell lung cancer to small-cell lung cancer: Molecular drivers and cells of origin. *Lancet Oncol.* **2015**, *16*, e165–e172. [CrossRef]
3. Satoh, Y.; Matsuo, Y.; Kuba, T.; Yamashita, K.; Sawano, M.; Tozaka, S.; Yamazaki, H.; Sonoda, D.; Mikubo, M.; Naito, M.; et al. EGFR mutation genotyping and ALK status determination in liquid-based cytology samples of non-small cell lung cancer. *Virchows Arch.* **2020**, *476*, 753–762. [CrossRef] [PubMed]
4. Rossi, A.; Di Maio, M.; Chiodini, P.; Rudd, R.M.; Okamoto, H.; Skarlos, D.V.; Fruh, M.; Qian, W.; Tamura, T.; Samantas, E.; et al. Carboplatin- or cisplatin-based chemotherapy in first-line treatment of small-cell lung cancer: The COCIS meta-analysis of individual patient data. *J. Clin. Oncol.* **2012**, *30*, 1692. [CrossRef] [PubMed]

5. Quelle, D.E.; Zindy, F.; Ashmun, R.A.; Sherr, C.J. Alternative reading frames of the INK4a tumor suppressor gene encode two unrelated proteins capable of inducing cell cycle arrest. *Cell* **1995**, *83*, 993–1000. [PubMed]
6. Maggi, L.B., Jr.; Winkeler, C.L.; Miceli, A.P.; Apicelli, A.J.; Brady, S.N.; Kuchenreuther, M.J.; Weber, J.D. ARF tumor suppression in the nucleolus. *Biochim. Biophys. Acta* **2014**, *1842*, 831–839. [CrossRef]
7. Basu, S.; Murphy, E. Genetic Modifiers of the p53 Pathway. *Cold Spring Harb. Perspect. Med.* **2016**, *6*, a026302. [CrossRef]
8. Carrasco-Garcia, E.; Moreno, M.; Moreno-Cugnon, L.; Mathew, A. Increased Arf/p53 activity in stem cells, aging and cancer. *Aging Cell* **2017**, *16*, 219–225. [CrossRef]
9. Inoue, K.; Fry, E.A. Aberrant expression of $p16^{INK4a}$ in human cancer—A new biomarker? *Cancer Rep. Rev.* **2018**, *2*, 1–7. [CrossRef]
10. Inoue, K.; Fry, E.A. Overexpression of ARF in human cancer—A new biomarker? *Tumor Microenviron.* **2018**, *1*, 37–44. [CrossRef]
11. Palmero, I.; Pantoja, C.; Serrano, M. p19ARF links the tumour suppressor p53 to ras. *Nature* **1998**, *395*, 125–126. [CrossRef] [PubMed]
12. Zindy, F.; Eischen, C.M.; Randle, D.H.; Kamijo, T.; Cleveland, J.L.; Sherr, C.J.; Roussel, M.F. Myc signaling via the ARF tumor suppressor regulates p53-dependent apoptosis and immortalization. *Genes Dev.* **1998**, *12*, 2424–2433. [CrossRef]
13. Bates, S.; Phillips, A.C.; Clarke, P.A.; Stott, F.; Peters, G.; Ludwig, R.L.; Vousden, K.H. p14ARF links tumor suppressors RB and *p53*. *Nature* **1998**, *395*, 124–125. [CrossRef]
14. Taneja, P.; Maglic, D.; Kai, F.; Sugiyama, T.; Kendig, R.D.; Frazier, D.P.; Willingham, M.C.; Inoue, K. Critical role of Dmp1 in HER2/neu-p53 signaling and breast carcinogenesis. *Cancer Res.* **2010**, *70*, 9084–9094. [CrossRef]
15. Sherr, C.J. Autophagy by ARF: A short story. *Mol. Cell* **2006**, *22*, 436–437. [CrossRef] [PubMed]
16. Quelle, D.E.; Cheng, M.; Ashmun, R.A.; Sherr, C.J. Cancer-associated mutations at the INK4a locus cancel cell cycle arrest by p16INK4a but not by the alternative reading frame protein p19ARF. *Proc. Natl. Acad. Sci. USA* **1997**, *94*, 669–673. [CrossRef] [PubMed]
17. Krimpenfort, P.; Quon, K.C.; Mooi, W.J.; Loonstra, A.; Berns, A. Loss of p16Ink4a confers susceptibility to metastatic melanoma in mice. *Nature* **2001**, *413*, 83–86. [CrossRef]
18. Krimpenfort, P.; Ijpenberg, A.; Song, J.Y.; van der Valk, M.; Nawijn, M.; Zevenhoven, J.; Berns, A. p15Ink4b is a critical tumour suppressor in the absence of p16Ink4a. *Nature* **2007**, *448*, 943–946. [CrossRef]
19. Kamijo, T.; Zindy, F.; Roussel, M.F.; Quelle, D.E.; Downing, J.R.; Ashmun, R.A.; Grosveld, G.; Sherr, C.J. Tumor suppression at the mouse INK4a locus mediated by the alternative reading frame product p19ARF. *Cell* **1997**, *91*, 649–655. [CrossRef]
20. Sharpless, N.E.; Bardeesy, N.; Lee, K.H.; Carrasco, D.; Castrillon, D.H.; Aguirre, A.J.; Wu, E.A.; Horner, J.W.; DePinho, R.A. Loss of p16Ink4a with retention of p19Arf predisposes mice to tumorigenesis. *Nature* **2001**, *413*, 86–91. [CrossRef]
21. Ruas, M.; Peters, G. The p16INK4a/CDKN2A tumor suppressor and its relatives. *Biochim. Biophys. Acta* **1998**, *1378*, F115–F177. [CrossRef]
22. Gil, J.; Peters, G. Regulation of the INK4b-ARF-INK4a tumour suppressor locus: All for one or one for all. *Nat. Rev. Mol. Cell Biol.* **2006**, *7*, 667–677. [CrossRef]
23. Kuo, M.L.; den Besten, W.; Bertwistle, D.; Roussel, M.F.; Sherr, C.J. N-terminal polyubiquitination and degradation of the Arf tumor suppressor. *Genes Dev.* **2004**, *18*, 1862–1874. [CrossRef] [PubMed]
24. Stott, F.J.; Bates, S.; James, M.C.; McConnell, B.B.; Starborg, M.; Brookes, S.; Palmero, I.; Ryan, K.; Hara, E.; Vousden, K.H.; et al. The alternative product from the human CDKN2A locus, p14(ARF), participates in a regulatory feedback loop with p53 and MDM2. *EMBO J.* **1998**, *17*, 5001–5014. [CrossRef] [PubMed]
25. Weber, J.D.; Taylor, L.J.; Roussel, M.F.; Sherr, C.J.; Bar-Sagi, D. Nucleolar Arf sequesters Mdm2 and activates p53. *Nat. Cell Biol.* **1999**, *1*, 20–26. [CrossRef] [PubMed]
26. Martelli, F.; Hamilton, T.; Silver, D.P.; Sharpless, N.E.; Bardeesy, N.; Rokas, M.; DePinho, R.A.; Livingston, D.M.; Grossman, S.R. p19ARF targets certain E2F species for degradation. *Proc. Natl. Acad. Sci. USA* **2001**, *98*, 4455–4460. [CrossRef] [PubMed]
27. Itahana, K.; Zhang, Y. Mitochondrial p32 is a critical mediator of ARF-induced apoptosis. *Cancer Cell* **2008**, *13*, 542–553. [CrossRef]
28. Chen, D.; Tavana, O.; Chu, B.; Erber, L.; Chen, Y.; Baer, R.; Gu, W. NRF2 is a major target of ARF in p53-independent tumor Suppression. *Mol. Cell* **2017**, *68*, 224–332. [CrossRef]

29. Hirai, H.; Sherr, C.J. Interaction of D-type cyclins with a novel myb-like transcription factor, DMP1. *Mol. Cell. Biol.* **1996**, *16*, 6457–6467. [CrossRef]
30. Inoue, K.; Sherr, C.J. Gene expression and cell cycle arrest mediated by transcription factor DMP1 is antagonized by D-type cyclins through a cyclin-dependent-kinase-independent mechanism. *Mol. Cell. Biol.* **1998**, *18*, 1590–1600. [CrossRef]
31. Inoue, K.; Sherr, C.J.; Shapiro, L.H. Regulation of the CD13/aminopeptidase N gene by DMP1, a transcription factor antagonized by D-type cyclins. *J. Biol. Chem.* **1998**, *273*, 29188–29194. [CrossRef] [PubMed]
32. Inoue, K.; Roussel, M.F.; Sherr, C.J. Induction of ARF tumor suppressor gene expression and cell cycle arrest by transcription factor DMP1. *Proc. Natl. Acad. Sci. USA* **1999**, *96*, 3993–3998. [CrossRef] [PubMed]
33. Bodner, S.M.; Naeve, C.W.; Rakestraw, K.M.; Jones, B.G.; Valentine, V.A.; Valentine, M.B.; Luthardt, F.W.; Willman, C.L.; Raimondi, S.C.; Downing, J.R.; et al. Cloning and chromosomal localization of the gene encoding human cyclin D-binding Myb-like protein (hDMP1). *Gene* **1999**, *229*, 223–229. [CrossRef]
34. Inoue, K.; Wen, R.; Rehg, J.E.; Adachi, M.; Cleveland, J.L.; Roussel, M.F.; Sherr, C.J. Disruption of the ARF transcriptional activator DMP1 facilitates cell immortalization, Ras transformation, and tumorigenesis. *Genes Dev.* **2000**, *14*, 1797–1809.
35. Inoue, K.; Zindy, F.; Randle, D.H.; Rehg, J.E.; Sherr, C.J. Dmp1 is haplo-insufficient for tumor suppression and modifies the frequencies of Arf and p53 mutations in Myc-induced lymphomas. *Genes Dev.* **2001**, *15*, 2934–2939. [CrossRef]
36. Sreeramaneni, R.; Chaudhry, A.; McMahon, M.; Sherr, C.J.; Inoue, K. Ras-Raf-Arf signaling critically depends on Dmp1 transcription factor. *Mol. Cell. Biol.* **2005**, *25*, 220–232. [CrossRef]
37. Mallakin, A.; Taneja, P.; Matise, L.A.; Willingham, M.C.; Inoue, K. Expression of Dmp1 in specific differentiated, nonproliferating cells and its repression by E2Fs. *Oncogene* **2006**, *25*, 7703–7713. [CrossRef]
38. Taneja, P.; Mallakin, A.; Matise, L.A.; Frazier, D.P.; Choudhary, M.; Inoue, K. Repression of Dmp1 and Arf transcription by anthracyclins: Critical roles of the NF-kappaB subunit p65. *Oncogene* **2007**, *26*, 7457–7466. [CrossRef]
39. Mallakin, A.; Sugiyama, T.; Taneja, P.; Matise, L.A.; Frazier, D.P.; Choudhary, M.; Hawkins, G.A.; D'Agostino, R.B., Jr.; Willingham, M.C.; Inoue, K. Mutually exclusive inactivation of DMP1 and ARF/p53 in lung cancer. *Cancer Cell* **2007**, *12*, 381–394. [CrossRef]
40. Mallakin, A.; Sugiyama, T.; Kai, F.; Taneja, P.; Kendig, R.D.; Frazier, D.P.; Maglic, D.; Matise, L.A.; Willingham, M.C.; Inoue, K. The Arf-inducing transcription factor Dmp1 encodes transcriptional activator of amphiregulin, thrombospondin-1, JunB and Egr1. *Int. J. Cancer* **2010**, *126*, 1403–1416. [CrossRef]
41. Frazier, D.P.; Kendig, R.D.; Kai, F.; Maglic, D.; Sugiyama, T.; Morgan, R.L.; Fry, E.A.; Lagedrost, S.J.; Sui, G.; Inoue, K. Dmp1 physically interacts with p53 and positively regulates p53's stabilization, nuclear localization, and function. *Cancer Res.* **2012**, *72*, 1740–1750. [CrossRef]
42. Maglic, D.; Zhu, S.; Taneja, P.; Fry, E.A.; Kai, F.; Kendig, R.D.; Sugiyama, T.; Miller, L.D.; Willingham, M.C.; Inoue, K. Prognostic value of the hDMP1-ARF-Hdm2-p53 pathway in breast cancer. *Oncogene* **2013**, *32*, 4120–4129. [CrossRef] [PubMed]
43. Zhu, S.; Mott, R.T.; Fry, E.A.; Taneja, P.; Kulik, G.; Sui, G.; Inoue, K. Cooperation between cyclin D1 expression and Dmp1-loss in breast cancer. *Am. J. Pathol.* **2013**, *183*, 1339–1350. [CrossRef]
44. Fry, E.A.; Taneja, P.; Maglic, D.; Zhu, S.; Sui, G.; Inoue, K. Dmp1α inhibits HER2/neu-induced mammary tumorigenesis. *PLoS ONE* **2013**, *8*, e77870. [CrossRef]
45. Maglic, D.; Stovall, D.B.; Cline, J.M.; Fry, E.A.; Mallakin, A.; Taneja, P.; Caudell, D.L.; Willingham, M.C.; Sui, G.; Inoue, K. DMP1β, a splice isoform of the tumor suppressor DMP1 locus, induces proliferation and progression of breast cancer. *J. Pathol.* **2015**, *236*, 90–102. [CrossRef] [PubMed]
46. Tschan, M.P.; Federzoni, E.A.; Haimovici, A.; Britschgi, C.; Moser, B.A.; Jin, J.; Reddy, V.A.; Sheeter, D.A.; Fischer, K.M.; Sun, P.; et al. Human DMTF1β antagonizes DMTF1α regulation of the p14(ARF) tumor suppressor and promotes cellular proliferation. *Biochim. Biophys. Acta* **2015**, *1849*, 1198–1208. [CrossRef]
47. Kendig, R.D.; Kai, F.; Fry, E.A.; Inoue, K. Stabilization of the p53-DNA complex by the nuclear protein Dmp1α. *Cancer Investig.* **2017**, *35*, 301–312. [CrossRef] [PubMed]
48. Inoue, K.; Fry, E.A.; Frazier, D.P. Transcription factors that interact with p53 and Mdm2. *Int. J. Cancer* **2016**, *138*, 1577–1585. [CrossRef]
49. Inoue, K.; Fry, E.A. Aberrant splicing of the DMP1-INK4a/ARF-MDM2-p53 pathway in cancer. *Int. J. Cancer* **2016**, *139*, 33–41. [CrossRef] [PubMed]

50. Fry, E.A.; Taneja, P.; Inoue, K. Oncogenic and tumor-suppressive mouse models for breast cancer employing HER2/neu. *Int. J. Cancer* **2017**, *140*, 495–503. [CrossRef]
51. Inoue, K.; Fry, E.A. Haplo-insufficient tumor suppressor genes. In *Advances in Medicine and Biology*; Nova Science Publishers, Inc.: Suite N Hauppauge, NY, USA, 2017; Volume 118, Chapter 6.
52. Qie, S.; Diehl, J.A. Cyclin D1, cancer progression, and opportunities in cancer treatment. *J. Mol. Med. (Berl.)* **2016**, *94*, 1313–1326. [CrossRef] [PubMed]
53. Scott, S.C.; Lee, S.S.; Abraham, J. Mechanisms of therapeutic CDK4/6 inhibition in breast cancer. *Semin. Oncol.* **2017**, *44*, 385–394. [CrossRef] [PubMed]
54. Sui, G.; Affar, B.; Shi, Y.; Brignone, C.; Wall, N.R.; Yin, P.; Donohoe, M.; Luke, M.P.; Calvo, D.; Grossman, S.R.; et al. Yin Yang 1 is a negative regulator of p53. *Cell* **2004**, *117*, 859–872. [CrossRef] [PubMed]
55. Zhang, Q.; Stovall, D.B.; Inoue, K.; Sui, G. The oncogenic role of Yin Yang 1. *Crit. Rev. Oncog.* **2011**, *16*, 163–197. [CrossRef]
56. Kratzke, R.A.; Greatens, T.M.; Rubins, J.B.; Maddaus, M.A.; Niewoehner, D.E.; Niehans, G.A.; Geradts, J. Rb and p16INK4a expression in resected non-small cell lung tumors. *Cancer Res.* **1996**, *56*, 3415–3420.
57. Zhou, J.X.; Niehans, G.A.; Shar, A.; Rubins, J.B.; Frizelle, S.P.; Kratzke, R.A. Mechanisms of G1 checkpoint loss in resected early stage non-small cell lung cancer. *Lung Cancer* **2001**, *32*, 27–38. [CrossRef]

International Journal of
Molecular Sciences

MDPI

Review

The Multifaced Role of STAT3 in Cancer and Its Implication for Anticancer Therapy

Manlio Tolomeo * and Antonio Cascio

Department of Health Promotion Sciences, Maternal and Infant Care, Internal Medicine and Medical Specialties, University of Palermo, via del Vespro 129, 90127 Palermo, Italy; antonio.cascio03@unipa.it
* Correspondence: manlio.tolomeo@policlinico.pa.it

Abstract: Signal transducer and activator of transcription (STAT) 3 is one of the most complex regulators of transcription. Constitutive activation of STAT3 has been reported in many types of tumors and depends on mechanisms such as hyperactivation of receptors for pro-oncogenic cytokines and growth factors, loss of negative regulation, and excessive cytokine stimulation. In contrast, somatic STAT3 mutations are less frequent in cancer. Several oncogenic targets of STAT3 have been recently identified such as c-myc, c-Jun, PLK-1, Pim1/2, Bcl-2, VEGF, bFGF, and Cten, and inhibitors of STAT3 have been developed for cancer prevention and treatment. However, despite the oncogenic role of STAT3 having been widely demonstrated, an increasing amount of data indicate that STAT3 functions are multifaced and not easy to classify. In fact, the specific cellular role of STAT3 seems to be determined by the integration of multiple signals, by the oncogenic environment, and by the alternative splicing into two distinct isoforms, STAT3α and STAT3β. On the basis of these different conditions, STAT3 can act both as a potent tumor promoter or tumor suppressor factor. This implies that the therapies based on STAT3 modulators should be performed considering the pleiotropic functions of this transcription factor and tailored to the specific tumor type.

Keywords: STAT3; cancer; tumor promoter; tumor suppressor

Citation: Tolomeo, M.; Cascio, A. The Multifaced Role of STAT3 in Cancer and Its Implication for Anticancer Therapy. *Int. J. Mol. Sci.* **2021**, *22*, 603. https://doi.org/10.3390/ijms22020603

Received: 17 November 2020
Accepted: 5 January 2021
Published: 9 January 2021

1. Introduction

Signal transducer and activator of transcription 3 (STAT3) is a multifunctional transcription factor involved in multiple biological processes. It was identified in 1993 by Wegenka et al. [1]. These authors observed that interleukin-6 (IL-6) rapidly induced the activation of a DNA-binding factor, termed acute-phase response factor (APRF), in rat liver and in human hepatoma cells. In 1994, Zhong et al. [2] described a 92 kD mouse protein from the signal transducer and activator of transcription (STAT) family that was activated as a DNA-binding protein by IL-6 and by epidermal growth factor (EGF). The partial amino acid sequence obtained from purified rat APRF and the use of a specific antiserum demonstrated that APRF was related to STAT3 [3].

STAT3 belongs to the STAT family of cytoplasmic transcription factors that mediate signal transduction from the plasma membrane to the nucleus in various cellular activities [4].

There are seven STAT proteins: STAT 1, 2, 3, 4, 5A, 5B, and 6 encoded by genes clustered on different chromosomes. Each of them consists of six regions: (1) a helical N-terminal domain (ND) for protein–protein interactions between adjacent STAT dimers on DNA; (2) a coiled-coil (CC) domain for interactions with regulatory proteins that positively or negatively modulate the transcriptional activity; (3) a DNA-binding domain (DBD) for the recognition of specific DNA-sequences of target genes; (4) a helical linker (LK) domain involved in nuclear export and DNA binding; (5) an Src homology 2 (SH2) domain for receptor binding and dimerization; (6) a C-terminal transactivation domain (TAD) that contains specific residues that are phosphorylated upon transcriptional activation [5,6].

STATs are activated in the cytoplasm by Janus kinases (JAKs), a family composed of four different intracellular non-receptor tyrosine kinases, that transduce cytokine-mediated signals [7,8]. JAKs are associated with two different types of receptors (type I and type II) that do not possess any catalytic activity but, after binding with the specific ligand, allow the auto-phosphorylation of JAK, which, in turn, phosphorylates and activates STAT proteins [9].

STATs recognize DNA motifs with a consensus sequence of 5′-TTCN3/4GAA-3′ that are located in promoter and enhancer regions and in the first introns of target genes [10].

STAT3 is one of the most complex transcription regulators and is involved in many biological functions such as cell proliferation, maturation, and survival. STAT3 is activated by the entire family of IL-6-type cytokines comprised of IL-6, IL-11, IL-22, IL-27, IL-31, oncostatin M (OSM), cardiotrophin 1 (CT-1), ciliary neurotrophic factor (CNTF), cardiotrophin-like cytokine factor 1 (CLCF1), and leukemia inhibitory factor (LIF) [11,12]. These cytokines are involved in embryonic development, immunologic activity, inflammation, hematopoiesis, cardiovascular physiology, liver, and neuronal regeneration.

A main role in the IL-6 cytokine family/STAT3 axis is played by a transmembrane—cytokine receptor associated—protein of 130 kD called gp130. In fact, this protein mediates, as a common signal transducer, the pleiotropic but overlapping functions of the IL-6 cytokine family. When a member of the IL-6-type cytokine family binds to its receptor, it induces the recruitment of gp130 and the formation of a complex of three proteins (IL-6-type cytokine/IL-6 member receptor/gp130). After complexing, gp130 is phosphorylated on tyrosine residues. The phosphorylation leads to the association with JAK, which, in turn, activates STAT3, which leads to the activation of downstream genes. Animals lacking gp130 are not viable, indicating the importance of this receptor associated protein [13].

While gp130 expression is relatively ubiquitous in a wide variety of tissues and organs, receptors of the IL6-cytokine family are more restricted and cell-type specific. These receptors include IL-6R (IL-6 receptor), IL-11R (IL-11 receptor), IL-27Rα (IL-27 receptor alpha), OSMR (OSM receptor), LIFR (LIF receptor), and CNTFRα (CNTF receptor alpha).

Many studies have demonstrated the critical role of aberrant STAT3 in malignant transformation, and several STAT3 oncogenic targets have been identified. However, recent evidence has shown that STAT3 can have opposite functions in cancer, and based on different conditions, STAT3 can act as both a potent tumor promoter and a tumor suppressor factor. This appears to depend on several factors such as the integration of multiple signals, the oncogenic environment, and the alternative splicing into different isoforms.

2. STAT3 and Tumorigenesis

STAT3 is currently considered an oncogene, and aberrant regulation of STAT3 has been reported in nearly 70% of cancers [14,15]. Aberrant activation of STAT3 in cancer cells causes the continuous transcription of cell growth factors and anti-apoptotic molecules that play a crucial role in maintaining cell growth and survival [16]. Moreover, STAT3 confers tumor great malignancy by promoting tumor invasion, migration, metastasis, and angiogenesis. This has led many researchers to develop STAT3 inhibitors for the treatment of different malignancies.

The first study demonstrating the involvement of STAT3 in cancer was published in 1995 by Yu et al. They showed that oncoprotein Src can activate the STAT3 signaling pathway, raising the possibility that this transcription factor contributes to oncogenesis by Src [17]. In 1996, Cao et al. reported that in all v-Src-transformed cell lines examined, STAT3 was constitutively activated, and phosphorylation of tyrosine on STAT3 was enhanced by the induction of v-Src expression [18]. They also showed that Src was associated with the presence of a constitutively activated STAT3 in vivo [18]. Although Src activation of STAT3 is the most well-characterized model, there was evidence that STAT3 can be activated by other members of the Src family kinases [19] and by other oncogenic tyrosine kinases, such as v-Ros, v-Fps, Etk/BMX, and v-Abl [20–25], or by viral proteins that directly or indirectly activate tyrosine kinase pathways, including human T lymphotropic virus (HTLV)1, polyomavirus middle T antigen, Epstein–Barr virus (EBV), and herpes virus saimiri [26–31].

It is noteworthy that Bromberg et al. reported that the substitution of two cysteine residues within the C-terminal loop of the SH2 domain of STAT3 produces a molecule that dimerizes spontaneously, binds to DNA, and activates transcription, causing cellular transformation, thus demonstrating that the activated STAT3 molecule by itself can mediate cellular transformation [32].

Hyperactivation of STAT3 has been reported in many types of tumors and can depend on the following mechanisms: (i) hyperactivation of receptors for pro-oncogenic cytokines and growth factors; (ii) elevated activity of cytoplasmic non-receptor tyrosine kinases, such as Src, Janus kinases (JAKs), and Abelson (Abl) kinase; (iii) loss of negative STAT3 regulation; (iv) excessive stimulation by cytokines such as IL-6 or EGF. These mechanisms can induce uncontrolled cell growth, malignant cell transformation, angiogenesis, metastasis, invasion, and immune escape. For example, the growth and survival of human myeloma cells is dependent on an IL-6 autocrine loop that induces a constitutive activation of STAT3 [33]. Likewise, IL-6 is an autocrine growth factor for human prostate cancer cells, and the effects of IL-6 on prostate cancer cell growth are mediated through the Jak/STAT3 signaling pathway [34]. Elevated epidermal growth factor receptor (EGFR)-mediated signaling or Src and Jak kinases' activities have been observed in breast cancer, prostate cancer, non-small cell lung cancer (NSCLC), melanoma, pancreatic cancer, and head and neck squamous carcinoma (HNSCC) cells [23,35–40].

In contrast, somatic STAT3 mutations are less frequent in cancer. For example, sixty percent of inflammatory hepatocellular adenomas have IL-6 signal transducer mutations that cause hyperactivation of IL-6/STAT3 signaling, while 12% of these tumors lacking IL-6 signal transducer mutations have somatic STAT3 mutations [41].

However, some types of cancers frequently have STAT3 mutations. For example, STAT3 mutations were found in 40% of patients with large granular lymphocytic leukemia. In this leukemia, all mutations were located in exon 21, encoding the Src homology 2 (SH2) domain, which mediates the dimerization and activation of STAT3 [42].

STAT3 activation in normal and tumor cells is dependent on the phosphorylation of a tyrosine residue (Y705), which is located between the SH2 domain and TAD (Figure 1). Phosphorylation of Y705 is crucial for STAT3 dimerization, nuclear translocation, and DNA binding. A second phosphorylation site is located on serine 727 (S727) in the C-terminal domain (Figure 1). Phosphorylation on S727 is required for maximal transcriptional activity. However, its functions are probably more complex since the S727 site is phosphorylated in response to various cellular stresses and through the interaction with transcriptional coactivators such as SRC, Cdk9, or CBP in the absence of Y705 phosphorylation [43,44]. In chronic lymphocytic leukemia, STAT3 is constitutively phosphorylated on S727 and not on Y705, and dephosphorylation of inducible tyrosine pSTAT3 does not affect STAT3-DNA binding [45]. It has also been reported that Ras-mediated transformation is significantly reduced when STAT3 is mutated on its S727 residue. Malignant transformation by activated Ras is impaired without STAT3, in spite of the inability of Ras to drive STAT3 tyrosine phosphorylation or nuclear translocation. The cooperation between Ras and STAT3 requires the serine phosphorylation site at the C-terminus. The mutation of S727 or deletion of the C-terminal domain abrogated cooperation between STAT3 and Ras [46]. Some observations suggest that phosphorylation on S727 could be connected to the genetic instability and DNA damage that occur early in tumor cells [47].

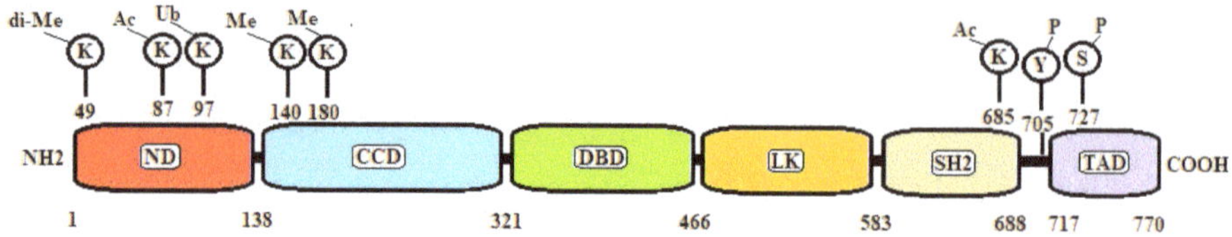

Figure 1. Structure of signal transducer and activator of transcription 3 (STAT3). Functional domains: ND, NH2-terminal domain; CCD, coiled-coil domain; DBD, DNA-binding domain; LK, linker domain; SH2, Src homology 2 domain; TAD, transactivation domain. STAT3 activation is dependent on the phosphorylation (P) of a tyrosine residue Y705, which is located between the SH2 domain and TAD. Phosphorylation on serine (S) 727 is required for maximal transcriptional activity. After cytokine treatment, STAT3 is acetylated (Ac) on lysine (K) K87 and 685. Other post-translational modifications are di-methylation (di-Me) of K49, monoubiquitination of K97, and methylation (Me) of K140 and K180.

After cytokine treatment, STAT3 is also acetylated on a single lysine (K) residue, K685 (Figure 1). Acetylation of K685 is critical for STAT3 to form stable dimers required for cytokine-stimulated DNA binding and transcriptional regulation. This site of acetylation also plays an important role in cancer growth. In fact, lysine acetylation of STAT3 is elevated in tumors such as melanoma and colorectal cancer. Genetically altering STAT3 at K685 reduces tumor growth, which is accompanied by demethylation and reactivation of several tumor-suppressor genes [48].

Other post-translational STAT3 modifications are p300-mediated acetylation of K87 at the STAT3 NH2 terminus, which is required for IL-6-induced target gene activation [49], dimethylation of K49 at the NH2 terminus, which is crucial for the expression of many IL-6-dependent genes [50], monoubiquitination at K97, which plays a role in STAT3 anti-apoptotic gene expression [51], methylation of K140, which is a negative regulatory event, because its blockade greatly increases the amount of activated STAT3 [52], and methylation of K180, which leads to enhanced STAT3 activity [53] (Figure 1).

STAT3 regulates multiple biological functions in the initiation of malignant transformation, and it is the convergence point of several major oncogenic signaling pathways. Moreover, STAT3 modulates host immunity against tumor cells leading to tumor-induced immunosuppression. This depends on the ability of STAT3 to suppress anticancer immune cells and to activate cancer promoting immune cells. Therefore, therapies based on STAT3 modulators can not only directly inhibit the growth of tumors, but also enhance antitumor immunity.

3. Role of STAT3 in Oncogenic Transformation

Activated STAT3 can upregulate the mRNA levels of many genes involved in cell growth and apoptosis such as cyclins D1, D2, D3, A, and B, Cdc25A, Cdc2, c-Myc, PLK1, Pim-1/2, Cten, survivin, Bcl-xL, IAPs, and Mcl-1. These upregulated genes cooperate in inducing the oncogenic transformation of cells. In addition, STAT3 downregulate genes encoding cell cycle checkpoint proteins such as p21, p27, and p53. Other genes, such as c-Jun, c-Fos, and FGFR, interact with STAT3 in inducing cancer transformation (Figure 2).

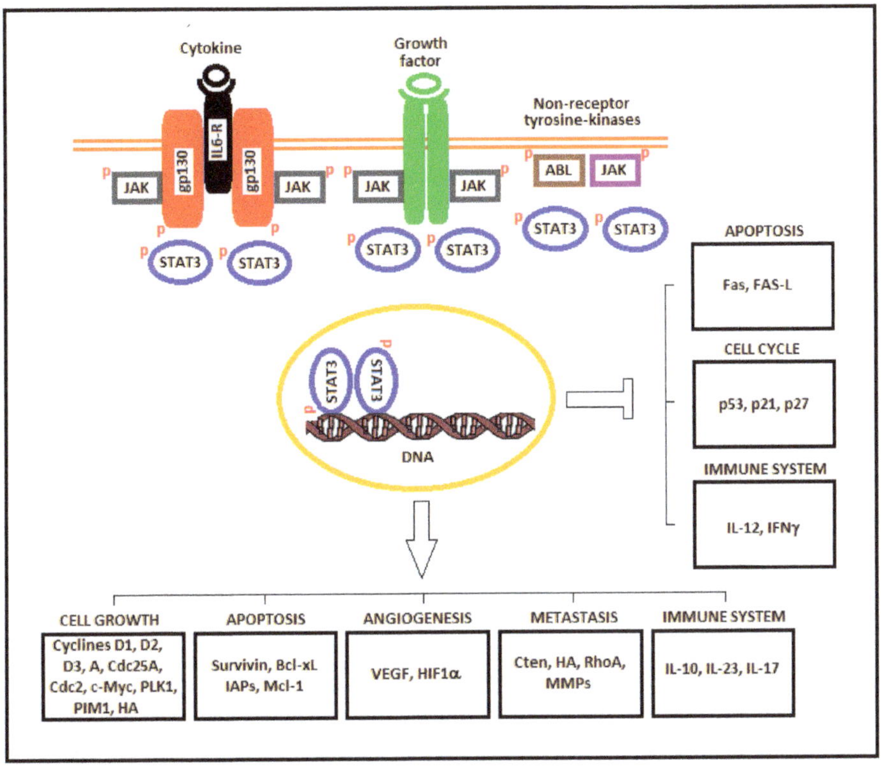

Figure 2. STAT3 upregulation and downregulation of factors involved in cell growth, apoptosis, angiogenesis, invasion, metastasis, and the immune system. STAT3 is activated through the interaction of cytokines and growth factors. Non-receptor tyrosine kinases have intrinsic kinase activity, whereas the receptors of ligands have associated JAK, which, when phosphorylated, acts as a platform for un-phosphorylated STAT3 to become activated. Phosphorylated STAT3 dimers translocate to the nucleus where they upregulate and downregulate a variety of genes that can contribute to tumorigenesis.

STAT3 plays a key role in the G1 to S phase cell cycle transition through the upregulation of cyclins D1, D2, D3, A, and Cdc25A and the concomitant downregulation of p21 and p27. Moreover, STAT3 participates in modulating the G2–M phase checkpoint by regulating gene expressions of cyclin B1 and Cdc2 via E2F [54]. Cyclin D1 mRNA levels are increased in primary rat-, mouse-, and human-derived cell lines expressing either the oncogenic variant of STAT3 (STAT3-C) or vSrc, which constitutively phosphorylates STAT3 [55]. Cyclin D1 mediates the progression of cells from the G1 to S phase of the cell cycle by phosphorylating the retinoblastoma protein (Rb). Phosphorylation of the Rb protein releases the E2F transcription factor, leading to transcription and S phase entry. Overexpression of cyclin D1 is found in many cancers and is sufficient to mediate mammary tumorigenesis [56,57]. Furthermore, cyclin D1-deficient animals are resistant to Ras- and Neu-mediated skin and breast tumorigenesis [58,59].

c-Myc is another important downstream effector of STAT3 signaling involved in cell growth and transformation. Disruption of STAT3 signaling by using dominant-negative STAT3β protein in NIH 3T3 fibroblasts suppresses c-Myc expression. Moreover, fibroblasts with c-Myc gene knockout are refractory to transformation by v-Src, and disruption of STAT3 signaling in normal cells inhibits PDGF-induced mitogenesis in a manner that is reversed by ectopic c-Myc expression [60].

Zhang et al. reported that there is an interaction between a region within c-Jun and specific sites within STAT3 [61]. c-Jun was the first oncogenic transcription factor discovered [62]. It is required for progression through the G1 phase of the cell cycle and protects cells from UV-induced apoptosis. Mutations in the contact region of STAT3 both reduce c-Jun–STAT3 protein interaction and disrupt the cooperation between these two proteins, which is required for maximal IL-6-dependent gene activation driven by the α2-macroglobulin enhancer [61,63]. c-Fos encodes a 62 kDa protein, which forms a heterodimer with c-Jun, resulting in the formation of the AP-1 (activator protein-1) complex, which binds DNA at AP-1 specific sites. It plays an important role in many cellular functions and has been found to be overexpressed in a variety of cancers. Overexpression of c-Jun and c-Fos strongly enhances STAT3-driven gene transactivation [64].

Two other factors implicated in cell growth and apoptosis and modulated by STAT3 are PLK-1 and Pim 1 and 2. Serine/threonine-protein kinase PLK1, also known as polo-like kinase 1 (PLK-1), is an enzyme that in humans promotes in cells the G2/M transition. It is considered a proto-oncogene, whose overexpression is often observed in tumor cells. The oncogenic properties of PLK1 are believed to be due to its role in driving cell cycle progression. In addition, PLK1 can inhibit the transactivation and pro-apoptotic functions of p53. STAT3 and PLK1 control each other's transcription in a positive feedback loop that contributes to the development of cancer such as esophageal squamous cell carcinoma [65]. The PIM family of serine/threonine kinases possess weak oncogenic abilities, but enhance other genes or chemical carcinogens to induce tumors. Pim-1 and Pim-2 are targets for STAT3 signal and the expression of a kinase-defective Pim-1 mutant attenuated STAT3-mediated cell proliferation. Furthermore, constitutive expression of Pim-1 together with c-Myc fully compensated for the loss of the STAT3-mediated cell cycle progression, antiapoptosis, and Bcl-2 expression [66].

STAT3 activation upregulates factors implicated in cell motility and malignant transformation such as COOH terminal tensin-like (Cten). Tensin is a signaling molecule that binds to actin filaments and localizes to focal adhesions. Cten is an atypical tensin family member lacking the actin-binding domain that promotes colon cancer tumorigenicity and cell motility [67,68]. In addition, its overexpression correlates with increasing tumor stage in thymomas, lung tumors, and gastric tumors [69–71].

A novel pathway that contributes to tumor growth, involving fibroblast growth factor receptors (FGFRs), STAT3, and hyaluronan (HA) synthesis, has been recently described by Bohrer et al. [72]. FGFR is a transmembrane receptor tyrosine kinase that is activated by FGF. Aberrant expression of these receptors was observed in different cancers such as cholangiocarcinoma, breast cancer, and prostate cancer [73]. Activation of FGFR induces the production of IL-6 family members, which activate the STAT3 pathway. FGFR-induced STAT3 activation contributes to the synthesis of hyaluronan (HA), a glycosaminoglycan, which, interacting with cancer, promotes proliferation and migration.

Finally, STAT3 inhibits apoptosis by modulating five important factors implicated in apoptosis regulation: survivin, Bcl-xL, IAPs, Mcl-1, and p53.

Survivin is an inhibitor of apoptosis. It acts by inhibiting caspase activation, thereby leading to the negative regulation of apoptosis. Survivin protein is overexpressed in most human tumors and fetal tissues. Survivin gene has been identified as a STAT3-regulated gene in breast cancer cells [74]. Direct inhibition of activated STAT3 signaling with antisense oligonucleotides inhibits survivin expression. Moreover, the block of constitutively active STAT3 or survivin expression in breast cancer cells by antisense oligonucleotides induces apoptosis. Both constitutive STAT3 activation and elevated survivin expression occur concurrently in high-risk breast tumors that are resistant to chemotherapy with docetaxel and doxorubicin [74].

Bcl-xL is another important anti-apoptotic factor, and its expression level is upregulated by STAT3 other than by other factors such as NF-κB and tumor necrosis factor-alpha (TNF-α). Inhibition of STAT3 leads to a decrease in the expression of Bcl-xL [75]. In a study designed to investigate the potential use of RNA interference (RNAi) to block STAT3 expression and activation on human breast cancer cells, Kunigal et al. showed that the knockdown of STAT3 expression by siRNA reduced the expression of Bcl-xL and survivin.

Of interest, knockdown of STAT3 caused also a Fas and Fas-L upregulation, induction of apoptosis, and tumor suppression. This study confirms that STAT3 contributes to the inhibition of apoptosis in tumor cells by acting on the expression of anti-apoptotic factors and downregulating the extrinsic apoptotic pathway [76].

Activation of STAT3 increases the transcription and expression of the IAP (inhibitor of apoptosis proteins) family proteins and Mcl-1 [77]. IAPs are a family of functionally and structurally related proteins that act as endogenous inhibitors of programmed cell death. They play a role in oncogenesis by suppressing apoptosis. IAPs inhibitors, such as LCL161, promote cancer cell death by antagonizing IAPs. Mcl-1 (myeloid cell leukemia-1) is a protein that belongs to the Bcl-2 family. Amplification and overexpression of Mcl-1 have been reported in various human tumors, including hematological malignancies and solid tumors [78].

STAT3 is able to modulate the expression of p53, one of the most important onco-suppressor proteins, which plays a crucial role in many tumors. In fact, the p53 gene is frequently mutated in cancer, but p53 mutations generally occur at a late stage in tumor development. In contrast, in the early stage, many clinically detectable cancers have only reduced p53 expression. STAT3 seems to play an important role in reducing p53 transcription by binding to the p53 promoter. Site-specific mutation of an STAT3 DNA-binding site in the p53 promoter partially abrogates STAT3-induced p53 inhibition, and the blocking of STAT3 in cancer cells upregulates expression of p53, leading to p53-mediated tumor cell apoptosis [79].

All these data indicate the complex role of STAT3 in modulating cell growth, apoptosis, and tumorigenesis. However, many tumors express pro-survival and growth genes in a non-STAT3-dependent manner. In other tumors, STAT3 activation alone appears insufficient to induce tumorigenesis, but acts by conferring greater malignancy to neoplastic cells. This was demonstrated in two different studies. In a mouse model of tumorigenesis study, STAT3 did not alter mammary tumor initiation but affected metastatic progression. In fact, only a few mice bearing STAT3-null tumors showed lung metastasis and lung lesions compared with the wild-type cohort. Tumors from STAT3-null mice were less vascularized than tumors from the wild-type animals and showed a significant reduction of the proangiogenic factor, VEGF. Moreover, a gene expression analysis showed that STAT3-null mice expressed lower levels of both cebpd and osmr, two genes that are induced by STAT3. These genes code for C/EBPδ and the oncostatin M receptor, two factors that are known to potentiate the acute-phase response of inflammation [80].

Similar results were obtained by Babieri et al. in mice expressing the activated rat ErbB-2 (neu) but lacking STAT3 in the mammary epithelium. STAT3 was apparently not required for neu-driven mammary tumorigenesis. However, STAT3 silencing in a neu-overexpressing tumor-derived cell line completely abolished both neu-driven anchorage-independent growth and lung metastasis [81].

These data indicate that STAT3 confers tumor great malignancy by promoting tumor invasion, migration, metastasis, and angiogenesis. Moreover, it plays an important role in cancer immune escape.

4. Invasion, Metastasis, and Angiogenesis

The formation of metastases depends on the ability of cancer cells to degrade the basal membrane and extracellular matrix and to move within. STAT3 regulates both of these two processes. RhoA-Rho-associated protein kinase (ROCK) is a kinase belonging to the AGC family of serine-threonine kinases. It is involved mainly in regulating the shape and movement of cells by acting on the cytoskeleton. The ROCK-myosin axis is the most well-known mechanism of cell contractility and is the major signaling pathway that induces amoeboid movement [82,83]. Amoeboid movement has been described for many cells including cancer cells [84]. STAT3 coordinates ameboid movement through activating RhoA, and inhibition of JAK/STAT3 activity suppresses the migration of malignant cells [85]. Invasion of the extracellular matrix is a key step in tumor metastasis formation, and several pieces of evidence indicate that STAT3 plays a crucial role also in this process by regulating the matrix metalloproteinases (MMPs), a family of zinc-dependent endoproteinases whose enzymatic activity is directed against components of the extracellular matrix such as collagens, laminins, and proteoglycans. STAT3 protein directly binds to the promoter of MMP genes, upregulating their expression and favoring the invasiveness of cancer cells [86–88]. On the contrary, STAT3 knockdown reduces MMP expression and cancer cell invasiveness in nude mice [89].

After degradation of the basement membrane, tumor cells enter the circulatory or lymphatic system, forming metastasis in lymph nodes and in distant organs. STAT3 activation has an important role in protecting metastatic cells from the immune system by modulating the secretion of various inflammatory factors such as IL6 and TNFα [90] and reducing the activity of NK cells [91].

Another essential step for tumor growth and metastasis formation is angiogenesis, which is mediated by the vascular endothelial growth factor (VEGF) [92]. STAT3 is a direct transcriptional activator of the VEGF gene [93]. Moreover, STAT3 induces the expression of hypoxia-inducible factor-1α (HIF1α), which is another key mediator of angiogenesis [94]. Both STAT3 and HIF1α bind simultaneously to the VEGF promoter, leading to its maximum transcriptional activation and angiogenesis [95].

5. Immune Escape

The immune system can suppress or promote cancer depending on the type of immune cells. Innate immune cells (such as macrophages, natural killer, and dendritic cells) and cells from the adaptive immune system (Th1 cells) can destroy tumor cells when activated [96]. In contrast, tumor-associated macrophages and MDSCs promote cancer through their ability to secrete angiogenetic, metastatic, and growth factors [97,98] or by suppressing anti-tumor immunity [99]. Evidence indicates that STAT3 is involved in the immune escape of tumor cells through the suppression of anticancer immune cells and activation of cancer promoting immune cells.

T helper cells (Th cells), also known as CD4+ cells, play a central role in cancer immune response by activating antigen-specific effector cells and recruiting cells of the innate immune system such as macrophages and mast cells. Once activated, Th cells proliferate rapidly and secrete cytokines to regulate immune response [97].

Th cells are activated and develop into effector T cells by two costimulatory signals from antigen presenting cells (APCs). The first costimulatory signal is the stimulation with antigens presented by major histocompatibility complex class-II (MHC-II) molecules, expressed on the surface of APCs. The second costimulatory signal is mediated by the linkage between CD80 or CD86 on the APCs and CD28 on the T cells.

Activated STAT3 in cancer cells induces the release of factors that block APCs maturation/activation and inhibits the generation of antigen-specific T cells, causing immune tolerance [91]. Furthermore, STAT3 activation in APCs determines immune tolerance by preventing the activation of T cells. The tolerance of T cells can be reversed if activation of STAT3 signaling is blocked in APCs [100].

Naive CD4 + T cells that develop into effector T cells differentiate into five major subtypes of cells known as Th1, Th2, Threg, Th17, and Th9.

Th1 cell differentiation is induced in the presence of IL-12. Th1 cells produce the proinflammatory cytokine IFN-γ and are associated with a good prognosis in several cancer types. They are primarily responsible for activating and regulating the development and persistence of cytotoxic T cells (CTLs). In addition, Th1 cells activate APC. Stat1 is the key transcription factor for the IFN-γ-mediated gene [101], and NF-kB is involved in the activation of IL-12 transcription [102]. STAT3 activation inhibits anti-tumor immune response by antagonizing the NF-κB- and STAT1-mediated expression of anti-tumor Th1 cytokines such as IL-12 and IFNγ [103,104]. The ablation of the STAT3 gene in natural killer cells and macrophages in mice increases the expression of Th1 cytokines, leading to STAT1 upregulation and increased anti-tumor immune responses [105].

Th2 and Th17 cells are two other major T cell subsets found within the tumor microenvironment. Th2 cells produce anti-inflammatory cytokines, whereas Th17 cells produce mainly proinflammatory cytokines including IL-17 and IL-22. These two cell types appear to be overall pro-tumoral.

The Th2 anti-inflammatory cytokine IL-10 is involved in cancer development and is transcriptionally upregulated by STAT3 in tumors. IL-6 promotes STAT3 recruitment in both colon cancer cells and T cells, which, in turn, upregulates IL-10 secretion [106]. IL-10 downregulates the expression of Th1 cytokines and MHC class II antigens and blocks NF-κB activity.

IL-6, transforming growth factor-β (TGFβ), and IL-23 are involved in the activation of Th17 immune response. In contrast to the tumor suppressor role of Th1 immune response, Th17 response and IL-23 play a cancer-promoting role. STAT3 is involved in Th17 development because it acts both as the major IL-6R-dependent transcription factor in T cells and as a transcriptional activator of Il-23a that encodes part of IL-23 [107]. Furthermore, the STAT3- and Th17-dependent pathways are involved in inflammation-induced cancer by a common human commensal bacterium [108].

Treg cells are characterized by the production of major suppressive cytokines, including IL-10 and TGF-β. The presence of Treg cells in the tumor correlates mainly with a bad prognosis. Treg cells secrete IL-10 and TGF-β. TGF-β in turn induces the expression of Forkhead box P3 (FOXP3), a mediator of Treg cells that converts naive CD4+ T cells into CD4+CD25+ FOXP3+ Treg cells [109,110]. STAT3 was demonstrated to transcriptionally upregulate the expression of FOXP3 in CD4+CD25+ Treg cells by binding to the promoter of FOXP3 [111].

Finally, STAT3 signaling is required for the immunosuppressive and tumor-promoting effects of myeloid-derived suppressor cells (MDSCs), tumor-associated macrophages [112], and in expansion of T regulatory cells, which promote tumor progression by inhibiting anti-tumor immune responses mediated by Th1-type CD4+ T cells and CD8+ T cells [113–116].

6. The Dual Role of STAT3 in Cancer

Although STAT3;s role as a tumor promoter has been widely demonstrated, several studies indicate that STAT3, under specific conditions, can act as a tumor suppressor factor in many different tumors (Table 1).

6.1. Glioblastomas

The first dual function of STAT3 was described in the nervous system during the development of cells along the glial lineage. The different cell types of the mammalian central nervous system (neurons, astrocytes, and oligodendrocytes) are derived from common neural precursor cells (NPCs) [117,118]. In the early stages of brain development, NPCs proliferate and generate neurons and glial cells in a sequential manner [119]. The differentiation of neural stem cells into astrocytes depends on the activation of two different factors: (i) the cytokine receptor leukemia inhibitory factor receptor β (LIFRβ—a receptor implicated in several cell differentiation processes) and (ii) STAT3 [120,121]. These factors induce the blocking of proliferative processes followed by cell differentiation. However, STAT3 is implicated also in the renewal capacity of NPCs, supporting the proliferation of these cells [122]. This dual role of STAT3 has raised questions about the relevance of STAT3 in glial tumorigenesis. In fact, STAT3 activation has been described in human gliomas [123]. At the same time, other authors have provided evidence of the absence of STAT3 activity in a large percentage of gliomas [124,125].

Table 1. Tumor suppressor and tumor promoter mechanisms of Signal transducer and activator of transcription 3 (STAT3) in different malignancies.

Tumor	Tumor Background	Tumor Suppressor Activity	Tumor Promoter Activity
Glioma	Normal PTEN expression	LIFRβ activation of STAT3 that inhibits IL-8 expression.	
	EGFRvIII expression		Formation of oncogenic EGFRvIII-STAT3 complex.
Prostate cancer		STAT3 activation of ARF-p53 pathway.	
Lung Adenocarcinomas	KRAS mutations	STAT3 dependent sequestration of NF-kB in the cytoplasm and inhibition of IL-8 expression.	
	EGFR mutations		STAT3 tumor promoter activity.
Colorectal cancer		STAT3 mediated degradation of Snail-1 and inhibition of epithelial-mesenchymal transition. STAT3β expression.	STAT3 mutations associated with APC mutations.
Thyroid cancer		STAT3 dependent suppression of HIF1α and aerobic glycolysis under hypoxic stress. STAT3 dependent activation of IGFBP7.	
Breast cancer	Primary tumor		STAT3 tumor promoter activity.
	Metastasis	STAT3 activation of dormancy-associated genes under hypoxic condition.	
Esophageal squamous cell carcinoma			STAT3α expression.
Esophageal squamous cell carcinoma with longer overall survival		STAT3β expression.	

These opposite functions could be determined by the interaction of STAT3 with different signals in the oncogenic environment. For example, STAT3 has a tumor suppressive activity in glioma cells with an intact PTEN function and an oncogenic role in EGFRvIII-expressing gliomas and glioblastomas [126].

PTEN is an oncosuppressor protein, and STAT3 plays a key role in the PTEN pathway to suppress malignant transformation of astrocytes. In this pathway, IL8 plays a critical role. IL8 is not expressed in normal brain but is expressed in human glioblastoma tumors where it promotes proliferation and invasiveness. Gene-profiling analyses of glioblastoma cells have shown that the IL8 gene is directly repressed by STAT3 [126].

In cells with intact PTEN function, the protein kinase AKT is directly inhibited by PTEN. AKT is an inhibitor of FOXO3, a transcription factor involved in several biological functions including apoptosis. This transcription factor activates the transcription of the LIFRβ gene, which, in turn, activates STAT3, causing the repression of the IL8 gene, thereby inhibiting glioma cell proliferation and invasiveness. Conversely, in PTEN-deficient tumor cells, Akt is constitutively active, thus inhibiting FOXO3 function and leading to LIFRβ downregulation. In this scenario, STAT3 is not active and cannot repress the IL8 gene. This leads to an overregulation of IL8, which drives glioma cell proliferation and invasiveness [126].

In glioblastoma tumors that are not deficient in PTEN, STAT3 can play an oncogenic role in response to the expression of the oncogenic truncated protein EGFRvIII [29].

EGFRvIII forms a complex with STAT3 in the nucleus, converting the tumor-suppressive form of STAT3 into an oncogenic protein [126].

These findings indicate that STAT3 plays distinct roles in cell transformation depending on the mutational background of the tumor.

6.2. Prostate Cancer

Hyperactivation of the IL-6/STAT3 axis is frequently observed in prostate cancer cell lines. However, the blocking of the IL-6/STAT3 axis did not result in a survival advantage in patients with advanced prostate cancer. About 70% of metastatic prostate cancers show deletions or mutations in the PTEN gene. The loss of PTEN function in these tumors is associated with the activation of the oncosuppressor ARF–p53 pathway, which likely acts as a oncosuppressor compensatory mechanism [127]. In fact, in normal cells, ARF (p14ARF in humans; p19ARF in mice) is readily degraded, but it is stabilized in cells in which PTEN is deleted or mutated. ARF promotes the degradation of MDM2, a negative regulator of p53. Thus, loss of PTEN causes the activation of p53, which induces cycle arrest and apoptosis or senescence. It has been observed that genetic inactivation of STAT3 in a PTEN-deficient prostate cancer mouse model accelerates cancer progression by down-modulating ARF. In accordance with this experimental data, loss of STAT3 and ARF expression in patients with prostate cancer correlated with increased risk of disease recurrence and metastasis. ARF is a direct target of STAT3, which is required for the activation of the ARF-p53 pathway. Hence, STAT3 acts as an indirect tumor suppressor in prostate cancer, and the loss of STAT3 signaling disrupts the ARF–Mdm2–p53 tumor suppressor axis, promoting the progression of cancer [127].

6.3. Lung Cancer

The genetic background of tumor plays an important role in determining the activity of STAT3 as a pro-oncogenic or anti-oncogenic factor also in lung cancer. STAT3 is considered to play an oncogenic role in lung cancer [128,129]; consequently, STAT3 has been considered an important target for therapeutic intervention. The most frequent genetic alterations in lung adenocarcinomas are missense mutations and amplifications of the Kirsten rat sarcoma viral oncogene (KRAS) and EGFR. STAT3 is considered to play a tumor-promoting role in EGFR mutant non-small-cell lung cancer (NSCLC). In contrast, STAT3 plays an unexpected tumor suppressive role in KRAS mutant lung adenocarcinoma. This depends on the ability of STAT3 to control NF-κB-induced IL-8 expression by sequestering NF-κB within the cytoplasm, thereby inhibiting IL-8-mediated MDSCs and tumor vascularization, and hence tumor progression. Therefore, KRAS mutations in lung adenocarcinoma should be carefully considered in the therapeutic approach with STAT3 inhibitors [130].

6.4. Colorectal Cancer

Another case in which the oncosuppressor role of STAT3 is not dependent on its transcription function but the interaction with specific factors is colorectal cancer. The development of human colon adenomas is initially induced by mutations in tumor suppressor protein APC (adenomatous polyposis coli). APC is involved in the β-catenin/Wnt signaling pathway. The transition from adenoma to adenocarcinoma occurs only if two other oncosuppressor genes are mutated: SMAD4 (small mother against decapentaplegic) and EPHB3 (ephrin type-B receptor 3) genes. SMAD4 is a transcription factor that acts as a mediator of TGF-β signal transduction. When the structure of SMAD4 is altered, the expression of the genes involved in cell growth is deregulated, and cells proliferate without any inhibition [131]. EPHB3 is a protein implicated in several tumor-suppressive signaling pathways [132]. It has been observed that STAT3 knockdown causes in Apcmin/+ mice the transition from adenoma to carcinoma with the acquisition of an invasive phenotype. Therefore, in Apcmin/+ mice, STAT3 acts as a tumor suppressor similarly to SMAD4 and EPHB3. In this case, the tumor suppressor activity of STAT3 depends on its ability to suppress the expression of Snail-1, a zinc finger protein that acts as an inductor of the epithelial-mesenchymal transition. This is a process in which the epithelial tumor cells lose their epithelial phenotype and acquire a mesenchymal phenotype displaying reduced intercellular interactions and increased motility and invasiveness. Mechanistically, STAT3 facilitates glycogen synthase kinase (GSK) 3β-mediated degradation of SNAI by regulating the phosphorylation of GSK3β. [133].

6.5. Thyroid Cancer

STAT3 is a modulator of glycolysis and mitochondrial respiration, and constitutive activation of STAT3 induces aerobic glycolysis and downregulates mitochondrial activity both in primary fibroblasts and in STAT3-dependent tumor cell lines [134]. STAT3 induced aerobic glycolysis is dependent on the transcriptional and posttranscriptional induction of HIF1α. However, in thyroid cancer-derived cell lines, the absence of STAT3 led to a growth advantage under hypoxic stress because of a metabolic reprogramming characterized by increased glucose consumption, lactate production, and a reduced rate of oxygen consumption. Moreover, knockdown of STAT3 increases the expression of genes encoding glycolytic enzymes such as PDK1, GLUT3, and HK2. These effects seem to depend on an increased expression of HIF1α and its transcriptional targets, suggesting a paradoxical role in thyroid cancer for STAT3 as a negative regulator of HIF1α and aerobic glycolysis under hypoxic stress [135].

These data are consistent with the observation of high levels of activated phospho-STAT3 (Y705) in benign follicular thyroid adenomas and human primary papillary thyroid carcinoma without evidence of distant metastasis. Furthermore, a positive correlation between the expression of activated phospho-STAT3 (Y705) and smaller human primary papillary thyroid carcinoma tumor sizes has been observed. Finally, STAT3 knockdown resulted in the downregulation of multiple transcripts, including the tumor suppressor insulin-like growth factor binding protein 7 (IGFBP7) [135]. All these data strongly suggest a growth-suppressive role for STAT3 in human primary papillary thyroid carcinoma.

6.6. Breast Cancer

STAT3 is a regulator of epithelial cell death during mammary gland involution. Removal of the suckling stimulus and the consequent milk stasis lead to an increased expression of leukemia inhibitory factor (LIF), which is the initial activator of STAT3 during gland involution [136]. Activated STAT3 does not induce apoptosis in epithelial cells but activates a lysosome-mediated programmed cell death pathway (LM-PCD) by upregulating the expression of cathepsins B and L [137,138].

In contrast to this function, STAT3 acts as a promoter of survival in breast cancer cells [139,140]. In fact, several works have shown that STAT3 is constitutively active in invasive breast cancer biopsies, but not in biopsies from benign breast tumors, indicating that this transcription factor is involved in breast cancer progression [139,141]. Thus, STAT3 seems to have two opposite functions in normal epithelial cells of mammary gland and in breast cancer cells [140,142].

However, an analysis of 346 node-negative breast cancer showed that the expression of nuclear activated phospho-STAT3 (Y705) in the biopsied samples was correlated with a significant improved short-term (five years) and long-term (20-year) survival [143].

By studying the distribution of mutations in breast cancer metastases matched with their primary lesions, it was observed that mutations in synchronous metastases (mainly lymph nodal) were similar to the those of the primary breast tumor. These mutations typically interested genes frequently mutated in breast cancer, such as TP53, PIK3CA, and GATA3. In contrast, metachronous distant metastases had additional mutations, in particular mutations that caused loss of JAK2 and STAT3 functions [144]. This suggests that the JAK2/STAT3 system may act as a tumor suppressor in these metastases, explaining the improved short- and long-term survival in patients expressing nuclear activated phospho-STAT3 (Y705) in the biopsied samples. There are two possible explanations for this apparent paradox: (i) The JAK/STAT pathway's inactivation could inhibit the immune response against metastases. This hypothesis is supported by the recent observation that JAK2 truncating mutations resulted in a lack of response to interferon gamma, including insensitivity to its antiproliferative effects on cancer cells [145]. (ii) JAK/STAT inactivation could induce cancer cells to exit from a quiescent state (G0/G1). In fact, under hypoxic conditions in the osteoblast niche, LIF receptor/STAT3 signaling confers a dormancy phenotype to disseminated breast cancer cells. Loss of the LIFR or STAT3 enables dormant breast cancer cells to downregulate dormancy-, quiescence-, and cancer stem cell-associated genes and to proliferate [146]. Moreover, STAT3 was identified as a dormancy-associated gene in estrogen receptor, ER-positive breast cancer cells [147].

6.7. Head and Neck Squamous Cell Cancers

Constitutive activation of STAT3 has been described in head and neck squamous cell cancers (HNSCCs), where it is involved in deregulation of the cell cycle, increased cell growth, and inhibition of apoptosis [147–150]. In these cells, constitutive activation of STAT3 is dependent on TGF alpha-induced activation of the EGFR [150]. However, although the anti-apoptotic and cell growth inducing activity of STAT3 has been well documented, high nuclear STAT3 expression levels were associated with a favorable outcome in HNSCCs. Survival analysis in a cohort of 102 patients with HNSCCs showed that high nuclear STAT3 expression was associated with longer progression-free survival. Moreover, a multivariable analysis including different prognostic variables including gender, TNM stage, tumor grade, and tumor site showed that only STAT3 was significant, revealing a lower risk of progression and death for patients with high nuclear STAT3-expressing tumors. Since in the cohorts of this study, the majority of tumors were of advanced stage, it is possible to suppose that in an advanced stage, tumors become independent of STAT3-mediated pro-oncogenic signaling. In this case, STAT3 could function as a tumor suppressor as described in breast cancer or in glioblastoma.

6.8. STAT3 Isoforms

STAT3 has four isoforms: STAT3α, STAT3β, STAT3γ, and STAT3δ. STAT3α, STAT3β are generated by alternative splicing and differ structurally in their C-terminal transactivation domains because the β isoform lacks the C-terminal transactivation domain [151]. STAT3γ is derived from STAT3α by proteolysis during granulocytic differentiation. Like STAT β, it lacks the C-terminal transactivating portion. STAT3δ is an isoform expressed during the early stage of granulocytic differentiation.

STAT3α is the full-length version of STAT3. It regulates essential STAT3 target genes involved in cell proliferation, migration, and survival, and the oncogenic functions of STAT3 have also been associated with constitutively active STAT3α [152].

Conversely, STAT3β inhibits cancer progression acting as a repressor of STAT3. Several studies have shown that STAT3β can inhibit cell proliferation by impairing STAT3α-driven activation of Bcl-xL, p21, and cyclin D1 [153–156]. Moreover, STAT3β upregulates TRAIL receptor 2 and Fas expression favoring apoptosis in tumor cells [157,158]. STAT3β also plays a role in chemoresistance; in fact, administration of the STAT3β plasmid reduces chemoresistance and inhibits invasion of gastric cancer cells [159]. STAT3β has been shown to cause tumor regression in vivo [160,161], promote programmed cell death in myeloma cells [33], and impaired proliferation of melanoma, ovarian, and breast cancer cells [162–164]. STAT3β expression is an independent protective prognostic marker in patients with esophageal squamous cell carcinoma, which strongly correlated with longer overall survival and recurrence-free survival [165]. Therefore, the evaluation of both STAT3 isoforms in cancer might have an important role to understand the prognosis and the usefulness of a treatment with STAT3 inhibitors.

7. Treatment of Cancer with STAT3 Inhibitors

Since constitutive activation of STAT3 plays a central role in many tumors, small and non-small-molecular STAT3 inhibitors have been developed as novel cancer therapeutics [166–168]. Currently, there are two different classes of STAT3 inhibitor compounds: (i) direct inhibitors, which directly block different domains of STAT3; (ii) indirect inhibitors, which inhibit the signal transduction functions of STAT3, for example, by inhibiting the function of JAKs, Src, and Abl. The former inhibitors can be divided into four types according to the different STAT3 target domains (SH2, DBD, ND, and TAD). Figure 3 shows the most important STAT3 inhibitors targeting the STAT3 structural domains or the signal transduction functions of STAT3.

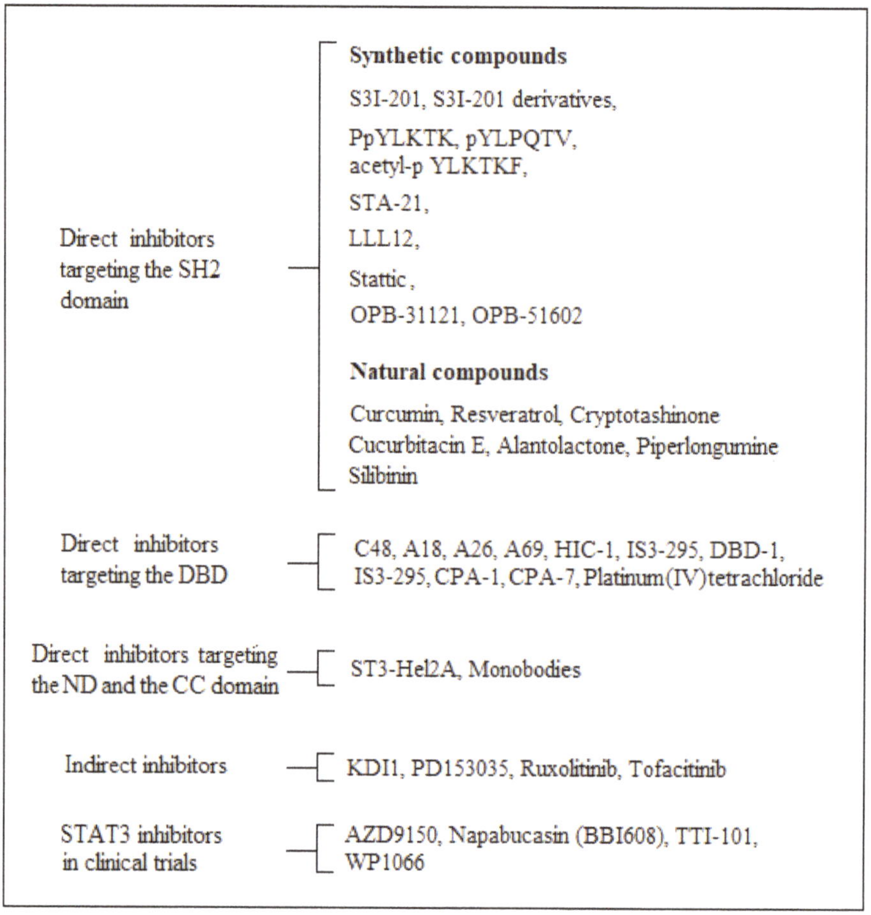

Figure 3. STAT3 inhibitors.

7.1. *Inhibitors Targeting the SH2 Domain*

SH2 domain interactions are critical for molecular activation and nuclear accumulation of phosphorylated STAT3 dimers to drive transcription. The SH2 domain mediates STAT3 dimerization via intermolecular phosphorylated tyrosine–SH2 interactions. Therefore, molecules able to block the SH2 domain of STAT3 have been considered for the treatment of different tumors. Many STAT3 inhibitors that block the SH2 domain were identified through a structure-activity relationship (SAR) study based on the previously identified inhibitor S3I-201. The key structural feature of these inhibitors is a salicylic acid moiety, which, by acting as a phosphotyrosine mimetic, facilitates binding to the STAT3 SH2 domain. S3I-201 and derivatives disrupt active STAT3:STAT3 dimers suppressing the expression levels of STAT3 target genes, such as Bcl-xL, survivin, cyclin D1, and MMP-9 and causing inhibition of cell growth, migration, and invasion of human cancer cells [169–171]. Other molecules able to block the SH2 domain of STAT3 include: (i) the STAT3 SH2 domain-binding peptides PpYLKTK [172], pYLPQTV [173], and acetyl-pYLKTKF [174]; (ii) STA-21 and its structurally optimized analog LLL12 [175]; (iii) stattic, a nonpeptidic small molecule [176].

OPB-31121 and OPB-51602 are two STAT3 inhibitors that interact with the SH2 domain with high affinity. Cell culture assays showed that OPB-31121 and OPB-51602 inhibited the phosphorylation of both Tyr705 and Ser727. OPB-31121 is a STAT3 inhibitor with potent antitumor effects against various human liver cancer cell lines [177]. In addition, OPB-31121 has shown the ability to decrease cell proliferation in both gastric cancer cells and in a xenograft model, induce apoptosis of gastric cancer cells, and inhibit the expression of antiapoptotic proteins [178]. Based on these findings, OPB-31121 has entered phase 1 and 2 clinical studies against hematopoietic and solid tumors, including hepatocellular carcinoma (HCC). However, in a phase 1 and pharmacological trial, OPB-31121 demonstrated insufficient antitumor activity for HCC and peripheral nervous system-related toxicities [179].

Another group of inhibitors targeting the SH2 domain of STAT3 are derivatives from natural compounds. Curcumin, a naturally-derived phytochemical from plants, has been shown to inhibit STAT3 phosphorylation and block the STAT3 signaling pathway in various tumor cell lines [180–182]. The stilbene compound resveratrol [183] and the natural compound cryptotanshinone, an extract from Salvia miltiorrhiza Bunge [184], are two other essential natural inhibitors of STAT3. Additionally, cucurbitacin E [185], alantolactone [186], piperlongumine [187], and silibinin [188] are natural STAT3 inhibitors that interact with the SH2 domain.

7.2. Inhibitors Targeting the DBD

DBD interacts with the promoter of STAT3 targeting genes and has a relatively high specificity. The small molecule C48 was the first inhibitor targeting the DBD. It acts by alkylating glutathione sulfhydryl on cysteine (C468) in the STAT3 DBD region. C48 blocks the accumulation of activated STAT3 in the nucleus in tumor cell lines that overexpress active STAT3, leading to impressive inhibition of tumor growth in mouse models [189]. Using an in silico screening approach, three STAT3 inhibitor, denominated A18, A26, and A69, targeting the DBD of STAT3, were identified. These compounds were found to inhibit STAT3-specific DNA binding activity, suppress the proliferation of cancer cells harboring aberrant STAT3 signaling, inhibit the migration and invasion of cancer cells, and inhibit the STAT3-dependent expression of downstream targets by blocking the binding of STAT3 to the promoter regions of responsive genes in cells. In addition, A18 was able to reduce tumor growth in a mouse xenograft model of lung cancer with little effect on body weight [190]. Other compounds targeting the DBD of STAT3 include HIC-1 [191] and DBD-1 [192].

Finally, several platinum compounds, including IS3-295, CPA-1, CPA-7, and platinum (IV) tetrachloride, have shown the ability to inhibit the STAT3 DBD, suggesting potential new applications for platinum complexes as modulators of the STAT3 pathway for cancer therapy [193,194].

7.3. Inhibitors Targeting the ND and the CC Domain

The N-terminal domain plays an important role in the interaction of dimers of STAT-3 and in the attachment of STAT-3 dimers to the sites of DNA. Therefore, in the last few years, compounds able to interact and block this domain have been developed. ST3-H2A2 is a highly selective STAT3 ND inhibitor. This compound induces apoptotic death in cancer cells associated with potent activation of proapoptotic genes. In particular, it activates the expression of the proapoptotic gene CHOP, causing apoptosis in tumor cells [195].

ST3-Hel2A is a synthetic peptide that selectively targets the N-terminal domain of STAT-3 and inhibits dimerization. In preclinical trials, ST3-Hel2A was found to inhibit the STAT-3 N-terminal domain, resulting in the blockage of STAT-3 dimerization and apoptosis in prostate cancer cell lines.

Synthetic antibody mimetics, termed monobodies, have been developed to interfere with STAT3 N-terminal and coiled-coil domain signaling. These monobodies are highly selective for STAT3 and bind to STAT3 with nanomolar affinity. The crystal structure of STAT3 in complex with monobody MS3-6 reveals bending of the coiled-coil domain, resulting in diminished DNA binding and nuclear translocation. MS3-6 expression strongly inhibits STAT3-dependent transcriptional activation and disrupts STAT3 interaction with the IL-22 receptor [196].

7.4. Indirect Inhibitors

STAT3 indirect inhibitors prevent STAT3 activation by inhibiting the tyrosine kinase events at the receptor level. Receptor-associated and non-receptor tyrosine kinases are critical upstream regulators of STAT3 activation, so targeting these kinases has attractive potential to block STAT3 activation. Molecules that indirectly target upstream activators of STAT3 include KDI1, a short peptide that specifically binds to the intracellular domain of the EGFR [197], PD153035, an RTK inhibitor [198], and several JAK and Src kinase inhibitors [199–205].

Numerous JAK2 inhibitors have been developed that are able to inhibit the JAK2/STAT signaling pathways, and ruxolitinib and tofacitinib have been approved by the FDA for the treatment of myelofibrosis and rheumatoid arthritis. However, apart from myelofibrosis, no satisfactory results have been obtained with JAK2 inhibitors in cancer patients, and clinical trials with JAK1/2 inhibitors and Src inhibitors have shown limited efficacy or excessive toxicity in advanced solid tumors [206].

7.5. Clinical Trials

Despite the large amount of STAT3 inhibitors studied in vitro, only a few compounds have entered clinical trials. AZD9150, napabucasin (BBI608), TTI-101, and WP1066 are the most studied compounds in different trials.

AZD9150 is a 16 oligonucleotide antisense molecule (ASO) targeting the 3′ untranslated part of STAT3 [207]. Clinical trials with AZD9150 in patients with advanced/metastatic carcinoma have shown the safety and anti-tumor activity of this compound. When used in combination with durvalumab, AZD9150 seems able to control advanced pancreatic, lung, and colorectal cancer.

Napabucasin (BBI608) is a newly-developed small molecule inhibitor of STAT3 [208] currently studied in several clinical trials in patients with metastatic pancreatic adenocarcinoma, colorectal cancer, gastric cancer, glioblastoma, hematologic malignancy, malignant pleural mesothelioma, and hepatocellular carcinoma.

TTI-101 is an orally bioavailable, binaphthol-sulfonamide-based inhibitor of STAT3 that specifically targets and binds to the phosphotyrosyl peptide binding site within the SH2 domain of STAT3 [209]. It is in a phase I trial in advanced cancers.

WP1066 has been designed to target the STAT3 pathway in tumor cells [210], and phase I trial studies have been started in patients with recurrent, or refractory, or progressive malignant brain tumors.

8. Conclusions

STAT3 is a transcription factor that has been extensively studied over the past 20 years due to its implication in many biological processes and diseases, including cancer. It is one of the most complex transcription regulators involved in multiple cellular processes, and its hyperactivation has been reported in many types of tumors. This has led to this transcription factor being considered an important target for the development of a new class of drugs. Despite a large number of STAT3 inhibitors having been developed and studied in vitro, only a few of them are currently in clinical trials, and at present, there are no clinically approved drugs directly targeting STAT3. This is, in part, caused by the serious toxic effects observed with the use of most STAT3 inhibitors. In fact, STAT3 is involved in many biological processes, and its inhibition can induce important alterations in normal cells. Furthermore, complete inhibition of the STAT3 signaling pathway may require very high drug concentrations as compared to inhibitors of ligands or cell surface receptors. In fact, cellular signaling transmitted by the latter is always amplified downstream, while STAT3 translates the signal without amplification. Finally, several studies have raised questions and doubts about the use of STAT3 inhibitors based on the tumor suppressor role of STAT3 in several cancers. This role does not depend on the transcription function of STAT3 but on its ability to interact with other biological factors. Therefore, the goal could be the discovery of high-quality targeted selective agents against STAT3 that do not affect the functions of healthy cells and the identification of patients who are likely to receive actual benefit from the treatment.

Funding: This research received no external funding.

Institutional Review Board Statement: Not applicable.

Informed Consent Statement: Not applicable.

Conflicts of Interest: The authors declare no conflict of interest.

References

1. Wegenka, U.M.; Buschmann, J.; Lütticken, C.; Heinrich, P.C.; Horn, F. Acute-phase response factor, a nuclear factor binding to acute-phase response elements, is rapidly activated by interleukin-6 at the posttranslational level. *Mol. Cell. Biol.* **1993**, *13*, 276–288. [CrossRef]
2. Zhong, Z.; Wen, Z.; Darnell, J.E. Stat3: A STAT family member activated by tyrosine phosphorylation in response to epidermal growth factor and interleukin-6. *Science* **1994**, *264*, 95–98. [CrossRef] [PubMed]
3. Hemmann, U.; Gerhartz, C.; Heesel, B.; Sasse, J.; Kurapkat, G.; Grötzingerl, J.; Wollmert, A.; Zhong, Z.; Darnell, J.E.; Graeve, L.; et al. Differential activation of acute phase response factor Stat3 and Stat1 via the cytoplasmic domain of the interleukin 6 signal transducer gp130. II. Src homology SH2 domains define the specificity of stat factor activation. *J. Biol. Chem.* **1996**, *271*, 2999–3007. [CrossRef] [PubMed]
4. Lim, C.P.; Cao, X. Structure, function, and regulation of STAT proteins. *Mol. BioSyst.* **2006**, *2*, 536–550. [CrossRef]
5. Quesnelle, K.M.; Boehm, A.L.; Grandis, J.R. STAT-mediated EGFR signaling in cancer. *J. Cell. Biochem.* **2007**, *102*, 311–319. [CrossRef]
6. Decker, T.; Kovarik, P. Serine phosphorylation of STATs. *Oncogene* **2000**, *19*, 2628–2637. [CrossRef]
7. Ihle, J.N. The janus kinase family and signaling through members of the cytokine receptor superfamily. *Proc. Soc. Exp. Biol. Med.* **1994**, *206*, 268–272. [CrossRef]
8. Darnell, J.E.; Kerr, I.M.; Stark, G.R. Jak-STAT pathways and transcriptional activation in response to IFNs and other extracellular signaling proteins. *Science* **1994**, *264*, 1415–1421. [CrossRef]
9. Wang, X.; Lupardus, P.; LaPorte, S.L.; Garcia, K.C. Structural biology of shared cytokine receptors. *Ann. Rev. Immunol.* **2009**, *27*, 29–60. [CrossRef]
10. Seidel, H.M.; Milocco, L.H.; Lamb, P.; Darnell, J.E.; Stein, R.B.; Rosen, J. Spacing of palindromic half sites as a determinant of selective STAT (signal transducers and activators of transcription) DNA binding and transcriptional activity. *Proc. Natl. Acad. Sci. USA* **1995**, *92*, 3041–3045. [CrossRef]
11. Garbers, C.; Hermanns, H.M.; Schaper, F.; Müller-Newen, G.; Grötzinger, J.; Rose-John, S.; Scheller, J. Plasticity and cross-talk of interleukin 6-type cytokines. *Cytokine Growth Factor Rev.* **2012**, *23*, 85–97. [CrossRef] [PubMed]
12. Jones, S.A.; Jenkins, B.J. Recent insights into targeting the IL-6 cytokine family in inflammatory diseases and cancer. *Nat. Rev. Immunol.* **2018**, *18*, 773–789. [CrossRef] [PubMed]
13. Betz, U.A.K.; Bloch, W.; Van Den Broek, M.; Yoshida, K.; Taga, T.; Kishimoto, T.; Addicks, K.; Rajewsky, K.; Müller, W. Postnatally induced inactivation of gp130 in mice results in neurological, cardiac, hematopoietic, immunological, hepatic, and pulmonary defects. *J. Exp. Med.* **1998**, *188*, 1955–1965. [CrossRef] [PubMed]
14. Turkson, J.; Jove, R. STAT proteins: Novel molecular targets for cancer drug discovery. *Oncogene* **2000**, *19*, 6613–6626. [CrossRef]
15. Levy, D.E.; Inghirami, G. STAT3: STAT3: A multifaceted oncogene. *Proc. Natl. Acad. Sci. USA* **2006**, *103*, 10151–10152. [CrossRef]
16. Pensa, S.; Regis, G.; Boselli, D.; Novelli, F.; Poli, V. STAT1 and STAT3 in tumorigenesis: Two sides of the same coin? In *JAK-STAT Pathway in Disease*; Landes Bioscience Books: Austin, TX, USA, 2009; pp. 100–121.
17. Yu, C.L.; Meyer, D.J.; Campbell, G.S.; Larner, A.C.; Carter-Su, C.; Schwartz, J.; Jove, R. Enhanced DNA-binding activity of a Stat3-related protein in cells transformed by the Src oncoprotein. *Science* **1995**, *269*, 81–83. [CrossRef]

18. Cao, X.; Tay, A.; Guy, G.R.; Tan, Y.H. Activation and association of Stat3 with Src in v-Src-transformed cell lines. *Mol. Cell. Biol.* **1996**, *16*, 1595–1603. [CrossRef]
19. Rane, S.G.; Reddy, E.P. JAKs, STATs and Src kinases in hematopoiesis. *Oncogene* **2002**, *21*, 3334–3358. [CrossRef]
20. Danial, N.N.; Rothman, P. Jak-STAT signaling induced by the v-abl oncogene. *Oncogene* **2000**, *269*, 1875–1877. [CrossRef]
21. Besser, D.; Bromberg, J.F.; Darnell, J.E.; Hanafusa, H. A single amino acid substitution in the v-Eyk intracellular domain results in activation of Stat3 and enhances cellular transformation. *Mol. Cell. Biol.* **1999**, *19*, 1401–1409. [CrossRef]
22. Zong, C.S.; Zeng, L.; Jiang, Y.; Sadowski, H.B.; Wang, L.H. Stat3 plays an important role in oncogenic Ros- and insulin-like growth factor I receptor-induced anchorage-independent growth. *J. Biol. Chem.* **1998**, *273*, 28065–28072. [CrossRef] [PubMed]
23. Garcia, R.; Yu, C.L.; Hudnall, A.; Catlett, R.; Nelson, K.L.; Smithgall, T.; Fujita, D.J.; Ethier, S.P.; Jove, R. Constitutive activation of Stat3 in fibroblasts transformed by diverse oncoproteins and in breast carcinoma cells. *Cell Growth Differ.* **1997**, *8*, 1267–1276. [PubMed]
24. Wen, X.; Lin, H.H.; Shih, H.M.; Kung, H.J.; Ann, D.K. Kinase activation of the non-receptor tyrosine kinase Etk/BMX alone is sufficient to transactivate STAT-mediated gene expression in salivary and lung epithelial cells. *J. Biol. Chem.* **1999**, *274*, 38204–38210. [CrossRef] [PubMed]
25. Lund, T.C.; Prator, P.C.; Medveczky, M.M.; Medveczky, P.G. The Lck binding domain of herpesvirus saimiri tip-484 constitutively activates Lck and STAT3 in T cells. *J. Virol.* **1999**, *73*, 1689–1694. [CrossRef] [PubMed]
26. Migone, T.S.; Lin, J.X.; Cereseto, A.; Mulloy, J.C.; O'Shea, J.J.; Franchini, G.; Leonard, W.J. Constitutively activated Jak-STAT pathway in T cells transformed with HTLV-I. *Science* **1995**, *269*, 79–81. [CrossRef] [PubMed]
27. Weber-Nordt, R.; Egen, C.; Wehinger, J.; Ludwig, W.; Gouilleux-Gruart, V.; Mertelsmann, R.; Finke, J. Constitutive activation of STAT proteins in primary lymphoid and myeloid leukemia cells and in Epstein-Barr virus (EBV)-related lymphoma cell lines. *Blood* **1996**, *88*, 809–816. [CrossRef]
28. Lund, T.C.; Garcia, R.; Medveczky, M.M.; Jove, R.; Medveczky, P.G. Activation of STAT transcription factors by herpesvirus Saimiri Tip-484 requires p56lck. *J. Virol.* **1997**, *71*, 6677–6682. [CrossRef]
29. Nepomuceno, R.R.; Snow, A.L.; Beatty, P.R.; Krams, S.M.; Martinez, O.M. Constitutive activation of Jak/STAT proteins in Epstein-Barr virus-infected B-cell lines from patients with post-transplant lymphoproliferative disorder. *Transplantation* **2002**, *74*, 396–402. [CrossRef]
30. Samaan, A.; Mahana, W. Constitutive and induced activation of JAK/Stat pathway in leukemogenic and asymptomatic human T-cell lymphoptropic virus type 1 (HTLV-1) transformed rabbit cell lines. *Immunol. Lett.* **2007**, *109*, 113–119. [CrossRef]
31. Tomita, M.; Kawakami, H.; Uchihara, J.N.; Okudaira, T.; Masuda, M.; Matsuda, T.; Tanaka, Y.; Ohshiro, K.; Mori, N. Inhibition of constitutively active Jak-Stat pathway suppresses cell growth of human T-cell leukemia virus type 1-infected T-cell lines and primary adult T-cell leukemia Cells. *Retrovirology* **2006**, *3*, 22. [CrossRef]
32. Bromberg, J.F.; Wrzeszczynska, M.H.; Devgan, G.; Zhao, Y.; Pestell, R.G.; Albanese, C.; Darnell, J.E. Stat3 as an oncogene. *Cell* **1999**, *98*, 295–303. [CrossRef]
33. Catlett-Falcone, R.; Landowski, T.H.; Oshiro, M.M.; Turkson, J.; Levitzki, A.; Savino, R.; Ciliberto, G.; Moscinski, L.; Fernández-Luna, J.L.; Nuñez, G.; et al. Constitutive activation of Stat3 signaling confers resistance to apoptosis in human U266 myeloma cells. *Immunity* **1999**, *10*, 105–115. [CrossRef]
34. Lou, W.; Ni, Z.; Dyer, K.; Tweardy, D.J.; Gao, A.C. Interleukin-6 induces prostate cancer cell growth accompanied by activation of stat3 signaling pathway. *Prostate* **2000**, *42*, 239–242. [CrossRef]
35. Niu, G.; Bowman, T.; Huang, M.; Shivers, S.; Reintgen, D.; Daud, A.; Chang, A.; Kraker, A.; Jove, R.; Yu, H. Roles of activated Src and Stat3 signaling in melanoma tumor cell growth. *Oncogene* **2002**, *21*, 7001–7007. [CrossRef] [PubMed]
36. Chan, K.S.; Carbajal, S.; Kiguchi, K.; Clifford, J.; Sano, S.; DiGiovanni, J. Epidermal growth factor receptor-mediated activation of Stat3 during multistage skin carcinogenesis. *Cancer Res.* **2004**, *64*, 2382–2389. [CrossRef] [PubMed]
37. Song, J.I.; Grandis, J.R. STAT signaling in head and neck cancer. *Oncogene* **2000**, *19*, 2489–2495. [CrossRef]
38. Song, L.; Turkson, J.; Karras, J.G.; Jove, R.; Haura, E.B. Activation of Stat3 by receptor tyrosine kinases and cytokines regulates survival in human non-small cell carcinoma cells. *Oncogene* **2003**, *22*, 4150–4165. [CrossRef]
39. Grandis, J.R.; Zeng, Q.; Drenning, S.D. Epidermal growth factor receptor—Mediated stat3 signaling blocks apoptosis in head and neck cancer. *Laryngoscope* **2000**, *119*, 868–874. [CrossRef]
40. Trevino, J.G.; Gray, M.J.; Nawrocki, S.T.; Summy, J.M.; Lesslie, D.P.; Evans, D.B.; Sawyer, T.K.; Shakespeare, W.C.; Watowich, S.S.; Chiao, P.J.; et al. Src activation of Stat3 is an independent requirement from NF-kappaB activation for constitutive IL-8 expression in human pancreatic adenocarcinoma cells. *Angiogenesis* **2006**, *9*, 101–110. [CrossRef]
41. Pilati, C.; Amessou, M.; Bihl, M.P.; Balabaud, C.; Van Nhieu, J.T.; Paradis, V.; Nault, J.C.; Izard, T.; Bioulac-Sage, P.; Couchy, G.; et al. Somatic mutations activating STAT3 in human inflammatory hepatocellular adenomas. *J. Exp. Med.* **2011**, *206*, 1359–1366. [CrossRef]
42. Koskela, H.L.M.; Eldfors, S.; Ellonen, P.; van Adrichem, A.J.; Kuusanmäki, H.; Andersson, E.I.; Lagström, S.; Clemente, M.J.; Olson, T.; Jalkanen, S.E.; et al. Somatic STAT3 mutations in large granular lymphocytic leukemia. *N. Engl. J. Med.* **2012**, *366*, 1905–1913. [CrossRef] [PubMed]
43. Giraud, S.; Hurlstone, A.; Avril, S.; Coqueret, O. Implication of BRG1 and cdk9 in the STAT3-mediated activation of the p21waf1. *Oncogene* **2004**, *23*, 7391–7398. [CrossRef] [PubMed]

44. Giraud, S.; Bienvenu, F.; Avril, S.; Gascan, H.; Heery, D.M.; Coqueret, O. Functional interaction of STAT3 transcription factor with the coactivator NcoA/SRC1a. *J. Biol. Chem.* **2002**, *277*, 8004–8011. [CrossRef] [PubMed]
45. Hazan-Halevy, I.; Harris, D.; Liu, Z.; Liu, J.; Li, P.; Chen, X.; Shanker, S.; Ferrajoli, A.; Keating, M.J.; Estrov, Z. STAT3 is constitutively phosphorylated on serine 727 residues, binds DNA, and activates transcription in CLL cells. *Blood* **2010**, *115*, 2852–2863. [CrossRef] [PubMed]
46. Gough, D.J.; Corlett, A.; Schlessinger, K.; Wegrzyn, J.; Larner, A.C.; Levy, D.E. Mitochondrial STAT3 supports Ras-dependent oncogenic transformation. *Science* **2009**, *324*, 1713–1716. [CrossRef]
47. Sellier, H.; Rébillard, A.; Guette, C.; Barré, B.; Coqueret, O. How should we define STAT3 as an oncogene and as a potential target for therapy? *JAK-STAT* **2013**, *2*, e24716. [CrossRef]
48. Lee, H.; Zhang, P.; Herrmann, A.; Yang, C.; Xin, H.; Wang, Z.; Hoon, D.S.B.; Forman, S.J.; Jove, R.; Riggs, A.D.; et al. Acetylated STAT3 is crucial for methylation of tumor-suppressor gene promoters and inhibition by resveratrol results in demethylation. *Proc. Natl. Acad. Sci. USA* **2012**, *109*, 7765–7769. [CrossRef]
49. Ray, S.; Boldogh, I.; Brasier, A.R. STAT3 NH2-terminal acetylation is activated by the hepatic acute-phase response and required for IL-6 induction of angiotensinogen. *Gastroenterology* **2005**, *129*, 1616–1632. [CrossRef]
50. Dasgupta, M.; Dermawan, J.K.; Willard, B.; Stark, G.R. STAT3-driven transcription depends upon the dimethylation of K49 by EZH2. *Proc. Natl. Acad. Sci USA* **2015**, *112*, 3985–3990. [CrossRef]
51. Ray, S.; Zhao, Y.; Jamaluddin, M.; Edeh, C.B.; Lee, C.; Brasier, A.R. Inducible STAT3 NH2 terminal mono-ubiquitination promotes BRD4 complex formation to regulate apoptosis. *Cell Signal.* **2014**, *26*, 1445–1455. [CrossRef]
52. Yang, J.; Huang, J.; Dasgupta, M.; Sears, N.; Miyagi, M.; Wang, B.; Chance, M.R.; Chen, X.; Du, Y.; Wang, Y.; et al. Reversible methylation of promoter-bound STAT3 by histone-modifying enzymes. *Proc. Natl. Acad. Sci. USA* **2010**, *107*, 21499–21504. [CrossRef] [PubMed]
53. Kim, E.; Kim, M.; Woo, D.H.; Shin, Y.; Shin, J.; Chang, N.; Oh, Y.T.; Kim, H.; Rheey, J.; Nakano, I.; et al. Phosphorylation of EZH2 activates STAT3 signaling via STAT3 methylation and promotes tumorigenicity of glioblastoma stem-like cells. *Cancer Cell.* **2013**, *23*, 839–852. [CrossRef] [PubMed]
54. Sun, J.; Du, Y.; Song, Q.; Nan, J.; Guan, P.; Guo, J.; Wang, X.; Yang, J.; Zhao, C. E2F is required for STAT3-mediated upregulation of cyclin B1 and Cdc2 expressions and contributes to G2-M phase transition. *Acta Biochim. Biophys. Sin.* **2019**, *51*, 313–322. [CrossRef] [PubMed]
55. Leslie, K.; Lang, C.; Devgan, G.; Azare, J.; Berishaj, M.; Gerald, W.; Kim, Y.B.; Paz, K.; Darnell, J.E.; Albanese, C.; et al. Cyclin D1 is transcriptionally regulated by and required for transformation by activated signal transducer and activator of transcription 3. *Cancer Res.* **2006**, *66*, 2544–2552. [CrossRef] [PubMed]
56. Alle, K.M.; Henshall, S.M.; Field, A.S.; Sutherland, R.L. Cyclin D1 protein is overexpressed in hyperplasia and intraductal carcinoma of the breast. *Clin. Cancer Res.* **1998**, *4*, 847–854. [PubMed]
57. Wang, T.C.; Cardiff, R.D.; Zukerberg, L.; Lees, E.; Arnold, A.; Schmidt, E.V. Mammary hyperplasia and carcinoma in MMTV-cyclin D1 transgenic mice. *Nature* **1994**, *369*, 669–671. [CrossRef]
58. Robles, A.I.; Rodriguez-Puebla, M.L.; Glick, A.B.; Trempus, C.; Hansen, L.; Sicinski, P.; Tennant, R.W.; Weinberg, R.A.; Yuspa, S.H.; Conti, C.J. Reduced skin tumor development in cyclin D1-deficient mice highlights the oncogenic ras pathway in vivo. *Genes Dev.* **1998**, *12*, 2469–2474. [CrossRef]
59. Yu, Q.; Geng, Y.; Sicinski, P. Specific protection against breast cancers by cyclin D1 ablation. *Nature* **2001**, *411*, 1017–1021. [CrossRef]
60. Bowman, T.; Broome, M.A.; Sinibaldi, D.; Wharton, W.; Pledger, W.J.; Sedivy, J.M.; Irby, R.; Yeatman, T.; Courtneidge, S.A.; Jove, R. Stat3-mediated Myc expression is required for Src transformation and PDGF-induced mitogenesis. *Proc. Natl. Acad. Sci. USA* **2001**, *98*, 7319–7324. [CrossRef]
61. Zhang, Y.; Du, X.; Wang, C.; Lin, D.; Ruan, X.; Feng, Y.; Huo, Y.; Peng, H.; Cui, J.; Zhang, T.; et al. Reciprocal activation between PLK1 and Stat3 contributes to survival and proliferation of esophageal cancer cells. *Gastroenterology* **2012**, *142*, 521–530. [CrossRef]
62. Vogt, P.K. Fortuitous convergences: The beginnings of JUN. *Nat. Rev. Cancer* **2002**, *2*, 465–469. [CrossRef] [PubMed]
63. Parri, E.; Kuusanmäki, H.; van Adrichem, A.J.; Kaustio, M.; Wennerberg, K. Identification of novel regulators of STAT3 activity. *PLoS ONE* **2020**, *142*, 521–530. [CrossRef] [PubMed]
64. Schuringa, J.J.; Timmer, H.; Luttickhuizen, D.; Vellenga, E.; Kruijer, W. c-Jun and c-Fos cooperate with STAT3 in IL-6-induced transactivation of the IL-6 respone element (IRE). *Cytokine* **2001**, *14*, 78–87. [CrossRef] [PubMed]
65. Liu, X.; Erikson, R.L. Polo-like kinase (Plk)1 depletion induces apoptosis in cancer cells. *Proc. Natl. Acad. Sci. USA* **2003**, *100*, 5789–5794. [CrossRef]
66. Shirogane, T.; Fukada, T.; Muller, J.M.M.; Shima, D.T.; Hibi, M.; Hirano, T. Synergistic roles for Pim-1 and c-Myc in STAT3-mediated cell cycle progression and antiapoptosis. *Immunity* **1999**, *11*, 709–719. [CrossRef]
67. Albasri, A.; Seth, R.; Jackson, D.; Benhasouna, A.; Crook, S.; Nateri, A.S.; Chapman, R.; Ilyas, M. C-terminal Tensin-like (CTEN) is an oncogene which alters cell motility possibly through repression of E-cadherin in colorectal cancer. *J. Pathol.* **2009**, *218*, 57–65. [CrossRef]
68. Liao, Y.C.; Chen, N.T.; Shih, Y.P.; Dong, Y.; Lo, S.H. Up-regulation of C-terminal tensin-like molecule promotes the tumorigenicity of colon cancer through β-catenin. *Cancer Res.* **2009**, *69*, 4563–4566. [CrossRef]

69. Sasaki, H.; Yukiue, H.; Kobayashi, Y.; Fukai, I.; Fujii, Y. Cten mRNA expression is correlated with tumor progression in thymoma. *Tumor Biol.* **2003**, *24*, 271–274. [CrossRef]
70. Sakashita, K.; Mimori, K.; Tanaka, F.; Kamohara, Y.; Inoue, H.; Sawada, T.; Hirakawa, K.; Mori, M. Prognostic relevance of Tensin4 expression in human gastric cancer. *Ann. Surg. Oncol.* **2008**, *15*, 2606–2613. [CrossRef]
71. Sasaki, H.; Moriyama, S.; Mizuno, K.; Yukiue, H.; Konishi, A.; Yano, M.; Kaji, M.; Fukai, I.; Kiriyama, M.; Yamakawa, Y.; et al. Cten mRNA expression was correlated with tumor progression in lung cancers. *Lung Cancer* **2003**, *40*, 151–155. [CrossRef]
72. Bohrer, L.R.; Chuntova, P.; Bade, L.K.; Beadnell, T.C.; Leon, R.P.; Brady, N.J.; Ryu, Y.; Goldberg, J.E.; Schmechel, S.C.; Koopmeiners, J.S.; et al. Activation of the FGFR-STAT3 pathway in breast cancer cells induces a hyaluronan-rich microenvironment that licenses tumor formation. *Cancer Res.* **2014**, *74*, 374–386. [CrossRef] [PubMed]
73. Wu, Y.M.; Su, F.; Kalyana-Sundaram, S.; Khazanov, N.; Ateeq, B.; Cao, X.; Lonigro, R.J.; Vats, P.; Wang, R.; Lin, S.F.; et al. Identification of targetable FGFR gene fusions in diverse cancers. *Cancer Discov.* **2013**, *3*, 636–647. [CrossRef] [PubMed]
74. Gritsko, T.; Williams, A.; Turkson, J.; Kaneko, S.; Bowman, T.; Huang, M.; Nam, S.; Eweis, I.; Diaz, N.; Sullivan, D.; et al. Persistent activation of Stat3 signaling induces survivin gene expression and confers resistance to apoptosis in human breast cancer cells. *Clin. Cancer Res.* **2006**, *12*, 11–19. [CrossRef] [PubMed]
75. Lee, T.-L.; Yeh, J.; Van Waes, C.; Chen, Z. Bcl-xL is regulated by NF-KappaB and STAT3 through p53 -dependent control in head and neck squamous cell carcinoma. *Cancer Res.* **2004**, *64* (Suppl. 7), 1115.
76. Kunigal, S.; Laka, S.S.; Sodadasu, P.K.; Estes, N.; Rao, J.S. Stat3-siRNA induces Fas-mediated apoptosis in vitro and in vivo in breast cancer. *Int. J. Oncol.* **2009**, *34*, 1209–1220.
77. Bhattacharya, S.; Ray, R.M.; Johnson, L.R. STAT3-mediated transcription of Bcl-2, Mcl-1 and C-IAP2 prevents apoptosis in polyamine-depleted cells. *Biochem. J.* **2005**, *392*, 335–344. [CrossRef]
78. Xiang, W.; Yang, C.Y.; Bai, L. MCL-1 inhibition in cancer treatment. *Onco. Targets. Ther.* **2018**, *11*, 7301–7314. [CrossRef]
79. Niu, G.; Wright, K.L.; Ma, Y.; Wright, G.M.; Huang, M.; Irby, R.; Briggs, J.; Karras, J.; Cress, W.D.; Pardoll, D.; et al. Role of Stat3 in Regulating p53 Expression and Function. *Mol. Cell. Biol.* **2005**, *25*, 7432–7440. [CrossRef]
80. Ranger, J.J.; Levy, D.E.; Shahalizadeh, S.; Hallett, M.; Muller, W.J. Identification of a Stat3-dependent transcription regulatory network involved in metastatic progression. *Cancer Res.* **2009**, *69*, 6823–6830. [CrossRef]
81. Barbieri, I.; Quaglino, E.; Maritano, D.; Pannellini, T.; Riera, L.; Cavallo, F.; Forni, G.; Musiani, P.; Chiarle, R.; Poli, V. Stat3 is required for anchorage-independent growth and metastasis but not for mammary tumor development downstream of the ErbB-2 oncogene. *Mol. Carcinog.* **2010**, *49*, 114–120. [CrossRef]
82. Sahai, E.; Marshall, C.J. Differing modes for tumour cell invasion have distinct requirements for Rho/ROCK signalling and extracellular proteolysis. *Nat. Cell Biol.* **2003**, *5*, 711–719. [CrossRef] [PubMed]
83. Wyckoff, J.B.; Pinner, S.E.; Gschmeissner, S.; Condeelis, J.S.; Sahai, E. ROCK- and myosin-dependent Matrix deformation enables protease-independent tumor-cell invasion in vivo. *Curr. Biol.* **2006**, *16*, 1515–1523. [CrossRef] [PubMed]
84. Friedl, P.; Alexander, S. Cancer invasion and the microenvironment: Plasticity and reciprocity. *Cell* **2011**, *147*, 992–1009. [CrossRef] [PubMed]
85. Pan, Y.R.; Chen, C.C.; Chan, Y.T.; Wang, H.J.; Chien, F.T.; Chen, Y.L.; Liu, J.L.; Yang, M.H. STAT3-coordinated migration facilitates the dissemination of diffuse large B-cell lymphomas. *Nat. Commun.* **2018**, *12*, 3696. [CrossRef] [PubMed]
86. Xie, T.X.; Wei, D.; Liu, M.; Gao, A.C.; Ali-Osman, F.; Sawaya, R.; Huang, S. Stat3 activation regulates the expression of matrix metalloproteinase-2 and tumor invasion and metastasis. *Oncogene* **2004**, *23*, 3550–3560. [CrossRef] [PubMed]
87. Dechow, T.N.; Pedranzini, L.; Leitch, A.; Leslie, K.; Gerald, W.L.; Linkov, I.; Bromberg, J.F. Requirement of matrix metalloproteinase-9 for the transformation of human mammary epithelial cells by Stat3-C. *Proc. Natl. Acad. Sci. USA* **2004**, *101*, 10602–10607. [CrossRef] [PubMed]
88. Itoh, M.; Murata, T.; Suzuki, T.; Shindoh, M.; Nakajima, K.; Imai, K.; Yoshida, K. Requirement of STAT3 activation for maximal collagenase-1 (MMP-1) induction by epidermal growth factor and malignant characteristics in T24 bladder cancer cells. *Oncogene* **2006**, *25*, 1195–1204. [CrossRef] [PubMed]
89. Li, H.; Huang, C.; Huang, K.; Wu, W.; Jiang, T.; Cao, J.; Feng, Z.; Qiu, Z. STAT3 knockdown reduces pancreatic cancer cell invasiveness and matrix metalloproteinase-7 expression in nude mice. *PLoS ONE* **2011**, *6*, e25941. [CrossRef]
90. Nguyen, D.X.; Bos, P.D.; Massagué, J. Metastasis: From dissemination to organ-specific colonization. *Nat. Rev. Cancer* **2009**, *9*, 274–284. [CrossRef]
91. Wang, T.; Niu, G.; Kortylewski, M.; Burdelya, L.; Shain, K.; Zhang, S.; Bhattacharya, R.; Gabrilovich, D.; Heller, R.; Coppola, D.; et al. Regulation of the innate and adaptive immune responses by Stat-3 signaling in tumor cells. *Nat. Med.* **2004**, *10*, 48–54. [CrossRef]
92. Grunstein, J.; Roberts, W.G.; Mathieu-Costello, O.; Hanahan, D.; Johnson, R.S. Tumor-derived expression of vascular endothelial growth factor is a critical factor in tumor expansion and vascular function. *Cancer Res.* **1999**, *59*, 1592–1598. [PubMed]
93. Chen, Z.; Zhong, C.H. STAT3: A critical transcription activator in angiogenesis. *Med. Res. Rev.* **2008**, *28*, 185–200. [CrossRef] [PubMed]
94. Semenza, G.L. Targeting HIF-1 for cancer therapy. *Nat. Rev. Cancer* **2003**, *3*, 721–732. [CrossRef] [PubMed]
95. Oh, M.K.; Park, H.J.; Kim, N.H.; Park, S.J.; Park, I.Y.; Kim, I.S. Hypoxia-inducible factor-1α enhances haptoglobin gene expression by improving binding of STAT3 to the promoter. *J. Biol. Chem.* **2011**, *286*, 8857–8865. [CrossRef]

96. Trinchieri, G. Interleukin-12 and the regulation of innate resistance and adaptive immunity. *Nat. Rev. Immunol.* **2003**, *3*, 133–146. [CrossRef]
97. Matsukawa, A.; Kudo, S.; Maeda, T.; Numata, K.; Watanabe, H.; Takeda, K.; Akira, S.; Ito, T. Stat3 in resident macrophages as a repressor protein of inflammatory response. *J. Immunol.* **2005**, *175*, 3354–3359. [CrossRef]
98. Pollard, J.W. Tumour-educated macrophages promote tumour progression and metastasis. *Nat. Rev. Cancer* **2004**, *4*, 71–78. [CrossRef]
99. Gabrilovich, D.I.; Nagaraj, S. Myeloid-derived suppressor cells as regulators of the immune system. *Nat. Rev. Immunol.* **2009**, *182*, 4499–4506. [CrossRef]
100. Cheng, F.; Wang, H.W.; Cuenca, A.; Huang, M.; Ghansah, T.; Brayer, J.; Kerr, W.G.; Takeda, K.; Akira, S.; Schoenberger, S.P.; et al. A critical role for Stat3 signaling in immune tolerance. *Immunity* **2003**, *19*, 425–436. [CrossRef]
101. Jove, R. Preface: Stat signaling. *Oncogene* **2000**, *19*, 2466–2467. [CrossRef]
102. Murphy, T.L.; Cleveland, M.G.; Kulesza, P.; Magram, J.; Murphy, K.M. Regulation of interleukin 12 p40 expression through an NF-kappa B half-site. *Mol. Cell. Biol.* **1995**, *15*, 5258–5267. [CrossRef] [PubMed]
103. Kortylewski, M.; Xin, H.; Kujawski, M.; Lee, H.; Liu, Y.; Harris, T.; Drake, C.; Pardoll, D.; Yu, H. Regulation of the IL-23 and IL-12 balance by Stat3 signaling in the tumor microenvironment. *Cancer Cell* **2009**, *15*, 114–123. [CrossRef] [PubMed]
104. Ho, H.H.; Ivashkiv, L.B. Role of STAT3 in type I interferon responses: Negative regulation of STAT1-dependent inflammatory gene activation. *J. Biol. Chem.* **2006**, *281*, 14111–14118. [CrossRef] [PubMed]
105. Takeda, K.; Clausen, B.E.; Kaisho, T.; Tsujimura, T.; Terada, N.; Förster, I.; Akira, S. Enhanced Th1 activity and development of chronic enterocolitis in mice devoid of stat3 in macrophages and neutrophils. *Immunity* **1999**, *10*, 39–49. [CrossRef]
106. Herbeuval, J.P.; Lelievre, E.; Lambert, C.; Dy, M.; Genin, C. Recruitment of STAT3 for production of IL-10 by colon carcinoma cells induced by macrophage-derived IL-6. *J. Immunol.* **2004**, *172*, 4630–4636. [CrossRef]
107. Chen, Z.; Laurence, A.; Kanno, Y.; Pacher-Zavisin, M.; Zhu, B.M.; Tato, C.; Yoshimura, A.; Hennighausen, L.; O'Shea, J.J. Selective regulatory function of Socs3 in the formation of IL-17-secreting T cells. *Proc. Natl. Acad. Sci. USA* **2006**, *103*, 8137–8142. [CrossRef]
108. Wu, S.; Rhee, K.J.; Albesiano, E.; Rabizadeh, S.; Wu, X.; Yen, H.R.; Huso, D.L.; Brancati, F.L.; Wick, E.; McAllister, F.; et al. A human colonic commensal promotes colon tumorigenesis via activation of T helper type 17 T cell responses. *Nat. Med.* **2009**, *15*, 1016–1022. [CrossRef]
109. Yu, H.; Kortylewski, M.; Pardoll, D. Crosstalk between cancer and immune cells: Role of STAT3 in the tumour microenvironment. *Nat. Rev. Immunol.* **2007**, *7*, 41–51. [CrossRef]
110. Chen, W.; Jin, W.; Hardegen, N.; Lei, K.J.; Li, L.; Marinos, N.; McGrady, G.; Wahl, S.M. Conversion of peripheral CD4+CD25- naive T cells to CD4+CD25+ regulatory T cells by TGF-beta induction of transcription factor Foxp3. *J. Exp. Med.* **2003**, *198*, 1875–1886. [CrossRef]
111. Zorn, E.; Nelson, E.A.; Mohseni, M.; Porcheray, F.; Kim, H.; Litsa, D.; Bellucci, R.; Raderschall, E.; Canning, C.; Soiffer, R.J.; et al. IL-2 regulates FOXP3 expression in human CD4+CD25+ regulatory T cells through a STAT-dependent mechanism and induces the expansion of these cells in vivo. *Blood* **2006**, *108*, 1571–1579. [CrossRef]
112. Cheng, P.; Corzo, C.A.; Luetteke, N.; Yu, B.; Nagaraj, S.; Bui, M.M.; Ortiz, M.; Nacken, W.; Sorg, C.; Vogl, T.; et al. Inhibition of dendritic cell differentiation and accumulation of myeloid-derived suppressor cells in cancer is regulated by S100A9 protein. *J. Exp. Med.* **2008**, *205*, 2235–2249. [CrossRef] [PubMed]
113. Chen, Z.; O'Shea, J.J. Th17 cells: A new fate for differentiating helper T cells. *Immunol. Res.* **2008**, *41*, 87–102. [CrossRef] [PubMed]
114. Kortylewski, M.; Kujawski, M.; Wang, T.; Wei, S.; Zhang, S.; Pilon-Thomas, S.; Niu, G.; Kay, H.; Mulé, J.; Kerr, W.G.; et al. Inhibiting Stat3 signaling in the hematopoietic system elicits multicomponent antitumor immunity. *Nat. Med.* **2005**, *11*, 1314–1321. [CrossRef] [PubMed]
115. Matsumura, Y.; Kobayashi, T.; Ichiyama, K.; Yoshida, R.; Hashimoto, M.; Takimoto, T.; Tanaka, K.; Chinen, T.; Shichita, T.; Wyss-Coray, T.; et al. Selective expansion of Foxp3-positive regulatory T cells and immunosuppression by suppressors of cytokine signaling 3-deficient dendritic cells. *J. Immunol.* **2007**, *179*, 2170–2179. [CrossRef]
116. Zou, W. Regulatory T cells, tumour immunity and immunotherapy. *Nat. Rev. Immunol.* **2006**, *6*, 295–307. [CrossRef]
117. Anderson, D.J. Stem cells and pattern formation in the nervous system: The possible versus the actual. *Neuron* **2001**, *30*, 19–35. [CrossRef]
118. Gage, F.H. Mammalian neural stem cells. *Science* **2000**, *287*, 1433–1438. [CrossRef]
119. Qian, X.; Shen, Q.; Goderie, S.K.; He, W.; Capela, A.; Davis, A.A.; Temple, S. Timing of CNS cell generation: A programmed sequence of neuron and glial cell production from isolated murine cortical stem cells. *Neuron* **2000**, *28*, 69–80. [CrossRef]
120. Ware, C.B.; Horowitz, M.C.; Renshaw, B.R.; Hunt, J.S.; Liggitt, D.; Koblar, S.A.; Gliniak, B.C.; McKenna, H.J.; Papayannopoulou, T.; Thoma, B.; et al. Targeted disruption of the low-affinity leukemia inhibitory factor receptor gene causes placental, skeletal, neural and metabolic defects and results in perinatal death. *Development* **1995**, *121*, 1283–1299.
121. Rajan, P.; McKay, R.D.G. Multiple routes to astrocytic differentiation in the CNS. *J. Neurosci.* **1998**, *18*, 3620–3629. [CrossRef]
122. Yoshimatsu, T.; Kawaguchi, D.; Oishi, K.; Takeda, K.; Akira, S.; Masuyama, N.; Gotoh, Y. Non-cell-autonomous action of STAT3 in maintenance of neural precursor cells in the mouse neocortex. *Development* **2006**, *133*, 2553–2563. [CrossRef]
123. Weissenberger, J.; Loeffler, S.; Kappeler, A.; Kopf, M.; Lukes, A.; Afanasieva, T.A.; Aguzzi, A.; Weis, J. IL-6 is required for glioma development in a mouse model. *Oncogene* **2004**, *23*, 3308–3316. [CrossRef]

124. Schaefer, L.K.; Ren, Z.; Fuller, G.N.; Schaefer, T.S. Constitutive of Stat3α in brain tumors: Localization to tumor endothelial cells and activation by the endothelial tyrosine kinase receptor (VEGFR-2). *Oncogene* **2002**, *21*, 2058–2065. [CrossRef]
125. Wang, H.; Wang, H.; Zhang, W.; Huang, H.J.; Liao, W.S.L.; Fuller, G.N. Analysis of the activation status of Akt, NFκB, and Stat3 in human diffuse gliomas. *Lab. Investig.* **2004**, *84*, 941–951. [CrossRef]
126. La Iglesia, N.; Puram, S.; Bonni, A. STAT3 Regulation of Glioblastoma Pathogenesis. *Curr. Mol. Med.* **2009**, *9*, 580–590. [CrossRef]
127. Pencik, J.; Schlederer, M.; Gruber, W.; Unger, C.; Walker, S.M.; Chalaris, A.; Marié, I.J.; Hassler, M.R.; Javaheri, T.; Aksoy, O.; et al. STAT3 regulated ARF expression suppresses prostate cancer metastasis. *Nat. Commun.* **2015**, *22*, 7736. [CrossRef]
128. Gao, S.P.; Mark, K.G.; Leslie, K.; Pao, W.; Motoi, N.; Gerald, W.L.; Travis, W.D.; Bornmann, W.; Veach, D.; Clarkson, B.; et al. Mutations in the EGFR kinase domain mediate STAT3 activation via IL-6 production in human lung adenocarcinomas. *J. Clin. Investig.* **2007**, *117*, 3846–3856. [CrossRef]
129. Haura, E.B.; Zheng, Z.; Song, L.; Cantor, A.; Bepler, G. Activated epidermal growth factor receptor-stat-3 signaling promotes tumor survival in vivo in non-small cell lung cancer. *Clin. Cancer Res.* **2005**, *11*, 8288–8294. [CrossRef]
130. Grabner, B.; Schramek, D.; Mueller, K.M.; Moll, H.P.; Svinka, J.; Hoffmann, T.; Bauer, E.; Blaas, L.; Hruschka, N.; Zboray, K.; et al. Disruption of STAT3 signalling promotes KRAS-induced lung tumorigenesis. *Nat. Commun.* **2015**, *6*, 6285. [CrossRef]
131. Takaku, K.; Oshima, M.; Miyoshi, H.; Matsui, M.; Seldin, M.F.; Taketo, M.M. Intestinal tumorigenesis in compound mutant mice of both Dpc4 (Smad4) and Apc genes. *Cell* **1998**, *6*, 645–656. [CrossRef]
132. Batlle, E.; Bacani, J.; Begthel, H.; Jonkeer, S.; Gregorieff, A.; Van De Born, M.; Malats, N.; Sancho, E.; Boon, E.; Pawson, T.; et al. EphB receptor activity suppresses colorectal cancer progression. *Nature* **2005**, *435*, 1126–1130. [CrossRef]
133. Lee, J.; Kim, J.C.K.; Lee, S.E.; Quinley, C.; Kim, H.R.; Herdman, S.; Corr, M.; Raz, E. Signal transducer and activator of transcription 3 (STAT3) protein suppresses adenoma-to-carcinoma transition in Apcmin/+ mice via regulation of Snail-1 (SNAI) protein stability. *J. Biol. Chem.* **2012**, *287*, 18182–18189. [CrossRef]
134. Demaria, M.; Misale, S.; Giorgi, C.; Miano, V.; Camporeale, A.; Campisi, J.; Pinton, P.; Poli, V. STAT3 can serve as a hit in the process of malignant transformation of primary cells. *Cell Death Differ.* **2012**, *19*, 1390–1397. [CrossRef]
135. Couto, J.P.; Daly, L.; Almeida, A.; Knauf, J.A.; Fagin, J.A.; Sobrinho-Simões, M.; Lima, J.; Máximo, V.; Soares, P.; Lyden, D.; et al. STAT3 negatively regulates thyroid tumorigenesis. *Proc. Natl. Acad. Sci. USA* **2012**, *109*, 2361–2370. [CrossRef]
136. Kritikou, E.A.; Sharkey, A.; Abell, K.; Came, P.J.; Anderson, E.; Clarkson, R.W.E.; Watson, C.J. A dual, non-redundant, role for LIF as a regulator of development and STAT3-mediated cell death in mammary gland. *Development* **2003**, *130*, 3459–3468. [CrossRef]
137. Kreuzaler, P.A.; Staniszewska, A.D.; Li, W.; Omidvar, N.; Kedjouar, B.; Turkson, J.; Poli, V.; Flavell, R.A.; Clarkson, R.W.E.; Watson, C.J. Stat3 controls lysosomal-mediated cell death in vivo. *Nat. Cell Biol.* **2011**, *13*, 303–309. [CrossRef]
138. Lloyd-Lewis, B.; Krueger, C.C.; Sargeant, T.J.; D'Angelo, M.E.; Deery, M.J.; Feret, R.; Howard, J.A.; Lilley, K.S.; Watson, C.J. Stat3-mediated alterations in lysosomal membrane protein composition. *J. Biol. Chem.* **2018**, *293*, 4244–4261. [CrossRef]
139. Resemann, H.K.; Watson, C.J.; Lloyd-Lewis, B. The stat3 paradox: A killer and an oncogene. *Mol. Cell. Endocrinol.* **2014**, *382*, 603–611. [CrossRef]
140. Wake, M.S.; Watson, C.J. STAT3 the oncogene—Still eluding therapy? *FEBS J.* **2015**, *282*, 2600–2611. [CrossRef]
141. Watson, C.J.; Miller, W.R. Elevated levels of members of the STAT family of transcription factors in breast carcinoma nuclear extracts. *Br. J. Cancer* **1995**, *71*, 840–844. [CrossRef]
142. Alvarez, J.V.; Febbo, P.G.; Ramaswamy, S.; Loda, M.; Richardson, A.; Frank, D.A. Identification of a genetic signature of activated signal transducer and activator of transcription 3 in human tumors. *Cancer Res.* **2005**, *65*, 5054–5062. [CrossRef]
143. Dolled-Filhart, M.; Camp, R.L.; Kowalski, D.P.; Smith, B.L.; Rimm, D.L. Tissue microarray analysis of signal transducers and activators of transcription 3 (Stat3) and phospho-Stat3 (Tyr705) in node-negative breast cancer shows nuclear localization is associated with a better prognosis. *Clin. Cancer Res.* **2003**, *9*, 594–600.
144. Yates, L.R.; Knappskog, S.; Wedge, D.; Farmery, J.H.R.; Gonzalez, S.; Martincorena, I.; Alexandrov, L.B.; Van Loo, P.; Haugland, H.K.; Lilleng, P.K.; et al. Genomic Evolution of Breast Cancer Metastasis and Relapse. *Cancer Cell* **2017**, *32*, 169–184. [CrossRef]
145. Zaretsky, J.M.; Garcia-Diaz, A.; Shin, D.S.; Escuin-Ordinas, H.; Hugo, W.; Hu-Lieskovan, S.; Torrejon, D.Y.; Abril-Rodriguez, G.; Sandoval, S.; Barthly, L.; et al. Mutations associated with acquired resistance to PD-1 blockade in melanoma. *N. Engl. J. Med.* **2016**, *375*, 819–829. [CrossRef]
146. Johnson, R.W.; Finger, E.C.; Olcina, M.M.; Vilalta, M.; Aguilera, T.; Miao, Y.; Merkel, A.R.; Johnson, J.R.; Sterling, J.A.; Wu, J.Y.; et al. Induction of LIFR confers a dormancy phenotype in breast cancer cells disseminated to the bone marrow. *Nat. Cell Biol.* **2016**, *18*, 1078–1089. [CrossRef]
147. Kim, R.S.; Avivar-Valderas, A.; Estrada, Y.; Bragado, P.; Sosa, M.S.; Aguirre-Ghiso, J.A.; Segall, J.E. Dormancy signatures and metastasis in estrogen receptor positive and negative breast cancer. *PLoS ONE* **2012**, *7*, e35569. [CrossRef]
148. Pectasides, E.; Egloff, A.M.; Sasaki, C.; Kountourakis, P.; Burtness, B.; Fountzilas, G.; Dafni, U.; Zaramboukas, T.; Rampias, T.; Rimm, D.; et al. Nuclear localization of signal transducer and activator of transcription 3 in head and neck squamous cell carcinoma is associated with a better prognosis. *Clin. Cancer Res.* **2010**, *16*, 2427–2434. [CrossRef]
149. Grandis, J.R.; Drenning, S.D.; Zeng, Q.; Watkins, S.C.; Melhem, M.F.; Endo, S.; Johnson, D.E.; Huang, L.; He, Y.; Kim, J.D. Constitutive activation of Stat3 signaling abrogates apoptosis in squamous cell carcinogenesis in vivo. *Proc. Natl. Acad. Sci. USA* **2000**, *97*, 4227–4232. [CrossRef]
150. Grandis, J.R.; Drenning, S.D.; Chakraborty, A.; Zhou, M.Y.; Zeng, Q.; Pitt, A.S.; Tweardy, D.J. Requirement of Stat3 but not Stat1 activation for epidermal growth factor receptor-mediated cell growth in vitro. *J. Clin. Investig.* **1998**, *102*, 1385–1392. [CrossRef]

151. Shao, H.; Quintero, A.J.; Tweardy, D.J. Identification and characterization of cis elements in the STAT3 gene regulating STAT3α and STAT3β messenger RNA splicing. *Blood* **2001**, *98*, 3853–3856. [CrossRef]
152. Tang, J.Z.; Kong, X.J.; Banerjee, A.; Muniraj, N.; Pandey, V.; Steiner, M.; Perry, J.K.; Zhu, T.; Liu, D.X.; Lobie, P.E. STAT3α is oncogenic for endometrial carcinoma cells and mediates the oncogenic effects of autocrine human growth hormone. *Endocrinology* **2010**, *151*, 4133–4145. [CrossRef]
153. Dewilde, S.; Vercelli, A.; Chiarle, R.; Poli, V. Of alphas and betas: Distinct and overlapping functions of STAT3 isoforms. *Front. Biosci.* **2008**, *13*, 6501–6514. [CrossRef]
154. Karni, R.; Jove, R.; Levitzki, A. Inhibition of pp60(c-Src) reduces Bcl-X(L) expression and reverses the transformed phenotype of cells overexpressing EGF and HER-2 receptors. *Oncogene* **1999**, *18*, 4654–4662. [CrossRef]
155. Sinibaldi, D.; Wharton, W.; Turkson, J.; Bowman, T.; Pledger, W.J.; Jove, R. Induction of p21(WAF1/CIP1) and cyclin D1 expression by the Src oncoprotein in mouse fibroblasts: Role of activated STAT3 signaling. *Oncogene* **2000**, *19*, 5419–5427. [CrossRef]
156. Turkson, J.; Bowman, T.; Garcia, R.; Caldenhoven, E.; De Groot, R.P.; Jove, R. Stat3 activation by Src induces specific gene regulation and is required for cell transformation. *Mol. Cell. Biol.* **1998**, *18*, 2545–2552. [CrossRef]
157. Ivanov, V.N.; Zhou, H.; Partridge, M.A.; Hei, T.K. Inhibition of ataxia telangiectasia mutated kinase activity enhances TRAIL-mediated apoptosis in human melanoma cells. *Cancer Res.* **2009**, *69*, 3510–3519. [CrossRef]
158. Ivanov, V.N.; Krasilnikov, M.; Ronai, Z. Regulation of Fas expression by STAT3 and c-Jun is mediated by phosphatidylinositol 3-kinase-AKT signaling. *J. Biol. Chem.* **2002**, *277*, 4932–4944. [CrossRef]
159. Banerjee, A.S.; Pal, A.D.; Banerjee, S. Epstein-Barr virus-encoded small non-coding RNAs induce cancer cell chemoresistance and migration. *Virology* **2013**, *443*, 294–305. [CrossRef]
160. Benekli, M.; Baer, M.R.; Baumann, H.; Wetzler, M. Signal transducer and activator of transcription proteins in leukemias. *Blood* **2003**, *101*, 2940–2954. [CrossRef]
161. Niu, G.; Shain, K.H.; Huang, M.; Ravi, R.; Bedi, A.; Dalton, W.S.; Jove, R.; Yu, H. Overexpression of a dominant-negative signal transducer and activator of transcription 3 variant in tumor cells leads to production of soluble factors that induce apoptosis and cell cycle arrest. *Cancer Res.* **2001**, *61*, 3276–3280.
162. Niu, G.; Heller, R.; Catlett-Falcone, R.; Coppola, D.; Jaroszeski, M.; Dalton, W.; Jove, R.; Yu, H. Gene therapy with dominant-negative Stat3 suppresses growth of the murine melanoma B16 tumor in vivo. *Cancer Res.* **1999**, *59*, 5059–5063.
163. Burke, W.M.; Jin, X.; Lin, H.J.; Huang, M.; Liu, R.; Reynolds, R.K.; Lin, J. Inhibition of constitutively active Stat3 suppresses growth of human ovarian and breast cancer cells. *Oncogene* **2001**, *20*, 7925–7934. [CrossRef]
164. Christine, R.; Sylvie, R.; Erik, B.; Geneviève, P.; Amélie, R.; Gérard, R.; Marc, B.; Christian, G.; Samir, A. Implication of STAT3 signaling in human colonic cancer cells during intestinal trefoil factor 3 (TFF3)—And vascular endothelial growth factor-mediated cellular invasion and tumor growth. *Cancer Res.* **2005**, *65*, 195–202.
165. Zhang, H.F.; Chen, Y.; Wu, C.; Wu, Z.Y.; Tweardy, D.J.; Alshareef, A.; Liao, L.D.; Xue, Y.J.; Wu, J.Y.; Chen, B.; et al. The opposing function of STAT3 as an oncoprotein and tumor suppressor is dictated by the expression status of STAT3β in esophageal squamous cell carcinoma. *Clin. Cancer Res.* **2016**, *22*, 691–703. [CrossRef]
166. Haura, E.B.; Turkson, J.; Jove, R. Mechanisms of disease: Insights into the emerging role of signal transducers and activators of transcription in cancer. *Nat. Clin. Pract. Oncol.* **2005**, *2*, 315–324. [CrossRef]
167. Debnath, B.; Xu, S.; Neamati, N. Small molecule inhibitors of signal transducer and activator of transcription 3 (stat3) protein. *J. Med. Chem.* **2012**, *55*, 6645–6668. [CrossRef]
168. Deng, J.; Grande, F.; Neamati, N. Small molecule inhibitors of Stat3 signaling pathway. *Curr. Cancer Drug Targets* **2007**, *7*, 91–107. [CrossRef]
169. Siddiquee, K.A.Z.; Gunning, P.T.; Glen, M.; Katt, W.P.; Zhang, S.; Schroeck, C.; Sebti, S.M.; Jove, R.; Hamilton, A.D.; Turkson, J. An oxazole-based small-molecule stat3 inhibitor modulates stat3 stability and processing and induces antitumor cell effects. *ACS Chem. Biol.* **2007**, *2*, 787–798. [CrossRef]
170. Zhang, X.; Sun, Y.; Pireddu, R.; Yang, H.; Urlam, M.K.; Lawrence, H.R.; Guida, W.C.; Lawrence, N.J.; Sebti, S.M. A novel inhibitor of STAT3 homodimerization selectively suppresses STAT3 activity and malignant transformation. *Cancer Res.* **2013**, *73*, 1922–1933. [CrossRef]
171. Fletcher, S.; Page, B.D.G.; Zhang, X.; Yue, P.; Li, Z.H.; Sharmeen, S.; Singh, J.; Zhao, W.; Schimmer, A.D.; Trudel, S.; et al. Antagonism of the Stat3-Stat3 protein dimer with salicylic acid based small molecules. *ChemMedChem* **2011**, *6*, 1459–1470. [CrossRef]
172. Turkson, J.; Ryan, D.; Kim, J.S.; Zhang, Y.; Chen, Z.; Haura, E.; Laudano, A.; Sebti, S.; Hamilton, A.D.; Jove, R. Phosphotyrosyl peptides block Stat3-mediated DNA binding activity, gene regulation, and cell transformation. *J. Biol. Chem.* **2001**, *276*, 45443–45455. [CrossRef] [PubMed]
173. Ren, Z.; Cabell, L.A.; Schaefer, T.S.; McMurray, J.S. Identification of a high-affinity phosphopeptide inhibitor of Stat3. *Bioorgan. Med. Chem. Lett.* **2003**, *13*, 633–636. [CrossRef]
174. Chen, J.; Nikolovska-Coleska, Z.; Yang, C.Y.; Gomez, C.; Gao, W.; Krajewski, K.; Jiang, S.; Roller, P.; Wang, S. Design and synthesis of a new, conformationally constrained, macrocyclic small-molecule inhibitor of STAT3 via "click chemistry. " *Bioorgan. Med. Chem. Lett.* **2007**, *17*, 3939–3942. [CrossRef]
175. Song, H.; Wang, R.; Wang, S.; Lin, J. A low-molecular-weight compound discovered through virtual database screening inhibits Stat3 function in breast cancer cells. *Proc. Natl. Acad. Sci. USA* **2005**, *102*, 4700–4705. [CrossRef]

176. Schust, J.; Sperl, B.; Hollis, A.; Mayer, T.U.; Berg, T. Stattic: A small-molecule inhibitor of STAT3 activation and dimerization. *Chem. Biol.* **2006**, *13*, 1235–1242. [CrossRef]
177. Hayakawa, F.; Sugimoto, K.; Harada, Y.; Hashimoto, N.; Ohi, N.; Kurahashi, S.; Naoe, T. A novel STAT inhibitor, OPB-31121, has a significant antitumor effect on leukemia with STAT-addictive oncokinases. *Blood Cancer J.* **2013**, *3*, e166. [CrossRef]
178. Kim, M.J.; Nam, H.J.; Kim, H.P.; Han, S.W.; Im, S.A.; Kim, T.Y.; Oh, D.Y.; Bang, Y.J. OPB-31121, a novel small molecular inhibitor, disrupts the JAK2/STAT3 pathway and exhibits an antitumor activity in gastric cancer cells. *Cancer Lett.* **2013**, *335*, 145–152. [CrossRef]
179. Okusaka, T.; Ueno, H.; Ikeda, M.; Mitsunaga, S.; Ozaka, M.; Ishii, H.; Yokosuka, O.; Ooka, Y.; Yoshimoto, R.; Yanagihara, Y.; et al. Phase 1 and pharmacological trial of OPB-31121, a signal transducer and activator of transcription-3 inhibitor, in patients with advanced hepatocellular carcinoma. *Hepatol. Res.* **2015**, *45*, 1283–1291. [CrossRef]
180. Alexandrow, M.G.; Song, L.J.; Altiok, S.; Gray, J.; Haura, E.B.; Kumar, N.B. Curcumin: A novel Stat3 pathway inhibitor for chemoprevention of lung cancer. *Eur. J. Cancer Prev.* **2012**, *21*, 407–412. [CrossRef]
181. Liu, Y.; Wang, X.; Zeng, S.; Zhang, X.; Zhao, J.; Zhang, X.; Chen, X.; Yang, W.; Yang, Y.; Dong, Z.; et al. The natural polyphenol curcumin induces apoptosis by suppressing STAT3 signaling in esophageal squamous cell carcinoma. *J. Exp. Clin. Cancer Res.* **2018**, *37*, 303. [CrossRef]
182. Kumar, A.; Bora, U. Molecular docking studies on inhibition of Stat3 dimerization by curcumin natural derivatives and its conjugates with amino acids. *Bioinformation* **2012**, *8*, 988–993. [CrossRef] [PubMed]
183. Wu, M.; Wang, L.; Li, F.; Hu, R.; Ma, J.; Zhang, K.; Cheng, X. Resveratrol downregulates STAT3 expression and astrocyte activation in primary astrocyte cultures of rat. *Neurochem. Res.* **2020**, *45*, 455–464. [CrossRef] [PubMed]
184. Shin, D.S.; Kim, H.N.; Shin, K.D.; Yoon, Y.J.; Kim, S.J.; Han, D.C.; Kwon, B.M. Cryptotanshinone inhibits constitutive signal transducer and activator of transcription 3 function through blocking the dimerization in DU145 prostate cancer cells. *Cancer Res.* **2009**, *69*, 193–202. [CrossRef] [PubMed]
185. Dong, Y.; Lu, B.; Zhang, X.; Zhang, J.; Lai, L.; Li, D.; Wu, Y.; Song, Y.; Luo, J.; Pang, X.; et al. Cucurbitacin E, a tetracyclic triterpenes compound from Chinese medicine, inhibits tumor angiogenesis through VEGFR2-mediated Jak2-STAT3 signaling pathway. *Carcinogenesis* **2010**, *31*, 2097–2104. [CrossRef] [PubMed]
186. Huang, W.; Dong, Z.; Chen, Y.; Wang, F.; Wang, C.J.; Peng, H.; He, Y.; Hangoc, G.; Pollok, K.; Sandusky, G.; et al. Small-molecule inhibitors targeting the DNA-binding domain of STAT3 suppress tumor growth, metastasis and STAT3 target gene expression in vivo. *Oncogene* **2016**, *35*, 783–792. [CrossRef]
187. Bharadwaj, U.; Eckols, T.K.; Kolosov, M.; Kasembeli, M.M.; Adam, A.; Torres, D.; Zhang, X.; Dobrolecki, L.E.; Wei, W.; Lewis, M.T.; et al. Drug-repositioning screening identified piperlongumine as a direct STAT3 inhibitor with potent activity against breast cancer. *Oncogene* **2015**, *34*, 1341–1353. [CrossRef] [PubMed]
188. Bosch-Barrera, J.; Menendez, J.A. Silibinin and STAT3: A natural way of targeting transcription factors for cancer therapy. *Cancer Treat. Rev.* **2015**, *41*, 540–546. [CrossRef]
189. Buettner, R.; Corzano, R.; Rashid, R.; Lin, J.; Senthil, M.; Hedvat, M.; Schroeder, A.; Mao, A.; Herrmann, A.; Yim, J.; et al. *Alkylation of cysteine* 468 in stat3 defines a novel site for therapeutic development. *Proc. ACS Chem. Biol.* **2011**, *6*, 432–443. [CrossRef]
190. Huang, W.; Dong, Z.; Wang, F.; Peng, H.; Liu, J.Y.; Zhang, J.T. A small molecule compound targeting STAT3 DNA-binding domain inhibits cancer cell proliferation, migration, and invasion. *ACS Chem. Biol.* **2014**, *9*, 1188–1196. [CrossRef]
191. Lin, Y.M.; Wang, C.M.; Jeng, J.C.; Leprince, D.; Shih, H.M. HIC1 interacts with and modulates the activity of STAT3. *Cell Cycle* **2013**, *12*, 2266–2276. [CrossRef]
192. Nagel-Wolfrum, K.; Buerger, C.; Wittig, I.; Butz, K.; Hoppe-Seyler, F.; Groner, B. The interaction of specific peptide aptamers with the DNA binding domain and the dimerization domain of the transcription factor Stat3 inhibits transactivation and induces apoptosis in tumor cells. *Mol. Cancer Res.* **2004**, *2*, 170–182. [PubMed]
193. Turkson, J.; Zhang, S.; Palmer, J.; Kay, H.; Stanko, J.; Mora, L.B.; Sebti, S.; Yu, H.; Jove, R. Inhibition of constitutive signal transducer and activator of transcription 3 activation by novel platinum complexes with potent antitumor activity. *Mol. Cancer Ther.* **2004**, *3*, 1533–1542. [PubMed]
194. Turkson, J.; Zhang, S.; Mora, L.B.; Burns, A.; Sebti, S.; Jove, R. A novel platinum compound inhibits constitutive Stat3 signaling and induces cell cycle arrest and apoptosis of malignant cells. *J. Biol. Chem.* **2005**, *280*, 32979–32988. [CrossRef] [PubMed]
195. Timofeeva, O.A.; Tarasova, N.I.; Zhang, X.; Chasovskikh, S.; Cheema, A.K.; Wang, H.; Brown, M.L.; Dritschilo, A. STAT3 suppresses transcription of proapoptotic genes in cancer cells with the involvement of its N-terminal domain. *Proc. Natl. Acad. Sci. USA* **2013**, *110*, 1267–1272. [CrossRef] [PubMed]
196. La Sala, G.; Michiels, C.; Kükenshöner, T.; Brandstoetter, T.; Maurer, B.; Koide, A.; Lau, K.; Pojer, F.; Koide, S.; Sexl, V.; et al. Selective inhibition of STAT3 signaling using monobodies targeting the coiled-coil and N-terminal domains. *Nat. Commun.* **2020**, *28*, 253–262. [CrossRef]
197. Buerger, C.; Nagel-Wolfrum, K.; Kunz, C.; Wittig, I.; Butz, K.; Hoppe-Seyler, F.; Groner, B. Sequence-specific peptide aptamers, interacting with the intracellular domain of the epidermal growth factor receptor, interfere with Stat3 activation and inhibit the growth of tumor cells. *J. Biol. Chem.* **2003**, *278*, 37610–37621. [CrossRef]
198. Ge, H.; Liu, H.; Fu, Z.; Sun, Z. Therapeutic and preventive effects of an epidermal growth factor receptor inhibitor on oral squamous cell carcinoma. *J. Int. Med. Res.* **2012**, *40*, 455–466. [CrossRef]
199. Aittomäki, S.; Pesu, M. Therapeutic targeting of the JAK/STAT pathway. *Basic Clin. Pharmacol. Toxicol.* **2014**, *114*, 18–23. [CrossRef]

200. Buchert, M.; Burns, C.J.; Ernst, M. Targeting JAK kinase in solid tumors: Emerging opportunities and challenges. *Oncogene* **2016**, *35*, 939. [CrossRef]
201. Plimack, E.R.; LoRusso, P.M.; McCoon, P.; Tang, W.; Krebs, A.D.; Curt, G.; Eckhardt, S.G. AZD1480: A Phase I study of a novel JAK2 inhibitor in solid tumors. *Oncologist* **2013**, *18*, 819–820. [CrossRef]
202. Furtek, S.L.; Backos, D.S.; Matheson, C.J.; Reigan, P. Strategies and approaches of targeting STAT3 for cancer treatment. *ACS Chem. Biol.* **2016**, *11*, 308–318. [CrossRef] [PubMed]
203. Puls, L.N.; Eadens, M.; Messersmith, W. Current status of Src inhibitors in solid tumor malignancies. *Oncologist* **2011**, *16*, 566–578. [CrossRef] [PubMed]
204. Nam, S.; Wen, W.; Schroeder, A.; Herrmann, A.; Yu, H.; Cheng, X.; Merz, K.H.; Eisenbrand, G.; Li, H.; Yuan, Y.C.; et al. Dual inhibition of Janus and Src family kinases by novel indirubin derivative blocks constitutively-activated Stat3 signaling associated with apoptosis of human pancreatic cancer cells. *Mol. Oncol.* **2013**, *7*, 369–378. [CrossRef]
205. Liu, L.; Gaboriaud, N.; Vougogianopoulou, K.; Tian, Y.; Wu, J.; Wen, W.; Skaltsounis, L.; Jove, R. MLS-2384, a new 6-bromoindirubin derivative with dual JAK/Src kinase inhibitory activity, suppresses growth of diverse cancer cells. *Cancer Biol. Ther.* **2014**, *15*, 178–184. [CrossRef] [PubMed]
206. Murakami, T.; Takigawa, N.; Ninomiya, T.; Ochi, N.; Yasugi, M.; Honda, Y.; Kubo, T.; Ichihara, E.; Hotta, K.; Tanimoto, M.; et al. Effect of AZD1480 in an epidermal growth factor receptor-driven lung cancer model. *Lung Cancer* **2014**, *83*, 30–36. [CrossRef]
207. Lau, Y.T.K.; Ramaiyer, M.; Johnson, D.E.; Grandis, J.R. Targeting STAT3 in cancer with nucleotide therapeutics. *Cancers* **2019**, *11*, 1681. [CrossRef]
208. Guha, P.; Gardell, J.; Darpolor, J.; Cunetta, M.; Lima, M.; Miller, G.; Espat, N.J.; Junghans, R.P.; Katz, S.C. STAT3 inhibition induces Bax-dependent apoptosis in liver tumor myeloid-derived suppressor cells. *Oncogene* **2019**, *38*, 533–548. [CrossRef]
209. Zhang, L.; Wang, Y.; Dong, Y.; Chen, Z.; Eckols, T.K.; Kasembeli, M.M.; Tweardy, D.J.; Mitch, W.E. Pharmacokinetics and pharmacodynamics of TTI-101, a STAT3 inhibitor that blocks muscle proteolysis in rats with chronic kidney disease. *Am. J. Physiol. Ren. Physiol.* **2020**, *319*, F84–F92. [CrossRef]
210. Verstovsek, S.; Manshouri, T.; Quintás-Cardama, A.; Harris, D.; Cortes, J.; Giles, F.J.; Kantarjian, H.; Priebe, W.; Estrov, Z. WP1066, a novel JAK2 inhibitor, suppresses proliferation and induces apoptosis in erythroid human cells carrying the JAK2 V617F mutation. *Clin. Cancer Res.* **2008**, *14*, 788–796. [CrossRef]

International Journal of
Molecular Sciences

MDPI

Review

Microbiota-Associated Therapy for Non-Alcoholic Steatohepatitis-Induced Liver Cancer: A Review

Yi-Hsun Chen [1], Wei-Kai Wu [2] and Ming-Shiang Wu [1,3,*]

1 College of Medicine, National Taiwan University, Taipei 100, Taiwan; u9423201@gmail.com
2 Division of Gastroenterology, National Taiwan University Hospital Bei-Hu Branch, Taipei 108, Taiwan; weikaiwu0115@gmail.com
3 Department of Internal Medicine, National Taiwan University Hospital, Taipei 100, Taiwan
* Correspondence: mingshiang@ntu.edu.tw; Tel.: +886-02-2312-3456

Received: 10 July 2020; Accepted: 19 August 2020; Published: 20 August 2020

Abstract: Even though advancement in medicine has contributed to the control of many diseases to date, cancer therapy continues to pose several challenges. Hepatocellular carcinoma (HCC) etiology is multifactorial. Recently, non-alcoholic fatty liver disease (NAFLD) has been considered as an important risk factor of HCC. NAFLD can be divided into non-alcoholic simple fatty liver (NAFL) and non-alcoholic steatohepatitis (NASH) based on histopathological features. Recently, studies have indicated that the gut microbiota is associated with NAFLD and HCC. Therefore, in this review, we have discussed the effects of gut microbiota-related mechanisms, including dysbiosis and gut barrier function, and gut microbiota-derived metabolites on NAFLD and HCC pathogenesis and the potential therapeutic strategies for NAFLD and HCC. With a better understanding of the gut microbiota composition and function, new and improved diagnostic, prognostic, and therapeutic strategies for common liver diseases can be developed.

Keywords: gut microbiota; NAFLD; HCC; dysbiosis; metabolites

---

1. Introduction

With advancements in medicine, many diseases can be controlled, but cancer continues to pose many challenges. Hepatocellular carcinoma (HCC) ranks fifth and second in highest cancer incidence and global cancer-related mortality, respectively [1,2]. Traditionally, chronic viral hepatitis caused by hepatitis B and C viruses (HBV and HCV, respectively) is a major risk factor of HCC [3]. In recent decades, the chronic viral hepatitis disease burden has been gradually controlled by universal implementation of HBV vaccination and dramatic improvement in anti-HBV and anti-HCV treatments [4]. Therefore, the incidence of viral hepatitis-related liver cirrhosis and HCC is expected to decline. Other important causative risk factors of HCC include non-alcoholic fatty liver disease (NAFLD), obesity, diabetes, and alcoholism [5]. Currently, NAFLD is the second most common cause of end-stage liver disease or liver cancer, which requires liver transplantation, in the United States [6–8]. It has been estimated that, from 2016 to 2030, the number of NAFLD-induced end-stage liver disease cases and related deaths worldwide will be doubled [9]. Therefore, NAFLD is likely to become the most important cause of HCC in the future.

Generally, NAFLD progression can be divided into four pathological stages, including non-alcoholic simple fatty liver (NAFL), non-alcoholic steatohepatitis (NASH), hepatic cirrhosis, and HCC [10] (Figure 1). Recent epidemiological studies showed that the average prevalence rate of NAFLD appeared to be higher in Asia (27%) than in the United States (24%) and Europe (23%) [7,8,11]. NAFLD-related health issues have attracted global attention since NAFLD is increasingly becoming the most common cause of hepatic cirrhosis and HCC [9,12]. NAFLD patients who remain in the

NAFL stage show a much lower risk of developing HCC than NASH patients. Chronic and repetitive hepatocyte damage and repair in steatohepatitis may lead to advanced fibrosis, cirrhosis, and HCC, thus causing mortality within a few decades. Moreover, no FDA approved medication for NAFLD treatment exists presently. Therefore, there is an urgent and unmet need for a better understanding of NAFLD pathogenesis and progression to HCC.

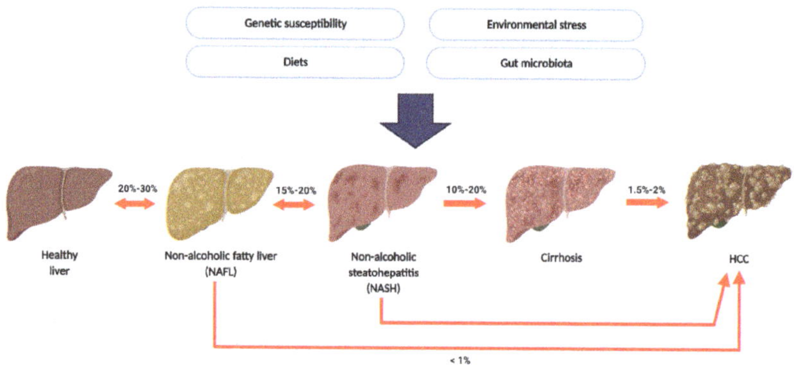

Figure 1. The natural history of non-alcoholic fatty liver disease (NAFLD) and its etiological risk factor. The prevalence of NAFL was around 20–30%. Around 15–20% of non-alcoholic simple fatty liver (NAFL) patients developed non-alcoholic steatohepatitis (NASH); 10–20% of NASH patients will further process to cirrhosis and Hepatocellular carcinoma (HCC). Several factors including genetic susceptibility, environmental stress, diet, and gut microbiota were considered as etiological risk factors.

Owing to this anatomical and functional connection, the liver and intestines maintain close functional and bidirectional communication which is subsumed in the "gut–liver axis". The liver is continually exposed to the products of digestion, absorption, and gut-derived factors through the portal vein [13]. On the other hand, the liver plays an important role in the regulation of the gut microbiota composition via bile acids (BAs) [14]. Recent studies have suggested that dysbiosis, the change in the gut microbiota, may be associated with liver diseases, including NAFLD and HCC [15–17]. Increased intestinal permeability, which is associated with dysbiosis, leads to the liver being exposed to intestinal toxic factors and bacteria such as lipopolysaccharides (LPS) [18,19]. Exposure of liver with these factors further induced hepatic inflammation and damage which contributed to HCC pathogenesis. In this review, we have discussed the pathophysiological roles of the gut microbiota and relevant molecules in NAFLD progression to HCC. Some potential diagnostic and therapeutic highlights, which could be implemented for future clinical applications, were included.

2. From NAFL to NASH: The Gut Microbiota-Associated Mechanisms

Although NAFL patients bear lower risk of developing HCC than NASH patients, approximately 10–20% of NAFL patients progress to NASH, with a significantly increasing risk of developing cirrhosis and HCC. Therefore, understanding the mechanistic roles of the gut microbiota that trigger the NAFLD inflammatory status may potentially help in discovering novel microbial and molecular pathways for preventing HCC development.

2.1. NASH and Dysbiosis

Many studies have indicated that gut dysbiosis is associated with NAFLD pathogenesis [20–26]. Compared to healthy individuals, NASH patients exhibit increased relative abundance of *Blautia*, *Dorea*, *Lactobacillus*, *Clostridium*, *Allisonella*, *Parabacteroides*, and *Escherichia* spp. [20,21,25–27] and decreased relative abundance of *Oscillospira*, *Coprococcus*, *Faecalibacterium*, and *Bifidobacterium* spp. [20,25–27]. A common finding in NASH patients, compared to NAFL patients, is increased relative abundance of

Blautia and *Bacteroides* spp. [23,28] and decreased relative abundance of *Prevotella* spp. [28]. Furthermore, gut dysbiosis includes not only compositional changes but also metabolic functional changes in the gut microbiome. For example, compositional changes in the gut microbiota lead to an altered short chain fatty acid (SCFA) profile, further affecting host energy absorption [20,25,29]. Gut dysbiosis can result in an increase in gut permeability, disruption of metabolic homeostasis, and changes in the microbiota-associated metabolites, thus eventually contributing to disease initiation and progression.

2.2. NASH and Leaky Gut

In previous clinical studies, NASH patients exhibited greater intestinal permeability than simple steatosis patients and healthy individuals [30]. Increased intestinal permeability is caused by decreased expression of zonula occludens-1 (ZO-1), a representative tight junction protein [30–32]. Several bacteria, including *Lactobacillus*, *Bifidobacterium*, Bacteroidetes, *Clostridiales*, *Oscillibacter*, *Desulfovibrio* spp., and *Akkermansia muciniphila*, have been associated with intestinal permeability in animal models. Among these bacteria, *Lactobacillus*, *Bifidobacterium*, Bacteroidetes, *Clostridiales* spp., and *Akkermansia muciniphila* are considered gut barrier-promoting microbes, while *Oscillibacter* and *Desulfovibrio* spp. are considered gut barrier-disrupting microbes [33]. *Lactobacillus*, *Bifidobacterium* spp. and *A. muciniphila* induce ZO-1 expression to promote the gut barrier [34–36]. On the other hand, *Desulfovibrio* spp. produce genotoxic hydrogen sulfide (H_2S), increasing the intestinal permeability [37]. Bacteroidetes, Verrucomicrobia, *Akkermansia*, and *Lactobacillus* spp. were positively correlated with increased expression levels of tight junction proteins, including ZO-1, occludin, and claudin-1, indicating that these bacteria maintained gut barrier function and improved hepatic inflammation and oxidative stress. On the other hand, Firmicutes, Proteobacteria, *Butyricimonas*, *Parabacteroides*, and *Bilophila* spp. exhibited the opposite effect [38]. Modulation of the gut microbiota by *Bifidobacterium bifidum* ATCC 2952 restored dextran sodium sulfate (DSS)-induced dysbiosis and up-regulated the expression of anti-inflammatory cytokines including interleukin (IL)-10, peroxisome proliferator-activated receptor (PPAR)-γ, and IL-6 in the gut, thereby indicating the important role of the gut microbiota [39]. A recent study, in which fecal microbial transplantation (FMT) from mice on high-fat diets (HFDs) to mice on standard diets was performed, showed the gut barrier damages in these mice, thereby indicating that altered gut microbiota was responsible for increased intestinal permeability [40] (Figure 2A).

Activation of the inflammasome by diverse microbial-, stress-, and danger-associated signals triggers pro-inflammatory cytokines including IL-1β and IL-18, thereby promoting innate immunity [41]. Previous studies have demonstrated that the intestinal epithelial nucleotide-binding oligomerization domain (NOD)-like receptor (NLR) family pyrin domain containing 6 (NLRP6) inflammasome maintains the intestinal barrier and microbial balance by regulating goblet cell mucus secretion [42] and anti-microbial peptide production [43]. NLRP6 is highly expressed in the epithelial cells of the small intestine, colon, and goblet cells and is co-expressed with apoptosis-associated speck-like protein containing a caspase recruitment domain (ASC) and caspase-1 in the intestinal epithelium [43]. A previous study indicated that fructose-fed mice exhibited impaired gut barrier and NLRP6 inflammasome [44]. NLRP6 activation induced the synthesis of anti-microbial peptides, including angiogenin-4, intelectin-1, and resistin-like molecule β, by gut epithelial cells [43]. Furthermore, NLRP6-deficient mice exhibited impaired anti-microbial peptides, resulting in dysbiosis, as indicated by the increased relative abundance of the *Prevotellaceae* spp. and members of the TM7 phylum and the decreased relative abundance of the *Lactobacillus* spp. and members of the Firmicutes phylum [45]. Therefore, the gut microbiota–NLRP6 axis plays an important role in maintaining the gut barrier function (Figure 2B).

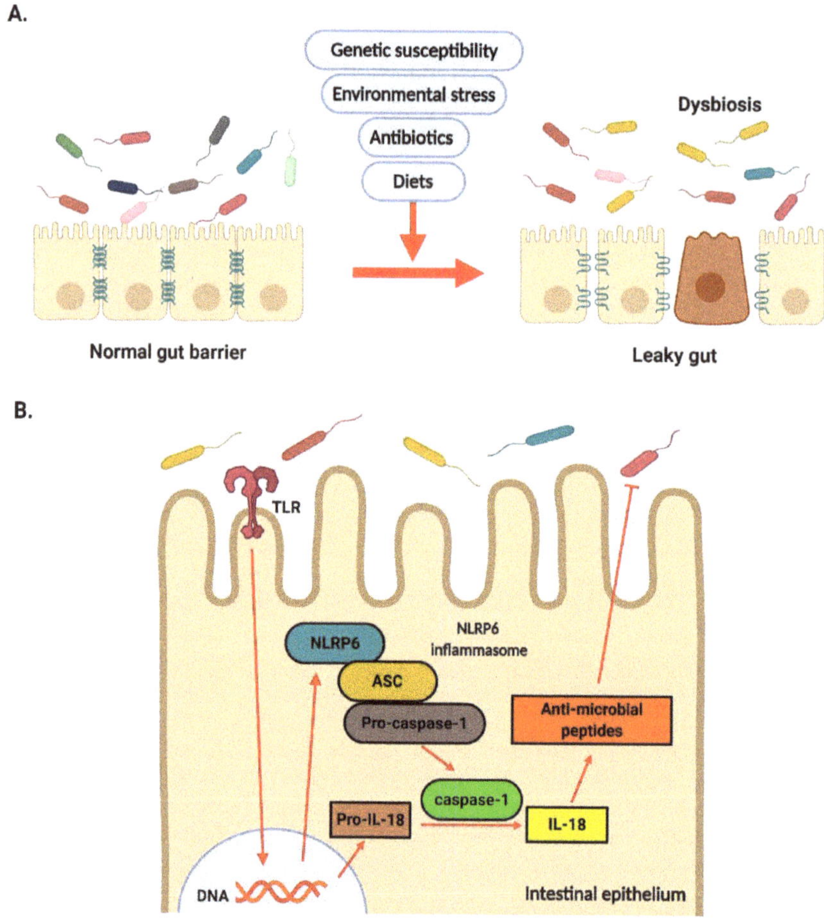

Figure 2. Dysbiosis and the role of NLRP6. (**A**) Dysbiosis induced by genetic susceptibility, environmental stress, diet, and gut microbiota results in disrupted tight junction and increased intestinal permeability. (**B**) Microbiota-derived factors activate NLRP6 inflammation via TLR signaling. Activation of NLRP6 results in induction of anti-microbial peptide synthesis and contributes to maintaining the homeostasis of gut microbiota.

2.3. Gut Microbiota and Hepatic Inflammation

The gut microbiota signals travel through the human body systemically via the liver. Both nutrients and microbe-derived molecules from the intestinal lumen converge in the liver through the portal vein. Modulation of intestinal permeability regulates the entry of microbe-derived molecules into the liver from the gut. Some of these molecules are harmful substances that can cause liver inflammation and induce the pathological process of NASH. For example, in JAM-A-deficient mice (genetically induced gut barrier dysfunction model) and a DSS-induced gut inflammation animal model, mice on high-fat, high-fructose, and cholesterol diets, compared to the control, showed LPS translocation and increased NASH severity [46,47]. LPS-triggered hepatic inflammation occurred through the activation of toll-like receptor 4 (TLR4) in several types of cells, including Kupffer cells, hepatocytes, hepatic stellate cells (HSCs), and liver sinusoidal endothelial cells (LSECs). In Kupffer cells, TLR4 signal activation via myeloid differentiation primary response 88 (MyD88) induced tumor necrosis factor (TNF)-α and reactive oxygen species (ROS), further enhancing hepatic inflammation. The LPS-triggered TLR4 on the HSCs induced the production of various chemokines and adhesion molecules, which in turn induced Kupffer cell chemotaxis. On the other hand, the activation of TLR4 on hepatocytes induced hepcidin production via the MyD88/c-Jun N-terminal kinase (JNK) pathway, which was associated with hepatic lipid accumulation [48]. These results were consistent with those of a previous human study that showed higher levels of antibodies against LPS, produced by Gram-negative bacteria, in NASH patients than in healthy individuals, and this increase paralleled disease severity [49]. In addition to LPS, other pathogen-associated molecular patterns (PAMPs), including peptidoglycan, flagellin, and bacterial RNA and DNA, can enter into the liver due to increased intestinal permeability and trigger inflammatory responses. TLR9 activation by bacterial DNA further induces IL-1β production in Kupffer cells, thus resulting in hepatic steatosis, inflammation, and fibrosis [50]. In addition to TLR, inflammasome proteins, which are activated by NLRs and assembled, recognize PAMPs, leading to IL-1 and IL-18 production and further triggering inflammation [51]. A previous study indicated that NLRP3 inflammasome components were significantly increased in NASH patients compared to in non-NASH NAFLD patients [52]. These results indicated the association between hepatic inflammation and NLRP3 inflammasome. Indeed, the lack of NLRP3 inflammasome attenuated hepatic injury, immune cell infiltration, and choline-deficient (CD) L-amino-defined (CDAA) diet-induced fibrosis, thereby confirming the important role of NLRP3 inflammasome [52]. The increased influx of different classes of lipotoxic lipids and insulin resistance-induced adipokines, in addition to PAMPs, into the liver due to a leaky gut can also trigger hepatic inflammation. Several lipid classes, including saturated non-esterified fatty acids (NEFAs), free cholesterol, sphingolipids, and sphingosine 1-phosphate (S1P), induce liver injury and inflammation [53,54]. For example, saturated NEFAs can bind to and activate hepatocyte death receptor TNF-related apoptosis-inducing ligand (TRAIL) receptor 2 (TRAIL-R2) and damage-associated molecular pattern (DAMP) receptors, such as TLR4, and further trigger downstream activation of the caspase cascade and execute hepatocyte apoptosis [55]. Additionally, accumulating ceramides, one type of sphingolipid, were observed in the HFD animals fed diets enriched with saturated fatty acids [56]. Increased ceramides contributed to ROS generation, and oxidative stress further induced apoptosis and inflammatory cell recruitment to the liver, thus resulting in worsening hepatic inflammation and damage [57]. Generally, lipotoxicity induces endoplasmic reticulum (ER) stress, mitochondrial dysfunction, inflammasome activation, and cell death [58] (Figure 3).

2.4. NASH and Gut Microbiota-Derived Metabolites

Dysbiosis includes changes in not only the gut microbiota composition but also in the microbiota-derived metabolites obtained from dietary nutrients, thereby affecting host metabolic homeostasis. The most important gut microbiota-derived metabolites are SCFAs. SCFAs are produced by the fermentation of dietary fibers by the gut microbiota, including *Roseburia*, *Ruminococcus*, *Salmonella*, *Blautia*, *Eubacterium*, *Anaerostipes*, *Coprococcus*, *Faecalibacterium*, *Marvinbryantia*, and *Megasphaera* spp. The relative abundance levels of acetate, propionate, and butyrate were the highest [59]. Propionate was associated with peptide-YY and glucagon-like peptide-1 (GLP-1) release using a primary cultured human colonic cell model and further demonstrated that increasing colonic propionate prevents weight gain and insulin resistance in overweight adult humans [60]. In an HFD-induced steatohepatitis mouse model, butyrate promoted CD4+ T cell differentiation into helper T 2 (Th2), Th22, or regulatory T (Treg) cells and inhibited CD4+ T cell differentiation into Th1 or Th17 cells, further preventing hepatic inflammation [61]. SCFAs exerted their biological functions mainly via G-protein coupled receptor (GPR) 41/43 activation or histone deacetylase (HDAC) inhibition. GPR41 and GPR43 were expressed in not only the gut but also the liver, and GPR41 and GPR43 activation attenuated host insulin resistance in murine models [62–64]. Butyrate could inhibit HDAC directly and regulate phosphorylation of cyclic adenosine monophosphate (cAMP) response element binding protein (CREB), which is involved in gluconeogenesis, further influencing the gut and liver metabolisms [65]. Thus, SCFAs affected not only the hepatic immune response but also hepatic metabolism.

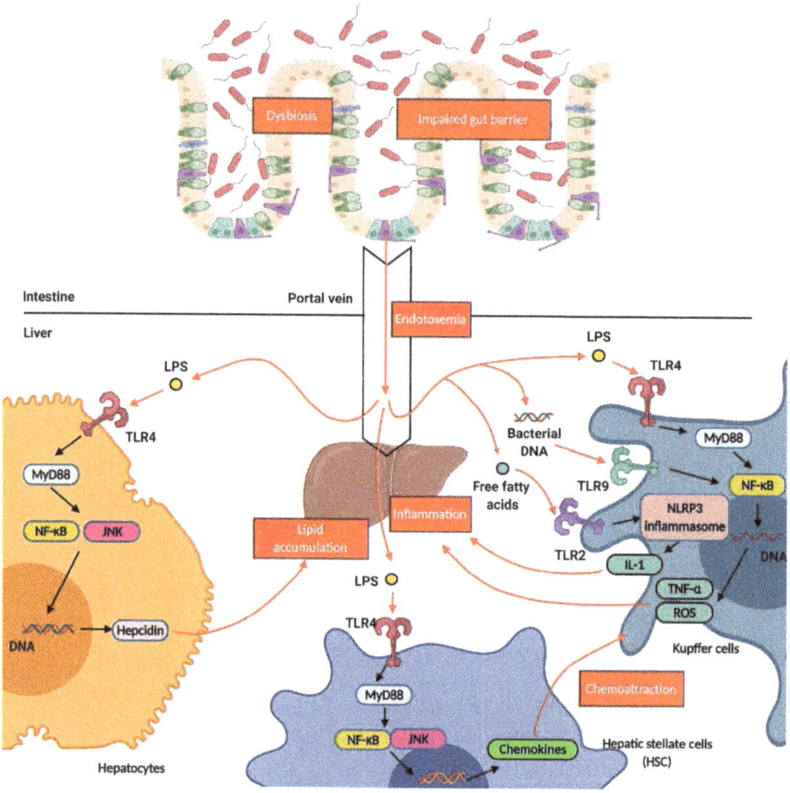

Figure 3. The mechanism of hepatic inflammation induced by gut microbiota. Dysbiosis results in impaired gut barrier and further induces the influx of bacterial DNA and LPS, which is termed endotoxemia, from gut to liver through portal vein. LPS further triggers TLR4 signaling in hepatocytes, Kupffer cells, and hepatic stellate cells (HSC). Activation of TLR4 and TLR9 in Kupffer cells induces the production of TNF-α and ROS, which further contributes to hepatic inflammation. Activation of TLR4 in hepatocytes induces the production of hepcidin, which further induces hepatic lipid accumulation. Activation of TLR4 in HSC induces the production of chemokines, which further contributes to chemoattraction for Kupffer cells. Additionally, influx of free fatty acid from gut to liver activates TLR2 signaling, which is termed lipotoxicity. Activation of TLR2 signaling results in activation of NLRP3 inflammation, which induces the production of IL-1. Increased IL-1 production leads to hepatic inflammation. LPS, Lipopolysaccharide; MyD88, myeloid differentiation primary response 88; NF-κB, nuclear factor kappa-light-chain-enhancer of activated B cells; JNK, c-Jun N-terminal kinase; TLR, toll-like receptors.

Amino acid imbalance is often found in NAFLD patients [66,67]. The ratio of branched-chain amino acids (BCAAs) to aromatic amino acids (AAAs) is a diagnostic marker for the severity of liver dysfunction. A decreasing ratio indicates severe liver dysfunction. BCAAs, including valine, leucine, and isoleucine, are essential amino acids for human beings and are involved in liver disease pathophysiology. Cohort studies have indicated that serum BCAA levels are positively correlated with insulin resistance and steatosis [67–69]. Further studies showed that *Prevotella copri* and *Bacteroides vulgatus* were the main species responsible for the association between BCCA biosynthesis and insulin resistance, and this finding was confirmed in the mouse model [70]. Although several studies indicated that BCAAs could inhibit triglyceride (TG) deposition in hepatocytes, reduce ER stress, and enhance gut barrier function by improving immune response, some inconsistencies in the results indicated that BCAAs caused hepatic damage, associated with abnormal lipolysis, in mice on HFDs [71–73]. Phenylacetic acid (PAA) is an AAA-derived metabolite which is produced by the gut microbiota. PAA has been found to induce hepatic steatosis by lowering protein kinase B (Akt) phosphorylation and affect BCAA metabolism by increasing acyl-CoA dehydrogenase short/branched chain (ACADSB) expression in primary hepatocytes and mice, indicating the causal role of PAA in NAFLD [67].

Recent studies have indicated that tryptophan metabolites may affect NAFLD development [59]. Indole and its derivatives, including indoleacrylic acid (IA), indole-3-acetic acid (IAA), indole-3-aldehyde (I3A), indole-3-propionic acid (IPA), and tryptamine, are the main tryptophan-derived gut bacterial products which are mainly produced by *Bacteroides, Eubacterium, and Clostridium* spp. [59]. Among them, tryptamine and I3A reduced hepatic fatty acid synthase (FAS) and sterol regulatory element-binding protein-1c (SREBP1c) expression via aryl-hydrocarbon receptor (AhR), further reducing Kupffer cell-induced hepatic inflammation [74]. Lower abundance of IPA in obese patients has been reported in previous studies, and IPA supplementation resulted in the reduction of weight gain in the antibiotic-induced dysbiosis animal model [75]. Additionally, IPA improved intestinal barrier function via pregnane X receptor (PXR), which in turn inhibited endotoxin-induced TLR4 signaling and improved tissue inflammation [76,77]. Therefore, tryptophan metabolites appear to be potential therapeutic targets.

Higher levels of ethanol in the blood and breath, accompanied with up-regulation of hepatic alcohol dehydrogenase, aldehyde dehydrogenase, and CYP2E1, were exhibited in NASH patients and ob/ob mice without alcohol consumption [78]. These results indicated that endogenous ethanol might be involved in NAFLD pathogenesis. Endogenous ethanol is obtained by carbohydrate fermentation by gut microbiota, and it stimulates oxidative stress and aggravates liver inflammation in NAFLD [79,80]. *Escherichia coli*, *Enterobacteriaceae* spp., and *Klebsiella pneumonia* have been identified as ethanol-producing bacteria and were found to be relatively abundant in NAFLD patients [79,81]. Ethanol can be further metabolized into acetaldehyde, which induces hepatic injury [82]. Therefore, increasing endogenous ethanol production may deteriorate hepatic inflammation. Ethanol exerts direct toxic effects on the liver, increasing intestinal permeability, which results in endotoxemia, and further triggers the inflammatory response, contributing to liver injury [83]. These findings indicate that endogenous ethanol might play a pivotal role in NASH pathogenesis. However, further investigation is required to determine the exact effects of endogenous ethanol on NAFLD and NASH.

BAs can be divided into primary and secondary BAs. Primary BAs, including cholic acid (CA) and chenodeoxycholic acid (CDCA), are produced using cholesterol in the liver. Primary BAs are converted into more than 20 secondary BAs, including deoxycholic acid (DCA) and lithocholic acid (LCA), by the gut microbiota [84,85]. Furthermore, distinct BA profiles were observed between the germ-free and conventional animals, thereby indicating the direct effects of the gut microbiota on BAs [86]. Adams et al. (2020) showed that increased DCA was associated with not only the increased relative abundance of specific bacterial groups, including *Bacteroidaceae* and *Lachnospiraceae* spp., but also advanced fibrosis in NAFLD [87]. At the molecular level, individual BAs act as agonists or antagonists for farnesoid X receptor (FXR) and takeda G-protein-coupled bile acid receptor 5 (TGR5) and affect energy, glucose, and lipoprotein metabolism, indicating that altered BA composition may affect host metabolism by modifying these signals [88]. For example, FXR activation by CA, CDCA, and obeticholic acid (OCA), which is derived from CDCA, stimulated production of fibroblast growth factor 15 (FGF15) in mice or FGF19 in humans. FGF15 and FGF19 further bound to FGF14 in the liver to inhibit BA synthesis, thereby altering the BA pool and exhibiting NASH improvement [89–92]. On the other hand, increasing CA or DCA may result in dysbiosis owing to their anti-microbial activity, further contributing to NAFLD pathogenesis [93,94]. Additionally, TGR5 activation in the intestine results in GLP-1 release from L cells, further promoting insulin release from pancreatic β-cells [95,96]. Taken together, these results indicate that the gut microbiota-induced alteration of the BA pool plays an important role in NAFLD pathogenesis.

Choline is metabolized to phosphatidylcholine, which is essential for very low-density lipoprotein (VLDL) production and hepatic lipid transfer, in the liver. Most phosphatidylcholine is derived from dietary choline. Therefore, hepatic lipid metabolism is affected by choline deficiency. CD diets are commonly used to induce NAFLD in animal models. Decreased VLDL levels and β-oxidation were observed in mice on CD diets, further resulting in fatty acid and cholesterol accumulation and increased oxidative stress and inflammation in the liver [20,97]. Three major bacterial phyla, including Proteobacteria, Firmicutes, and Actinobacteria, are associated with choline metabolism. These bacteria metabolize dietary choline to trimethylamine (TMA), which is further metabolized to trimethylamine-N-oxide (TMAO) by flavin-containing monooxygenases, in the liver [98]. High circulating TMAO levels have been reported to increase the risk of atherosclerosis and cardiovascular disease [99]. Previous studies also indicated that NAFLD patients exhibited elevated serum TMAO levels, which were positively correlated with the pathological progression of NAFLD [100]. Further research indicated that TMAO modulated BA metabolism and FXR signaling inhibition, contributing to NAFLD pathogenesis [101]. The effects of the gut microbiota-derived metabolites on NAFLD are shown in Figure 4 and Table 1.

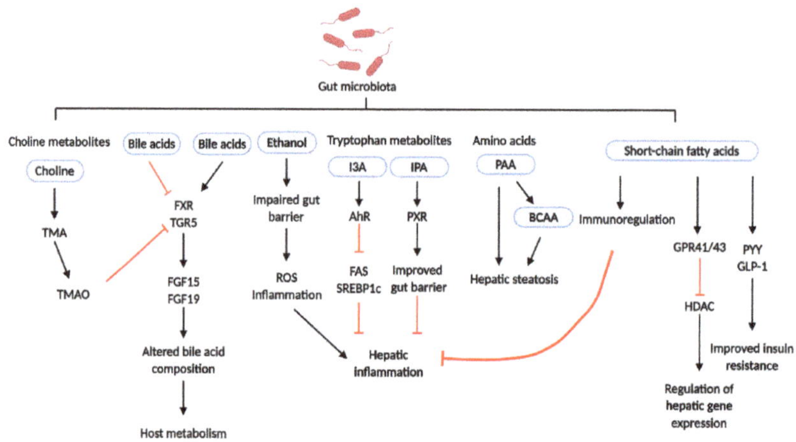

Figure 4. The roles of gut microbiota-derived metabolites. Short-chain fatty acids (SCFAs) are associated with the release of PYY and GLP-1, which further ameliorate insulin resistance. SCFAs also activate GPR41/43, performing the regulation of hepatic gene expression via inhibition of HDAC. Additionally, SCFAs also perform immunoregulation to inhibit hepatic inflammation. PAA can induce hepatic steatosis by itself or via affecting BCAA metabolism. IPA improves intestinal barrier function via PXR, which improves tissue inflammation. I3A reduces hepatic FAS and SREBP1 expression via AhR, further reducing hepatic inflammation. Increasing endogenous ethanol production may deteriorate hepatic inflammation. Specific BAs act as agonists or antagonists of FXR and TGR5 which affect the composition of BAs and further affect host metabolism. TMAO, which is derived from TMA and choline, modulates BA metabolism and FXR signaling inhibition, contributing to NAFLD pathogenesis. TMA, trimethylamine; TMAO, trimethylamine-N-oxide; FXR, farnesoid X receptor; TGR5, Takeda G protein-coupled bile acid receptor 5; FGF, fibroblast growth factor; I3A, indole-3-aldehyde; IPA, indole-3-propionic acid; AhR, aryl-hydrocarbon receptor; PXR, pregnane X receptor; FAS, fatty acid synthase; SREBP1c, sterol regulatory element-binding protein-1c; PAA, phenylacetic acid; BCAAs, branched-chain amino acids; GPR, G-protein coupled receptor; GLP-1, glucagon-like peptide-1; PYY, peptide-YY; HDAC, histone deacetylase.

Table 1. Overview of metabolite-relative effects on host in in vitro and animal models.

Metabolites	Effects	References
Acetate	HDAC inhibition.	[65]
Propionate	Induces PYY and GLP-1 release. HDAC inhibition.	[60,65]
Butyrate	Promotes Th2, Th22, or Treg cell differentiation, further preventing hepatic inflammation. HDAC inhibition.	[61,65]
PAA	Induces steatosis.	[67]
BCAA	Alleviates hepatic steatosis and liver injury by suppressing FAS gene expression and protein levels. Suppresses the progression of NASH by reducing oxidative stress.	[71,72]
	Exacerbates hepatic oxidative stress, increases hepatic apoptosis.	[73]
Tryptamine	Reduces production of pro-inflammatory cytokines and migration of macrophages.	[74]
I3A	Reduces production of pro-inflammatory cytokines and migration of macrophages. Reduces the expression of FAS and SREBP1c.	[74]
IPA	Reduces weight gain. Improves intestinal barrier function.	[75,76]
Ethanol	Transfer of high-alcohol-producing Klebsiella pneumoniae by oral gavage into mice induces NAFLD.	[81]
Obeticholic acid (OCA)	Decreases hepatic inflammation by inhibition of pro-inflammatory cytokines. Decreases fibrogenesis by inhibition of pro-fibrotic cytokines. Inhibits LSEC and Kupffer cell activation.	[92]
Cholic acid (CA) Deoxycholic acid(DCA)	Changes in the composition of gut microbiota.	[94]
TMAO	Increases hepatic TG accumulation and lipogenesis. Shifts hepatic BA composition toward FXR-antagonistic activity.	[101]
	Human Studies	
Metabolites	**Effects**	**References**
Propionate	Prevents weight gain and insulin resistance.	[60]
BCAA	Positive correlation with insulin resistance and steatosis.	[67–69]
IPA	Negative correlation with obesity.	[75,76]
Ethanol	Positive correlation with NASH.	[79]
DCA	Associated with fibrosis in NAFLD.	[87]
OCA	Reduction in ALP, ALT and GGT.	[91]
TMAO	Positively correlated with NAFLD. Positively correlated with the serum levels of total BA and hepatic CYP7A1 mRNA.	[100,101]

3. From NASH to HCC: The Gut Microbiota-Associated Mechanisms

NASH progression to HCC shows mechanisms similar to those of NAFL progression to NASH. Dysbiosis and a leaky gut result in PAMP and gut microbiota-derived metabolite influx into the liver, thereby further triggering hepatic inflammation and disrupting metabolism homeostasis. Several groups of bacteria were associated with HCC. A previous study showed *E. coli* overgrowth in the intestines of HCC and cirrhosis patients [102]. Another study indicated that HCC patients, compared to cirrhosis patients, exhibited increased levels of *Bacteroides* and *Ruminococcaceae* spp. [103,104] and decreased levels of *Akkermensia* and *Bifidobacterium* spp. [104]. At the molecular level, PAMPs, such as LPS, activated signaling of TLRs, including TLR4 and TLR9, and induced cytokine and chemokine production, further inducing immune cell infiltration into the liver. PAMPs also activated HSCs via TLR activation to senescence-associated secretory phenotype (SASP) and induced epiregulin production, further promoting fibrosis development [105,106] (Figure 5). Dysbiosis affects metabolic functions via the gut microbiota-derived metabolites in NASH progression to HCC, like in NAFL progression to NASH. Primary BA conversion to secondary BAs by the gut microbiota is involved in HCC pathogenesis. Dysbiosis promotes HCC by inhibiting primary BA production, further inhibiting LSEC activation. Inhibition of LSEC activation results in chemokine ligand 6 (CXCL6) down-regulation, CXCL6-mediated natural killer T cell recruitment, and further loss of liver tumor growth control [107].

On the other hand, secondary BAs promote HCC development by activating HSC SASP and the hepatic mTOR pathway [108]. Thus, controlling the production of secondary BAs using antibiotics reduces HCC development [109] (Figure 5).

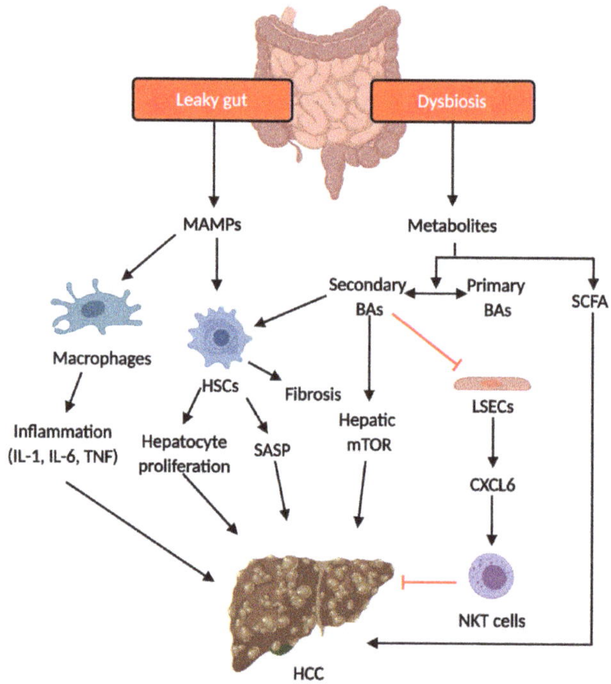

Figure 5. The mechanism of gut microbiota on the pathogenesis of HCC. Increased hepatic exposure to microbiota-derived metabolites and MAMPs results from dysbiosis and a leaky gut. Changes in BA pool (the ratio of primary BAs and secondary BAs) alter LSEC- and CXCL16-dependent NKT recruitment as well as HSC SASP. MAMPs induce the activation of macrophages, resulting in the production of pro-inflammatory cytokines including IL-1, IL-6, and TNF. Increased pro-inflammatory cytokines further contribute hepatic inflammation, which may also promote HCC development. MAMPs, microbiota-associated molecular patterns; HSCs, hepatic stellate cells; SASP, senescence-associated secretory phenotype; BAs, bile acids; LSECs, liver sinusoidal cells; SCFAs, short chain fatty acids; NKT cells, nature killer cells.

4. Potential Therapeutic Strategies and Non-Invasive Diagnosis

Owing to the higher daily calorie intake and sedentary lifestyle of NAFLD patients, the first step of NAFLD treatment includes weight loss by lifestyle modifications, including diet restriction and increased physical activity [110]. However, hepatic fat accumulation, inflammation, and necrosis are significantly improved only when more than 10% of the body weight is reduced [111,112]. Thus, lifestyle therapy appears to be insufficient for resolving NASH.

In addition to lifestyle interventions, potential NAFLD therapeutic strategies based on the gut microbiota and gut-liver axis have attracted attention in recent years. Antibiotic, prebiotic, and probiotic use can be applied to modulate the gut microbiota and prevent hepatocarcinogenesis. For example, a preclinical mouse model indicated that chronic oral administration of antibiotics decreased secondary bile acid levels, hepatic lipid accumulation, and attenuated hepatic inflammation and fibrosis via modulating the composition of gut microbiota [113,114]. On the contrary, Mahana D. et al. showed different results which indicated that mice treated with antibiotics exhibited severe insulin resistance and NAFLD and the composition of the gut microbiota was shifted from Firmicutes to Bifidobacterium, S24-7, and Prevotella [115]. The function of the gut microbiota is based on community, and a "healthy" microbiome has not been defined yet [116]. Therefore, these inconsistent results may arise from the complex community of gut microbiota. Otherwise, antibiotics may eliminate important species associated with healthy status and the risk of antibiotic resistance poses a larger concern, thereby reducing the efficiency of antibiotic use as a therapeutic strategy. Food ingredients which improve beneficial bacterial growth in the gut are termed prebiotics. In humans, supplementation with prebiotics such as oligofructose decreases the levels of hepatic inflammatory markers [117]. A previous study indicated that prebiotic treatment was negatively associated with endotoxin levels [118]. In addition to human studies, several animal studies have revealed the therapeutic potential of prebiotics. For example, prebiotic treatment reduced hepatic TG accumulation via the inhibition of expression of genes such as FAS, which is involved in the lipogenesis pathway [114]. On the other hand, probiotics are live bacteria which are beneficial to the host. For example, *Lactobacillus* and *Bifidobacterium* spp. have been reported to reduce gut inflammation and improve gut barrier function by remodeling the gut microbiota [119]. In humans, administration of *Lactobacillus acidophilus* reduced AST and ALT levels in NASH patients [120] and several clinical trials of probiotics were reported in other reviews [121]. However, most of the molecular mechanisms by which probiotics exert their functions remain unclear.

FMT is a novel therapeutic strategy which is defined as the transplantation of functional gut microbiota in healthy human feces into patients to alter the recipient's gut microbiota directly and normalize microbiota composition for therapeutic benefit [122]. Remarkable effectiveness of FMT was shown in patients with recurrent and refractory *Clostridium difficile* infection and has been confirmed as a clinical technique for treatment according to the 2013 guidelines [123–125]. FMT application, as a treatment strategy for extra-gastrointestinal diseases, has been evaluated in recent years. A previous study indicated that mice on HFDs showed decreased hepatic lipid accumulation and pro-inflammatory cytokine levels after FMT [126]. Additionally, FMT elevated the relative abundance of the beneficial bacterial species of *Christensenellaceae* and *Lactobacillus*, improved gut barrier function, and increased butyrate production, thereby further ameliorating endotoxemia [126]. In a human study, FMT from lean donors to individuals with metabolic syndrome temporarily increases insulin sensitivity [127]. A phase I clinical study has indicated that FMT with oral capsule restores microbial diversity and function and further reduced the recurrence of hepatic encephalopathy [128,129]. Although these human studies were not targeted toward the therapy of NAFLD, these findings still indicate the therapeutic potential of FMT. However, only a few control trials of FMT have been enrolled to date, and the role of FMT must be further examined because FMT application involves the risk of developing other pathogenic infections [130].

In addition to potential treatments, effective and non-invasive methods for diagnosing NAFLD are another important key to preventing HCC. Unfortunately, broadly applicable and non-invasive methods for diagnosing NAFLD are not available as yet. A recent study by Oh T.G. et al. demonstrated that a core gut microbiome signature can identify cirrhosis across separated cohorts, independent of disease etiology, host genetic, and environmental factors [131]. The identified disease microbiome included the elevated relative abundance of *Veillonella parvula*, *Veillonella atypica*, *Ruminococcus gnavus*, *Clostridium bolteae*, and *Acidaminococcus* sp. D21 and decreased abundance of *Eubacterium eligens*, *Eubacterium rectale*, and *Faecalibacterium prausnitzii* [131]. Although the results indicated the improved diagnostic accuracy in several cohorts, the authors claimed that these diagnostic methods need multi-center studies and well-phenotyped patients in order to be validated. However, it still is a promising non-invasive diagnostic method for NAFLD.

5. Conclusions

In general, the current NAFLD therapeutic strategies based on the gut microbiota and gut–liver axis mainly include prebiotic, probiotic, and FMT application. These therapeutic strategies improve NAFLD and HCC by recovering gut homeostasis from a state of dysbiosis, thereby improving gut barrier function to prevent endotoxemia, promoting anti-inflammatory effects and modulating gut microbiota-derived metabolite production. However, a huge gap in the development of therapies by targeting specific gut microbiota species or gut microbiota-derived metabolites remains. Although high-throughput sequencing including 16S rRNA and metagenomic sequencing help the researcher to identify gut microbiotas that are present in a sample without the need for culturing, the results only indicate the correlation of gut microbiota with diseases. Moving from association to causation remains a significant challenge. Specific species of gut microbiota may need to be cultured in order to conduct the causation test. Therefore, there is a strong demand for a culturomic technique. On the other hand, due to the complex community of gut microbiota, multi-omics analysis including transcriptomics, proteomics, and metabolomics may give us a glimpse of the entire disease picture and further contribute to the development of precision medicine. Therefore, advances in the understanding of the gut microbiota will allow the development of improved diagnostic, prognostic, and therapeutic strategies for liver diseases.

Author Contributions: Writing—original draft preparation, Y.-H.C.; writing—review and editing, W.-K.W., M.-S.W. All authors have read and agreed to the published version of the manuscript.

Funding: This research received no external funding.

Acknowledgments: We apologize to the authors whose work could not be cited due to space constraints.

Conflicts of Interest: The authors declare no conflict of interest.

Abbreviations

HCC	hepatocellular carcinoma
NAFLD	non-alcoholic fatty liver disease
NASH	non-alcoholic steatohepatitis
HBV	hepatitis B viruses
HCV	hepatitis C viruses
SCFA	short chain fatty acid
ZO-1	zonula occludens-1
DSS	dextran sodium sulfate
IL10	interleukin-10
IL6	interleukin-6
PPAR	peroxisome proliferator-activated receptor
FMT	fecal microbial transplantation
HFDs	high-fat diets
NLRP6	nucleotide-binding oligomerization domain -like receptor family pyrin domain containing 6
ASC	apoptosis-associated speck-like protein containing a caspase recruitment domain
TLR4	toll-like receptor 4
HSCs	hepatic stellate cells
LSECs	liver sinusoidal endothelial cells
MyD88	myeloid differentiation primary response 88
TNF-α	tumor necrosis factor-α
ROS	reactive oxygen species
JNK	c-Jun N-terminal kinase
PAMPs	pathogen-associated molecular patterns
NEFAs	non-esterified fatty acids
S1P	sphingosine 1-phosphate
TRAIL	TNF-related apoptosis-inducing ligand
DAMP	damage-associated molecular pattern
GPR	G-protein coupled receptor
HDAC	histone deacetylase
CREB	cyclic adenosine monophosphate (cAMP) response element binding protein
BCAAs	branched-chain amino acids

AAAs	aromatic amino acids
ACADSB	acyl-CoA dehydrogenase short/branched chain
IA	indoleacrylic acid
IAA	indole-3-acetic acid
I3A	indole-3-aldehyde
IPA	indole-3-propionic acid
SREBP1c	sterol regulatory element-binding protein-1c
AhR	aryl-hydrocarbon receptor
PXR	pregnane X receptor
CA	cholic acid
CDCA	chenodeoxycholic acid
DCA	deoxycholic acid
LCA	lithocholic acid
FXR	farnesoid X receptor
TGR5	takeda G-protein-coupled bile acid receptor 5
OCA	obeticholic acid
FGF	fibroblast growth factor
VLDL	very low-density lipoprotein
TMA	trimethylamine
TMAO	trimethylamine-N-oxide
CXCL6	chemokine ligand 6

References

1. Sia, D.; Villanueva, A.; Friedman, S.L.; Llovet, J.M. Liver cancer cell of origin, molecular class, and effects on patient prognosis. *Gastroenterology* **2017**, *152*, 745–761. [CrossRef] [PubMed]
2. Bray, F.; Ferlay, J.; Soerjomataram, I.; Siegel, R.L.; Torre, L.A.; Jemal, A. Global cancer statistics 2018: GLOBOCAN estimates of incidence and mortality worldwide for 36 cancers in 185 countries. *CA Cancer J. Clin.* **2018**, *68*, 394–424. [CrossRef]
3. Yang, J.D.; Hainaut, P.; Gores, G.J.; Amadou, A.; Plymoth, A.; Roberts, L.R. A global view of hepatocellular carcinoma: Trends, risk, prevention and management. *Nat. Rev. Gastroenterol. Hepatol.* **2019**, *16*, 589–604. [CrossRef] [PubMed]
4. Thomas, D.L. Global elimination of chronic hepatitis. *N. Engl. J. Med.* **2019**, *380*, 2041–2050. [CrossRef] [PubMed]
5. Ghouri, Y.A.; Mian, I.; Rowe, J.H. Review of hepatocellular carcinoma: Epidemiology, etiology, and carcinogenesis. *J. Carcinog.* **2017**, *16*, 1.
6. Sanyal, A.J. Past, present and future perspectives in nonalcoholic fatty liver disease. *Nat. Rev. Gastroenterol. Hepatol.* **2019**, *16*, 377–386. [CrossRef] [PubMed]
7. Younossi, Z.; Anstee, Q.M.; Marietti, M.; Hardy, T.; Henry, L.; Eslam, M.; George, J.; Bugianesi, E. Global burden of NAFLD and NASH: Trends, predictions, risk factors and prevention. *Nat. Rev. Gastroenterol. Hepatol.* **2018**, *15*, 11. [CrossRef] [PubMed]
8. Fan, J.-G.; Kim, S.-U.; Wong, V.W.-S. New trends on obesity and NAFLD in Asia. *J. Hepatol.* **2017**, *67*, 862–873. [CrossRef]
9. Estes, C.; Anstee, Q.M.; Arias-Loste, M.T.; Bantel, H.; Bellentani, S.; Caballeria, J.; Colombo, M.; Craxi, A.; Crespo, J.; Day, C.P.; et al. Modeling nafld disease burden in china, france, germany, italy, japan, spain, united kingdom, and united states for the period 2016–2030. *J. Hepatol.* **2018**, *69*, 896–904. [CrossRef]
10. Ferolla, S.M.; Armiliato, G.N.; Couto, C.A.; Ferrari, T.C. The role of intestinal bacteria overgrowth in obesity-related nonalcoholic fatty liver disease. *Nutrients* **2014**, *6*, 5583–5599. [CrossRef]
11. Rinella, M.; Charlton, M. The globalization of nonalcoholic fatty liver disease: Prevalence and impact on world health. *Hepatology* **2016**, *64*, 19–22. [CrossRef] [PubMed]
12. Estes, C.; Razavi, H.; Loomba, R.; Younossi, Z.; Sanyal, A.J. Modeling the epidemic of nonalcoholic fatty liver disease demonstrates an exponential increase in burden of disease. *Hepatology* **2018**, *67*, 123–133. [CrossRef] [PubMed]

13. O'Hara, S.P.; Karlsen, T.H.; LaRusso, N.F. Cholangiocytes and the environment in primary sclerosing cholangitis: Where is the link? *Gut* **2017**, *66*, 1873–1877. [CrossRef] [PubMed]
14. Brandl, K.; Kumar, V.; Eckmann, L. Gut-liver axis at the frontier of host-microbial interactions. *Am. J. Physiol. Gastrointest. Liver Physiol.* **2017**, *312*, G413–G419. [CrossRef]
15. Gupta, H.; Youn, G.S.; Shin, M.J.; Suk, K.T. Role of gut microbiota in hepatocarcinogenesis. *Microorganisms* **2019**, *7*, 121. [CrossRef]
16. Giau, V.V.; Wu, S.Y.; Jamerlan, A.; An, S.S.A.; Kim, S.; Hulme, J. Gut microbiota and their neuroinflammatory implications in Alzheimer's disease. *Nutrients* **2018**, *10*, 1765. [CrossRef]
17. Nishida, A.; Inoue, R.; Inatomi, O.; Bamba, S.; Naito, Y.; Andoh, A. Gut microbiota in the pathogenesis of inflammatory bowel disease. *Clin. J. Gastroenterol.* **2018**, *11*, 1–10. [CrossRef] [PubMed]
18. Ponziani, F.R.; Zocco, M.A.; Cerrito, L.; Gasbarrini, A.; Pompili, M. Bacterial translocation in patients with liver cirrhosis: Physiology, clinical consequences, and practical implications. *Expert Rev. Gastroenterol. Hepatol.* **2018**, *12*, 641–656. [CrossRef]
19. Carotti, S.; Guarino, M.P.L.; Vespasiani-Gentilucci, U.; Morini, S. Starring role of toll-like receptor-4 activation in the gut-liver axis. *World J. Gastrointest. Pathophysiol.* **2015**, *6*, 99. [CrossRef] [PubMed]
20. Del Chierico, F.; Nobili, V.; Vernocchi, P.; Russo, A.; De Stefanis, C.; Gnani, D.; Furlanello, C.; Zandonà, A.; Paci, P.; Capuani, G.; et al. Gut microbiota profiling of pediatric nonalcoholic fatty liver disease and obese patients unveiled by an integrated meta-omics-based approach. *Hepatology* **2017**, *65*, 451–464. [CrossRef]
21. Mouzaki, M.; Wang, A.Y.; Bandsma, R.; Comelli, E.M.; Arendt, B.M.; Zhang, L.; Fung, S.; Fischer, S.E.; McGilvray, I.G.; Allard, J.P. Bile acids and dysbiosis in non-alcoholic fatty liver disease. *PLoS ONE* **2016**, *11*, e0151829. [CrossRef] [PubMed]
22. Boursier, J.; Mueller, O.; Barret, M.; Machado, M.; Fizanne, L.; Araujo-Perez, F.; Guy, C.D.; Seed, P.C.; Rawls, J.F.; David, L.A. The severity of nonalcoholic fatty liver disease is associated with gut dysbiosis and shift in the metabolic function of the gut microbiota. *Hepatology* **2016**, *63*, 764–775. [CrossRef] [PubMed]
23. Shen, F.; Zheng, R.-D.; Sun, X.-Q.; Ding, W.-J.; Wang, X.-Y.; Fan, J.-G. Gut microbiota dysbiosis in patients with non-alcoholic fatty liver disease. *Hepatobiliary Pancreat. Dis. Int.* **2017**, *16*, 375–381. [CrossRef]
24. Loomba, R.; Seguritan, V.; Li, W.; Long, T.; Klitgord, N.; Bhatt, A.; Dulai, P.S.; Caussy, C.; Bettencourt, R.; Highlander, S.K.; et al. Gut microbiome-based metagenomic signature for non-invasive detection of advanced fibrosis in human nonalcoholic fatty liver disease. *Cell Metab.* **2017**, *25*, 1054–1062.e5. [CrossRef] [PubMed]
25. Da Silva, H.E.; Teterina, A.; Comelli, E.M.; Taibi, A.; Arendt, B.M.; Fischer, S.E.; Lou, W.; Allard, J.P. Nonalcoholic fatty liver disease is associated with dysbiosis independent of body mass index and insulin resistance. *Sci. Rep.* **2018**, *8*, 1–12.
26. Duarte, S.M.B.; Stefano, J.T.; Miele, L.; Ponziani, F.; Souza-Basqueira, M.; Okada, L.; de Costa, F.B.; Toda, K.; Mazo, D.; Sabino, E.; et al. Gut microbiome composition in lean patients with NASH is associated with liver damage independent of caloric intake: A prospective pilot study. *Nutr. Metab. Cardiovasc. Dis.* **2018**, *28*, 369–384. [CrossRef] [PubMed]
27. Nobili, V.; Putignani, L.; Mosca, A.; Del Chierico, F.; Vernocchi, P.; Alisi, A.; Stronati, L.; Cucchiara, S.; Toscano, M.; Drago, L. Bifidobacteria and lactobacilli in the gut microbiome of children with non-alcoholic fatty liver disease: Which strains act as health players? *Arch. Med Sci. AMS* **2018**, *14*, 81. [CrossRef]
28. Million, M.; Maraninchi, M.; Henry, M.; Armougom, F.; Richet, H.; Carrieri, P.; Valero, R.; Raccah, D.; Vialettes, B.; Raoult, D. Obesity-associated gut microbiota is enriched in Lactobacillus reuteri and depleted in Bifidobacterium animalis and Methanobrevibacter smithii. *Int. J. Obes.* **2012**, *36*, 817–825. [CrossRef] [PubMed]
29. Jumpertz, R.; Le, D.S.; Turnbaugh, P.J.; Trinidad, C.; Bogardus, C.; Gordon, J.I.; Krakoff, J. Energy-balance studies reveal associations between gut microbes, caloric load, and nutrient absorption in humans. *Am. J. Clin. Nutr.* **2011**, *94*, 58–65. [CrossRef] [PubMed]
30. Luther, J.; Garber, J.J.; Khalili, H.; Dave, M.; Bale, S.S.; Jindal, R.; Motola, D.L.; Luther, S.; Bohr, S.; Jeoung, S.W.; et al. Hepatic injury in nonalcoholic steatohepatitis contributes to altered intestinal permeability. *Cell. Mol. Gastroenterol. Hepatol.* **2015**, *1*, 222–232.e2. [CrossRef] [PubMed]
31. Giorgio, V.; Miele, L.; Principessa, L.; Ferretti, F.; Villa, M.P.; Negro, V.; Grieco, A.; Alisi, A.; Nobili, V. Intestinal permeability is increased in children with non-alcoholic fatty liver disease, and correlates with liver disease severity. *Dig. Liver Dis.* **2014**, *46*, 556–560. [CrossRef] [PubMed]

32. Miele, L.; Valenza, V.; La Torre, G.; Montalto, M.; Cammarota, G.; Ricci, R.; Masciana, R.; Forgione, A.; Gabrieli, M.L.; Perotti, G.; et al. Increased intestinal permeability and tight junction alterations in nonalcoholic fatty liver disease. *Hepatology* **2009**, *49*, 1877–1887. [CrossRef]
33. Rohr, M.W.; Narasimhulu, C.A.; Rudeski-Rohr, T.A.; Parthasarathy, S. Negative effects of a high-fat diet on intestinal permeability: A review. *Adv. Nutr.* **2020**, *11*, 77–91. [CrossRef] [PubMed]
34. Wu, W.; Lv, L.; Shi, D.; Ye, J.; Fang, D.; Guo, F.; Li, Y.; He, X.; Li, L. Protective effect of Akkermansia muciniphila against immune-mediated liver injury in a mouse model. *Front. Microbiol.* **2017**, *8*, 1804. [CrossRef] [PubMed]
35. Lam, Y.Y.; Ha, C.W.; Campbell, C.R.; Mitchell, A.J.; Dinudom, A.; Oscarsson, J.; Cook, D.I.; Hunt, N.H.; Caterson, I.D.; Holmes, A.J.; et al. Increased gut permeability and microbiota change associate with mesenteric fat inflammation and metabolic dysfunction in diet-induced obese mice. *PLoS ONE* **2012**, *7*, e34233. [CrossRef] [PubMed]
36. Ling, X.; Linglong, P.; Du Weixia, W.H. Protective effects of bifidobacterium on intestinal barrier function in LPS-induced enterocyte barrier injury of Caco-2 monolayers and in a rat NEC model. *PLoS ONE* **2016**, *11*, e0161635. [CrossRef] [PubMed]
37. Lam, Y.Y.; Ha, C.W.; Hoffmann, J.M.; Oscarsson, J.; Dinudom, A.; Mather, T.J.; Cook, D.I.; Hunt, N.H.; Caterson, I.D.; Holmes, A.J.; et al. Effects of dietary fat profile on gut permeability and microbiota and their relationships with metabolic changes in mice. *Obesity* **2015**, *23*, 1429–1439. [CrossRef] [PubMed]
38. Xia, T.; Zhang, B.; Li, S.; Fang, B.; Duan, W.; Zhang, J.; Song, J.; Wang, M. Vinegar extract ameliorates alcohol-induced liver damage associated with the modulation of gut microbiota in mice. *Food Funct.* **2020**, *11*, 2898–2909. [CrossRef] [PubMed]
39. Din, A.U.; Hassan, A.; Zhu, Y.; Zhang, K.; Wang, Y.; Li, T.; Wang, Y.; Wang, G. Inhibitory effect of Bifidobacterium bifidum ATCC 29521 on colitis and its mechanism. *J. Nutr. Biochem.* **2020**, *79*, 108353. [CrossRef]
40. Mouries, J.; Brescia, P.; Silvestri, A.; Spadoni, I.; Sorribas, M.; Wiest, R.; Mileti, E.; Galbiati, M.; Invernizzi, P.; Adorini, L.; et al. Microbiota-driven gut vascular barrier disruption is a prerequisite for non-alcoholic steatohepatitis development. *J. Hepatol.* **2019**, *71*, 1216–1228. [CrossRef]
41. Schroder, K.; Tschopp, J. The inflammasomes. *Cell* **2010**, *140*, 821–832. [CrossRef]
42. Wlodarska, M.; Thaiss, C.A.; Nowarski, R.; Henao-Mejia, J.; Zhang, J.-P.; Brown, E.M.; Frankel, G.; Levy, M.; Katz, M.N.; Philbrick, W.M.; et al. NLRP6 inflammasome orchestrates the colonic host-microbial interface by regulating goblet cell mucus secretion. *Cell* **2014**, *156*, 1045–1059. [CrossRef]
43. Levy, M.; Thaiss, C.A.; Zeevi, D.; Dohnalova, L.; Zilberman-Schapira, G.; Mahdi, J.A.; David, E.; Savidor, A.; Korem, T.; Herzig, Y.; et al. Microbiota-modulated metabolites shape the intestinal microenvironment by regulating NLRP6 inflammasome signaling. *Cell* **2015**, *163*, 1428–1443. [CrossRef] [PubMed]
44. Li, J.-M.; Yu, R.; Zhang, L.-P.; Wen, S.-Y.; Wang, S.-J.; Zhang, X.-Y.; Xu, Q.; Kong, L.-D. Dietary fructose-induced gut dysbiosis promotes mouse hippocampal neuroinflammation: A benefit of short-chain fatty acids. *Microbiome* **2019**, *7*, 98. [CrossRef] [PubMed]
45. Elinav, E.; Strowig, T.; Kau, A.L.; Henao-Mejia, J.; Thaiss, C.A.; Booth, C.J.; Peaper, D.R.; Bertin, J.; Eisenbarth, S.C.; Gordon, J.I.; et al. NLRP6 inflammasome regulates colonic microbial ecology and risk for colitis. *Cell* **2011**, *145*, 745–757. [CrossRef] [PubMed]
46. Gäbele, E.; Dostert, K.; Hofmann, C.; Wiest, R.; Schölmerich, J.; Hellerbrand, C.; Obermeier, F. DSS induced colitis increases portal LPS levels and enhances hepatic inflammation and fibrogenesis in experimental NASH. *J. Hepatol.* **2011**, *55*, 1391–1399. [CrossRef] [PubMed]
47. Rahman, K.; Desai, C.; Iyer, S.S.; Thorn, N.E.; Kumar, P.; Liu, Y.; Smith, T.; Neish, A.S.; Li, H.; Tan, S.; et al. Loss of junctional adhesion molecule a promotes severe steatohepatitis in mice on a diet high in saturated fat, fructose, and cholesterol. *Gastroenterology* **2016**, *151*, 733–746.e12. [CrossRef]
48. Kolodziejczyk, A.A.; Zheng, D.; Shibolet, O.; Elinav, E. The role of the microbiome in NAFLD and NASH. *EMBO Mol. Med.* **2019**, *11*, e9302. [CrossRef]
49. Verdam, F.J.; Rensen, S.S.; Driessen, A.; Greve, J.W.; Buurman, W.A. Novel evidence for chronic exposure to endotoxin in human nonalcoholic steatohepatitis. *J. Clin. Gastroenterol.* **2011**, *45*, 149–152. [CrossRef]
50. Miura, K.; Kodama, Y.; Inokuchi, S.; Schnabl, B.; Aoyama, T.; Ohnishi, H.; Olefsky, J.M.; Brenner, D.A.; Seki, E. Toll-like receptor 9 promotes steatohepatitis by induction of interleukin-1β in mice. *Gastroenterology* **2010**, *139*, 323–334. [CrossRef]

51. Szabo, G.; Petrasek, J. Inflammasome activation and function in liver disease. *Nat. Rev. Gastroenterol. Hepatol.* **2015**, *12*, 387. [CrossRef] [PubMed]
52. Wree, A.; McGeough, M.D.; Peña, C.A.; Schlattjan, M.; Li, H.; Inzaugarat, M.E.; Messer, K.; Canbay, A.; Hoffman, H.M.; Feldstein, A.E. NLRP3 inflammasome activation is required for fibrosis development in NAFLD. *J. Mol. Med.* **2014**, *92*, 1069–1082. [CrossRef] [PubMed]
53. Musso, G.; Cassader, M.; Paschetta, E.; Gambino, R. Bioactive lipid species and metabolic pathways in progression and resolution of nonalcoholic steatohepatitis. *Gastroenterology* **2018**, *155*, 282–302.e8. [CrossRef] [PubMed]
54. Yue, F.; Xia, K.; Wei, L.; Xing, L.; Wu, S.; Shi, Y.; Lam, S.M.; Shui, G.; Xiang, X.; Russell, R.; et al. Effects of constant light exposure on sphingolipidomics and progression of NASH in high-fat-fed rats. *J. Gastroenterol. Hepatol.* **2020**. [CrossRef] [PubMed]
55. Musso, G.; Cassader, M.; Gambino, R. Non-alcoholic steatohepatitis: Emerging molecular targets and therapeutic strategies. *Nat. Rev. Drug Discov.* **2016**, *15*, 249. [CrossRef]
56. Kaufman, R.J. Orchestrating the unfolded protein response in health and disease. *J. Clin. Investig.* **2002**, *110*, 1389–1398. [CrossRef]
57. Hirsova, P.; Ibrabim, S.H.; Gores, G.J.; Malhi, H. Lipotoxic lethal and sublethal stress signaling in hepatocytes: Relevance to NASH pathogenesis. *J. Lipid Res.* **2016**, *57*, 1758–1770. [CrossRef]
58. Ibrahim, S.H.; Hirsova, P.; Gores, G.J. Non-alcoholic steatohepatitis pathogenesis: Sublethal hepatocyte injury as a driver of liver inflammation. *Gut* **2018**, *67*, 963–972. [CrossRef]
59. Canfora, E.E.; Meex, R.C.; Venema, K.; Blaak, E.E. Gut microbial metabolites in obesity, NAFLD and T2DM. *Nat. Rev. Endocrinol.* **2019**, *15*, 261–273. [CrossRef]
60. Chambers, E.S.; Viardot, A.; Psichas, A.; Morrison, D.J.; Murphy, K.G.; Zac-Varghese, S.E.; MacDougall, K.; Preston, T.; Tedford, C.; Finlayson, G.S.; et al. Effects of targeted delivery of propionate to the human colon on appetite regulation, body weight maintenance and adiposity in overweight adults. *Gut* **2015**, *64*, 1744–1754. [CrossRef]
61. Zhou, D.; Pan, Q.; Liu, X.L.; Yang, R.X.; Chen, Y.W.; Liu, C.; Fan, J.G. Clostridium butyricum B1 alleviates high-fat diet-induced steatohepatitis in mice via enterohepatic immunoregulation. *J. Gastroenterol. Hepatol.* **2017**, *32*, 1640–1648. [CrossRef] [PubMed]
62. Brown, A.J.; Goldsworthy, S.M.; Barnes, A.A.; Eilert, M.M.; Tcheang, L.; Daniels, D.; Muir, A.I.; Wigglesworth, M.J.; Kinghorn, I.; Fraser, N.J. The Orphan G protein-coupled receptors GPR41 and GPR43 are activated by propionate and other short chain carboxylic acids. *J. Biol. Chem.* **2003**, *278*, 11312–11319. [CrossRef]
63. Le Poul, E.; Loison, C.; Struyf, S.; Springael, J.-Y.; Lannoy, V.; Decobecq, M.-E.; Brezillon, S.; Dupriez, V.; Vassart, G.; Van Damme, J.; et al. Functional characterization of human receptors for short chain fatty acids and their role in polymorphonuclear cell activation. *J. Biol. Chem.* **2003**, *278*, 25481–25489. [CrossRef] [PubMed]
64. Li, B.-Y.; Xu, X.-Y.; Gan, R.-Y.; Sun, Q.-C.; Meng, J.-M.; Shang, A.; Mao, Q.-Q.; Li, H.-B. Targeting Gut Microbiota for the Prevention and Management of Diabetes Mellitus by Dietary Natural Products. *Foods* **2019**, *8*, 440. [CrossRef] [PubMed]
65. Schilderink, R.; Verseijden, C.; de Jonge, W.J. Dietary inhibitors of histone deacetylases in intestinal immunity and homeostasis. *Front. Immunol.* **2013**, *4*, 226. [CrossRef]
66. Iida, A.; Kuranuki, S.; Yamamoto, R.; Uchida, M.; Ohta, M.; Ichimura, M.; Tsuneyama, K.; Masaki, T.; Seike, M.; Nakamura, T. Analysis of amino acid profiles of blood over time and biomarkers associated with non-alcoholic steatohepatitis in STAM mice. *Exp. Anim.* **2019**, *68*, 417–428. [CrossRef]
67. Hoyles, L.; Fernandez-Real, J.-M.; Federici, M.; Serino, M.; Abbott, J.; Charpentier, J.; Heymes, C.; Luque, J.L.; Anthony, E.; Barton, R.H.; et al. Molecular phenomics and metagenomics of hepatic steatosis in non-diabetic obese women. *Nat. Med.* **2018**, *24*, 1070–1080. [CrossRef]
68. Wang, T.J.; Larson, M.G.; Vasan, R.S.; Cheng, S.; Rhee, E.P.; McCabe, E.; Lewis, G.D.; Fox, C.S.; Jacques, P.F.; Fernandez, C. Metabolite profiles and the risk of developing diabetes. *Nat. Med.* **2011**, *17*, 448. [CrossRef]
69. Menni, C.; Fauman, E.; Erte, I.; Perry, J.R.; Kastenmüller, G.; Shin, S.-Y.; Petersen, A.-K.; Hyde, C.; Psatha, M.; Ward, K.J.; et al. Biomarkers for type 2 diabetes and impaired fasting glucose using a nontargeted metabolomics approach. *Diabetes* **2013**, *62*, 4270–4276. [CrossRef]

70. Pedersen, H.K.; Gudmundsdottir, V.; Nielsen, H.B.; Hyotylainen, T.; Nielsen, T.; Jensen, B.A.; Forslund, K.; Hildebrand, F.; Prifti, E.; Falony, G.; et al. Human gut microbes impact host serum metabolome and insulin sensitivity. *Nature* **2016**, *535*, 376–381. [CrossRef]
71. Tanaka, H.; Fukahori, S.; Baba, S.; Ueno, T.; Sivakumar, R.; Yagi, M.; Asagiri, K.; Ishii, S.; Tanaka, Y. Branched-Chain Amino Acid–Rich Supplements Containing Microelements Have Antioxidant Effects on Nonalcoholic Steatohepatitis in Mice. *J. Parenter. Enter. Nutr.* **2016**, *40*, 519–528. [CrossRef] [PubMed]
72. Honda, T.; Ishigami, M.; Luo, F.; Lingyun, M.; Ishizu, Y.; Kuzuya, T.; Hayashi, K.; Nakano, I.; Ishikawa, T.; Feng, G.-G.; et al. Branched-chain amino acids alleviate hepatic steatosis and liver injury in choline-deficient high-fat diet induced NASH mice. *Metabolism* **2017**, *69*, 177–187. [CrossRef] [PubMed]
73. Zhang, F.; Zhao, S.; Yan, W.; Xia, Y.; Chen, X.; Wang, W.; Zhang, J.; Gao, C.; Peng, C.; Yan, F.; et al. Branched chain amino acids cause liver injury in obese/diabetic mice by promoting adipocyte lipolysis and inhibiting hepatic autophagy. *EBioMedicine* **2016**, *13*, 157–167. [CrossRef] [PubMed]
74. Krishnan, S.; Ding, Y.; Saedi, N.; Choi, M.; Sridharan, G.V.; Sherr, D.H.; Yarmush, M.L.; Alaniz, R.C.; Jayaraman, A.; Lee, K. Gut microbiota-derived tryptophan metabolites modulate inflammatory response in hepatocytes and macrophages. *Cell Rep.* **2018**, *23*, 1099–1111. [CrossRef] [PubMed]
75. Konopelski, P.; Konop, M.; Gawrys-Kopczynska, M.; Podsadni, P.; Szczepanska, A.; Ufnal, M. Indole-3-propionic acid, a tryptophan-derived bacterial metabolite, reduces weight gain in rats. *Nutrients* **2019**, *11*, 591. [CrossRef]
76. Jennis, M.; Cavanaugh, C.; Leo, G.; Mabus, J.; Lenhard, J.; Hornby, P. Microbiota-derived tryptophan indoles increase after gastric bypass surgery and reduce intestinal permeability in vitro and in vivo. *Neurogastroenterol. Motil.* **2018**, *30*, e13178. [CrossRef]
77. Venkatesh, M.; Mukherjee, S.; Wang, H.; Li, H.; Sun, K.; Benechet, A.P.; Qiu, Z.; Maher, L.; Redinbo, M.R.; Phillips, R.S.; et al. Symbiotic bacterial metabolites regulate gastrointestinal barrier function via the xenobiotic sensor PXR and Toll-like receptor 4. *Immunity* **2014**, *41*, 296–310. [CrossRef]
78. Cope, K.; Risby, T.; Diehl, A.M. Increased gastrointestinal ethanol production in obese mice: Implications for fatty liver disease pathogenesis. *Gastroenterology* **2000**, *119*, 1340–1347. [CrossRef]
79. Zhu, L.; Baker, S.S.; Gill, C.; Liu, W.; Alkhouri, R.; Baker, R.D.; Gill, S.R. Characterization of gut microbiomes in nonalcoholic steatohepatitis (NASH) patients: A connection between endogenous alcohol and NASH. *Hepatology* **2013**, *57*, 601–609. [CrossRef]
80. Fuster, D.; Samet, J.H. Alcohol use in patients with chronic liver disease. *N. Engl. J. Med.* **2018**, *379*, 1251–1261. [CrossRef]
81. Yuan, J.; Chen, C.; Cui, J.; Lu, J.; Yan, C.; Wei, X.; Zhao, X.; Li, N.; Li, S.; Xue, G. Fatty liver disease caused by high-alcohol-producing Klebsiella pneumoniae. *Cell Metab.* **2019**, *30*, 675–688. [CrossRef] [PubMed]
82. Baker, S.S.; Baker, R.D.; Liu, W.; Nowak, N.J.; Zhu, L. Role of alcohol metabolism in non-alcoholic steatohepatitis. *PLoS ONE* **2010**, *5*, e9570. [CrossRef] [PubMed]
83. Ghoshal, U.C.; Goel, A.; Quigley, E.M. Gut microbiota abnormalities, small intestinal bacterial overgrowth, and non-alcoholic fatty liver disease: An emerging paradigm. *Indian J. Gastroenterol.* **2020**, *39*, 9–21. [CrossRef] [PubMed]
84. Sipe, L.M.; Chaib, M.; Pingili, A.K.; Pierre, J.F.; Makowski, L. Microbiome, bile acids, and obesity: How microbially modified metabolites shape anti-tumor immunity. *Immunol. Rev.* **2020**, *295*, 220–239. [CrossRef]
85. Gérard, P. Metabolism of cholesterol and bile acids by the gut microbiota. *Pathogens* **2014**, *3*, 14–24. [CrossRef]
86. Sayin, S.I.; Wahlström, A.; Felin, J.; Jäntti, S.; Marschall, H.-U.; Bamberg, K.; Angelin, B.; Hyötyläinen, T.; Orešič, M.; Bäckhed, F. Gut microbiota regulates bile acid metabolism by reducing the levels of tauro-beta-muricholic acid, a naturally occurring FXR antagonist. *Cell Metab.* **2013**, *17*, 225–235. [CrossRef]
87. Adams, L.A.; Wang, Z.; Liddle, C.; Melton, P.E.; Ariff, A.; Chandraratna, H.; Tan, J.; Ching, H.; Coulter, S.; de Boer, B.; et al. Bile acids associate with specific gut microbiota, low-level alcohol consumption and liver fibrosis in patients with non-alcoholic fatty liver disease. *Liver Int.* **2020**, *40*, 1356–1365. [CrossRef]
88. Jiang, C.; Xie, C.; Li, F.; Zhang, L.; Nichols, R.G.; Krausz, K.W.; Cai, J.; Qi, Y.; Fang, Z.-Z.; Takahashi, S.; et al. Intestinal farnesoid X receptor signaling promotes nonalcoholic fatty liver disease. *J. Clin. Investig.* **2015**, *125*, 386–402. [CrossRef]
89. Jahn, D.; Rau, M.; Hermanns, H.M.; Geier, A. Mechanisms of enterohepatic fibroblast growth factor 15/19 signaling in health and disease. *Cytokine Growth Factor Rev.* **2015**, *26*, 625–635. [CrossRef]

90. De Wit, N.; Derrien, M.; Bosch-Vermeulen, H.; Oosterink, E.; Keshtkar, S.; Duval, C.; Bosch, J.; Kleerebezem, M.; Müller, M.; van der Meer, R. Saturated fat stimulates obesity and hepatic steatosis and affects gut microbiota composition by an enhanced overflow of dietary fat to the distal intestine. *Am. J. Physiol. Gastrointest. Liver Physiol.* **2012**, *303*, G589–G599. [CrossRef]
91. Hirschfield, G.M.; Mason, A.; Luketic, V.; Lindor, K.; Gordon, S.C.; Mayo, M.; Kowdley, K.V.; Vincent, C.; Bodhenheimer, H.C., Jr.; Parés, A.; et al. Efficacy of obeticholic acid in patients with primary biliary cirrhosis and inadequate response to ursodeoxycholic acid. *Gastroenterology* **2015**, *148*, 751–761.e8. [CrossRef] [PubMed]
92. Verbeke, L.; Mannaerts, I.; Schierwagen, R.; Govaere, O.; Klein, S.; Elst, I.; Windmolders, P.; Farre, R.; Wenes, M.; Mazzone, M.; et al. FXR agonist obeticholic acid reduces hepatic inflammation and fibrosis in a rat model of toxic cirrhosis. *Sci. Rep.* **2016**, *6*, 1–12.
93. Devkota, S.; Wang, Y.; Musch, M.W.; Leone, V.; Fehlner-Peach, H.; Nadimpalli, A.; Antonopoulos, D.A.; Jabri, B.; Chang, E.B. Dietary-fat-induced taurocholic acid promotes pathobiont expansion and colitis in Il10−/− mice. *Nature* **2012**, *487*, 104–108. [CrossRef] [PubMed]
94. Islam, K.S.; Fukiya, S.; Hagio, M.; Fujii, N.; Ishizuka, S.; Ooka, T.; Ogura, Y.; Hayashi, T.; Yokota, A. Bile acid is a host factor that regulates the composition of the cecal microbiota in rats. *Gastroenterology* **2011**, *141*, 1773–1781. [CrossRef] [PubMed]
95. Katsuma, S.; Hirasawa, A.; Tsujimoto, G. Bile acids promote glucagon-like peptide-1 secretion through TGR5 in a murine enteroendocrine cell line STC-1. *Biochem. Biophys. Res. Commun.* **2005**, *329*, 386–390. [CrossRef] [PubMed]
96. Alarcon, C.; Wicksteed, B.; Rhodes, C. Exendin 4 controls insulin production in rat islet beta cells predominantly by potentiation of glucose-stimulated proinsulin biosynthesis at the translational level. *Diabetologia* **2006**, *49*, 2920–2929. [CrossRef]
97. Smallwood, T.; Allayee, H.; Bennett, B.J. Choline metabolites: Gene by diet interactions. *Curr. Opin. Lipidol.* **2016**, *27*, 33. [CrossRef]
98. Romano, K.A.; Vivas, E.I.; Amador-Noguez, D.; Rey, F.E. Intestinal microbiota composition modulates choline bioavailability from diet and accumulation of the proatherogenic metabolite trimethylamine-N-oxide. *MBio* **2015**, 6. [CrossRef]
99. Wu, W.-K.; Chen, C.-C.; Liu, P.-Y.; Panyod, S.; Liao, B.-Y.; Chen, P.-C.; Kao, H.-L.; Kuo, H.-C.; Kuo, C.-H.; Chiu, T.H.; et al. Identification of TMAO-producer phenotype and host–diet–gut dysbiosis by carnitine challenge test in human and germ-free mice. *Gut* **2019**, *68*, 1439–1449. [CrossRef]
100. Chen, Y.-M.; Liu, Y.; Zhou, R.-F.; Chen, X.-L.; Wang, C.; Tan, X.-Y.; Wang, L.-J.; Zheng, R.-D.; Zhang, H.-W.; Ling, W.-H.; et al. Associations of gut-flora-dependent metabolite trimethylamine-N-oxide, betaine and choline with non-alcoholic fatty liver disease in adults. *Sci. Rep.* **2016**, *6*, 1–9.
101. Tan, X.; Liu, Y.; Long, J.; Chen, S.; Liao, G.; Wu, S.; Li, C.; Wang, L.; Ling, W.; Zhu, H. Trimethylamine N-Oxide Aggravates Liver Steatosis through Modulation of Bile Acid Metabolism and Inhibition of Farnesoid X Receptor Signaling in Nonalcoholic Fatty Liver Disease. *Mol. Nutr. Food Res.* **2019**, *63*, 1900257.
102. Grąt, M.; Wronka, K.; Krasnodębski, M.; Lewandowski, Z.; Kosińska, I.; Grąt, K.; Stypułkowski, J.; Rejowski, S.; Wasilewicz, M.; Gałęcka, M. Profile of gut microbiota associated with the presence of hepatocellular cancer in patients with liver cirrhosis. In *Transplantation Proceedings*; Elsevier: Amsterdam, The Netherlands, 2016; pp. 1687–1691.
103. Ni, J.; Huang, R.; Zhou, H.; Xu, X.; Li, Y.; Cao, P.; Zhong, K.; Ge, M.; Chen, X.; Hou, B.; et al. Analysis of the relationship between the degree of dysbiosis in gut microbiota and prognosis at different stages of primary hepatocellular carcinoma. *Front. Microbiol.* **2019**, *10*, 1458. [CrossRef]
104. Ponziani, F.R.; Bhoori, S.; Castelli, C.; Putignani, L.; Rivoltini, L.; Del Chierico, F.; Sanguinetti, M.; Morelli, D.; Sterbini, F.P.; Petito, V.; et al. Hepatocellular carcinoma is associated with gut microbiota profile and inflammation in nonalcoholic fatty liver disease. *Hepatology* **2019**, *69*, 107–120. [PubMed]
105. Dapito, D.H.; Mencin, A.; Gwak, G.-Y.; Pradere, J.-P.; Jang, M.-K.; Mederacke, I.; Caviglia, J.M.; Khiabanian, H.; Adeyemi, A.; Bataller, R.; et al. Promotion of hepatocellular carcinoma by the intestinal microbiota and TLR4. *Cancer Cell* **2012**, *21*, 504–516. [PubMed]
106. Schwabe, R.F.; Greten, T.F. Gut microbiome in HCC–Mechanisms, diagnosis and therapy. *J. Hepatol.* **2020**, *72*, 230–238. [CrossRef]

107. Ma, C.; Han, M.; Heinrich, B.; Fu, Q.; Zhang, Q.; Sandhu, M.; Agdashian, D.; Terabe, M.; Berzofsky, J.A.; Fako, V. Gut microbiome–mediated bile acid metabolism regulates liver cancer via NKT cells. *Science* **2018**, *360*, eaan5931. [CrossRef]
108. Yamada, S.; Takashina, Y.; Watanabe, M.; Nagamine, R.; Saito, Y.; Kamada, N.; Saito, H. Bile acid metabolism regulated by the gut microbiota promotes non-alcoholic steatohepatitis-associated hepatocellular carcinoma in mice. *Oncotarget* **2018**, *9*, 9925.
109. Francescone, R.; Hou, V.; Grivennikov, S.I. Microbiome, inflammation and cancer. *Cancer J.* **2014**, *20*, 181. [CrossRef]
110. Gerber, L.; Otgonsuren, M.; Mishra, A.; Escheik, C.; Birerdinc, A.; Stepanova, M.; Younossi, Z. Non-alcoholic fatty liver disease (NAFLD) is associated with low level of physical activity: A population-based study. *Aliment. Pharmacol. Ther.* **2012**, *36*, 772–781. [CrossRef]
111. Promrat, K.; Kleiner, D.E.; Niemeier, H.M.; Jackvony, E.; Kearns, M.; Wands, J.R.; Fava, J.L.; Wing, R.R. Randomized controlled trial testing the effects of weight loss on nonalcoholic steatohepatitis. *Hepatology* **2010**, *51*, 121–129.
112. Vilar-Gomez, E.; Martinez-Perez, Y.; Calzadilla-Bertot, L.; Torres-Gonzalez, A.; Gra-Oramas, B.; Gonzalez-Fabian, L.; Friedman, S.L.; Diago, M.; Romero-Gomez, M. Weight loss through lifestyle modification significantly reduces features of nonalcoholic steatohepatitis. *Gastroenterology* **2015**, *149*, 367–378.e5. [CrossRef] [PubMed]
113. Janssen, A.W.; Houben, T.; Katiraei, S.; Dijk, W.; Boutens, L.; Van Der Bolt, N.; Wang, Z.; Brown, J.M.; Hazen, S.L.; Mandard, S.; et al. Modulation of the gut microbiota impacts nonalcoholic fatty liver disease: A potential role for bile acids. *J. Lipid Res.* **2017**, *58*, 1399–1416. [CrossRef] [PubMed]
114. Safari, Z.; Gérard, P. The links between the gut microbiome and non-alcoholic fatty liver disease (NAFLD). *Cell. Mol. Life Sci.* **2019**, *76*, 1541–1558. [PubMed]
115. Mahana, D.; Trent, C.M.; Kurtz, Z.D.; Bokulich, N.A.; Battaglia, T.; Chung, J.; Müller, C.L.; Li, H.; Bonneau, R.A.; Blaser, M.J. Antibiotic perturbation of the murine gut microbiome enhances the adiposity, insulin resistance, and liver disease associated with high-fat diet. *Genome Med.* **2016**, *8*, 1–20.
116. McBurney, M.I.; Davis, C.; Fraser, C.M.; Schneeman, B.O.; Huttenhower, C.; Verbeke, K.; Walter, J.; Latulippe, M.E. Establishing what constitutes a healthy human gut microbiome: State of the science, regulatory considerations, and future directions. *J. Nutr.* **2019**, *149*, 1882–1895.
117. Pachikian, B.D.; Essaghir, A.; Demoulin, J.B.; Catry, E.; Neyrinck, A.M.; Dewulf, E.M.; Sohet, F.M.; Portois, L.; Clerbaux, L.A.; Carpentier, Y.A.; et al. Prebiotic approach alleviates hepatic steatosis: Implication of fatty acid oxidative and cholesterol synthesis pathways. *Mol. Nutr. Food Res.* **2013**, *57*, 347–359.
118. Daubioul, C.; Horsmans, Y.; Lambert, P.; Danse, E.; Delzenne, N.M. Effects of oligofructose on glucose and lipid metabolism in patients with nonalcoholic steatohepatitis: Results of a pilot study. *Eur. J. Clin. Nutr.* **2005**, *59*, 723–726.
119. Kumar, M.; Verma, V.; Nagpal, R.; Kumar, A.; Gautam, S.K.; Behare, P.V.; Grover, C.R.; Aggarwal, P.K. Effect of probiotic fermented milk and chlorophyllin on gene expressions and genotoxicity during AFB1-induced hepatocellular carcinoma. *Gene* **2011**, *490*, 54–59. [CrossRef]
120. Monem, S.M.A. Probiotic therapy in patients with nonalcoholic steatohepatitis in Zagazig University hospitals. *Euroasian J. Hepatogastroenterol.* **2017**, *7*, 101. [CrossRef]
121. Meroni, M.; Longo, M.; Dongiovanni, P. The role of probiotics in nonalcoholic fatty liver disease: A new insight into therapeutic strategies. *Nutrients* **2019**, *11*, 2642.
122. Gupta, A.; Khanna, S. Fecal microbiota transplantation. *JAMA* **2017**, *318*, 102. [CrossRef] [PubMed]
123. Khoruts, A.; Dicksved, J.; Jansson, J.K.; Sadowsky, M.J. Changes in the composition of the human fecal microbiome after bacteriotherapy for recurrent Clostridium difficile-associated diarrhea. *J. Clin. Gastroenterol.* **2010**, *44*, 354–360. [CrossRef] [PubMed]
124. Van Nood, E.; Vrieze, A.; Nieuwdorp, M.; Fuentes, S.; Zoetendal, E.G.; de Vos, W.M.; Visser, C.E.; Kuijper, E.J.; Bartelsman, J.F.; Tijssen, J.G.; et al. Duodenal infusion of donor feces for recurrent Clostridium difficile. *N. Engl. J. Med.* **2013**, *368*, 407–415. [PubMed]
125. Surawicz, C.M.; Brandt, L.J.; Binion, D.G.; Ananthakrishnan, A.N.; Curry, S.R.; Gilligan, P.H.; McFarland, L.V.; Mellow, M.; Zuckerbraun, B.S. Guidelines for Diagnosis, Treatment, and Prevention of Clostridium difficileInfections. *Am. J. Gastroenterol.* **2013**, *108*, 478–498. [CrossRef] [PubMed]

126. Zhou, D.; Pan, Q.; Shen, F.; Cao, H.-X.; Ding, W.-J.; Chen, Y.-W.; Fan, J.-G. Total fecal microbiota transplantation alleviates high-fat diet-induced steatohepatitis in mice via beneficial regulation of gut microbiota. *Sci. Rep.* **2017**, *7*, 1–11.
127. Vrieze, A.; Van Nood, E.; Holleman, F.; Salojärvi, J.; Kootte, R.S.; Bartelsman, J.F.; Dallinga–Thie, G.M.; Ackermans, M.T.; Serlie, M.J.; Oozeer, R. Transfer of intestinal microbiota from lean donors increases insulin sensitivity in individuals with metabolic syndrome. *Gastroenterology* **2012**, *143*, 913–916.e7. [CrossRef]
128. Bajaj, J.S.; Salzman, N.H.; Acharya, C.; Sterling, R.K.; White, M.B.; Gavis, E.A.; Fagan, A.; Hayward, M.; Holtz, M.L.; Matherly, S.; et al. Fecal Microbial Transplant Capsules Are Safe in Hepatic Encephalopathy: A Phase 1, Randomized, Placebo-Controlled Trial. *Hepatology* **2019**, *70*, 1690–1703. [CrossRef]
129. Bajaj, J.S.; Kakiyama, G.; Savidge, T.; Takei, H.; Kassam, Z.A.; Fagan, A.; Gavis, E.A.; Pandak, W.M.; Nittono, H.; Hylemon, P.B.; et al. Antibiotic-associated disruption of microbiota composition and function in cirrhosis is restored by fecal transplant. *Hepatology* **2018**, *68*, 1549–1558. [CrossRef]
130. Leylabadlo, H.E.; Ghotaslou, R.; Kafil, H.S.; Feizabadi, M.M.; Moaddab, S.Y.; Farajnia, S.; Sheykhsaran, E.; Sanaie, S.; Shanehbandi, D.; Baghi, H.B. Non-alcoholic fatty liver diseases: From role of gut microbiota to microbial-based therapies. *Eur. J. Clin. Microbiol. Infect. Dis.* **2020**, *39*, 613–627. [CrossRef]
131. Oh, T.G.; Kim, S.M.; Caussy, C.; Fu, T.; Guo, J.; Bassirian, S.; Singh, S.; Madamba, E.V.; Bettencourt, R.; Richards, L.; et al. A Universal Gut-Microbiome-Derived Signature Predicts Cirrhosis. *Cell Metab.* **2020**. [CrossRef]

International Journal of
Molecular Sciences

MDPI

Review

Targeted Therapies in Early Stage NSCLC: Hype or Hope?

Alex Friedlaender [1,†], Alfredo Addeo [1,†], Alessandro Russo [2], Vanesa Gregorc [3], Diego Cortinovis [4] and Christian D. Rolfo [5,*]

1 Oncology Department, University Hospital of Geneva, 1205 Genève, Switzerland; alex.friedlaender@hcuge.ch (A.F.); alfredo.addeo@hcuge.ch (A.A.)
2 Medical Oncology Unit, A.O. Papardo, 98158 Messina, Italy; ale.russo1986@gmail.com
3 Department of Oncology, Division of Experimental Medicine, IRCCS San Raffaele, 20132 Milan, Italy; vanesa.gregorc@hsr.it
4 SC Medical Oncology, Azienda Socio Sanitaria Territoriale (ASST) H S Gerardo Monza, 20900 Monza, Italy; d.cortinovis@asst-monza.it
5 Thoracic Oncology Department and Early Phase Clinical Trials Section, School of Medicine, University of Maryland, Baltimore, MD 20742, USA
* Correspondence: christian.rolfo@umm.edu; Tel.: +410-328-7904; Fax: +410-328-9160
† They are co-first authors.

Received: 14 August 2020; Accepted: 26 August 2020; Published: 31 August 2020

Abstract: Non-small-cell lung cancer (NSCLC) represents roughly 85% of lung cancers, with an incidence that increases yearly across the world. The introduction in clinical practice of several new and more effective molecules has led to a consistent improvement in survival and quality of life in locally advanced and metastatic NSCLC. In particular, oncogenic drivers have indeed transformed the therapeutic algorithm for NSCLC. Nearly 25% of patients are diagnosed in an early stage when NSCLC is still amenable to radical surgery. In spite of this, five-year survival rates for fully resected early stage remains rather disappointing. Adjuvant chemotherapy has shown a modest survival benefit depending on the stage, but more than half of patients relapse. Given this need for improvement, over the last years different targeted therapies have been evaluated in early-stage NSCLC with no survival benefit in unselected patients. However, the identification of reliable predictive biomarkers to these agents in the metastatic setting, the design of molecularly-oriented studies, and the availability of novel potent and less toxic agents opened the way for a novel era in early stage NSCLC treatment. In this review, we will discuss the current landscape of targeted therapeutic options in early NSCLC.

Keywords: NSCLC; targeted therapy; early stage; EGFR; ALK; osimertinib

1. Introduction

Non-small-cell lung cancer (NSCLC) represents roughly 85% of lung cancers, with an incidence that keeps rising across the globe [1]. The treatment landscape of metastatic NSCLC has considerably improved over the last two decades due to a better understanding of cancer biology [2]. The introduction in clinical practice of several new and more effective molecules has led to a consistent improvement in overall survival (OS) and quality of life (QoL) in advanced/metastatic disease [3].

Oncogenic drivers have indeed reshaped the therapeutic algorithm for metastatic NSCLC, making us wonder whether they could also play a role in early disease. Roughly a quarter of patients are diagnosed with disease amenable to potentially radical surgery [3]

In spite of this, five-year survival rates for fully resected stage I disease range from 50 to 70% while lying between 10 and 30% for stage IIIA NSCLC [4].

Adjuvant chemotherapy has shown a modest survival benefit (with an absolute increase in survival of 4% at five years), but, depending on the stage [5,6], more than half of patients still relapse [7]. Similarly, a neoadjuvant approach yields a 5% absolute benefit on five-year survival [8], leaving room for improvement. The question of whether targeted therapy could fill this unmet medical need is far from new. Over the last several years, different targeted therapies have been evaluated in early stage NSCLC with no survival benefit in unselected patients [9–12]. However, the identification of reliable predictive biomarkers to these agents in the metastatic setting, the design of molecularly-oriented studies, and the availability of novel potent and less toxic agents paved the way for a novel era in early stage NSCLC treatment (Figure 1). In this review, we will discuss the current landscape of targeted therapeutic options being investigated in early NSCLC.

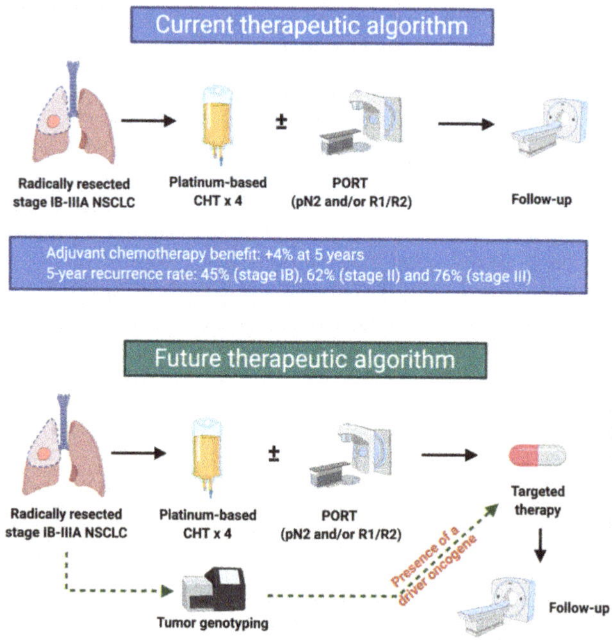

Figure 1. Potential role of targeted therapies in the adjuvant setting. Post-operative platinum-based chemotherapy (CHT) in stage IB-IIIA resected non-small-cell lung cancer (NSCLC) has been associated with a 4% survival benefit at 5 years [7]. According to the Lung Adjuvant Cisplatin Evaluation (LACE) study, 5-year recurrence rates range from 45% in stage IB to 76% in stage III after adjuvant chemotherapy [8]. Post-operative radiotherapy (PORT) is currently recommended in patients with pathologic N2 (pN2) disease and in those with microscopic (R1) or macroscopic (R2) residual disease.

2. Epidermal Growth Factor Receptor (EGFR) mutations

In EGFR mutant NSCLC, the most common oncogenic driver targeted today, the question of adding a tyrosine kinase inhibitor (TKI) or replacing chemotherapy with one is not new. Before EGFR mutations were identified as the major determinants of efficacy to first-generation EGFR TKIs in 2004 [13,14], two randomized phase III trials assessed the impact of erlotinib (RADIANT) [9] and gefitinib (BR19) [10] in unselected stage IB-IIIA NSCLCs. The RADIANT trial enrolled patients with fully resected stage IB to IIIA NSCLC and confirmed tumor EGFR expression by immunohistochemistry (IHC) or fluorescent in situ hybridization (FISH). Patients were randomized in a 2:1 ratio to receive daily erlotinib for two years or placebo. When indicated, patients received adjuvant chemotherapy before starting the study therapy. The primary endpoint of improved DFS was not met. It is noteworthy that only 163 out of 973 patients recruited in the study harbored an exon 19 deletion or L858R EGFR mutation. In this specific subgroup of EGFR-driven diseases, DFS was superior with erlotinib (HR, 0.61; 95% CI, 0.384–0.981; $p = 0.0391$). This difference could not be retained as statistically significant, given the hierarchical testing that allowed assessment of secondary endpoints only if the primary endpoint was statistically significant. The follow-up was too short at the time of the analysis to properly assess survival differences. The phase 3 BR19 study assessed the role of gefitinib as adjuvant therapy for up to two years versus placebo in patients with completed resected stage IB to IIIA NSCLC. At a median follow-up of 4.7 years among 503 patients, there was neither DFS (1.22, 95% CI 0.90–1.71) nor OS (HR 1.24, 95% CI 0.94–1.64) benefit in the experimental arm. The trial was closed early. No benefit was also observed in the small subgroup of patients harboring EGFR mutations in terms of both DFS (HR, 1.84; 95% CI, 0.44 to 7.73; $p = 0.395$) and OS (HR, 3.16; 95% CI, 0.61 to 16.45; $p = 0.15$) [10].

Following the successful experience in advanced/metastatic setting with multiple EGFR TKIs approved in molecularly selected patients [15–19] different studies have sought to demonstrate a survival advantage with the use of these agents as adjuvant therapy in EGFR-mutated radically resected NSCLCs (Table 1). Nevertheless the role of co-mutations remains unclear [20].

The SELECT trial is a single-arm phase 2 trial and was the first to test the efficacy of adjuvant erlotinib in resected EGFR-mutated NSCLC. One hundred patients with stage IA to IIIA EGFR mutant NSCLC were given erlotinib for up to two years after completing standard adjuvant therapy. The 2-year DFS was 88%, and the authors concluded that this was an improvement compared to historical matched controls, which had a two-year DFS of 76%. Furthermore, the median time to recurrence was 25 months after discontinuing erlotinib [11]. Since then, a number of randomized adjuvant targeted therapy trials have been published.

The CTONG1104/ADJUVANT trial is a phase 3 study that compared the standard cisplatin-vinorelbine chemotherapy to gefitinib, a first-generation EGFR TKI in fully resected stage II to IIIA EGFR mutant NSCLC [12]. Chemotherapy was administered for four cycles, while gefitinib was given until progression for up to two years. The primary endpoint was disease-free survival (DFS), while OS was a secondary endpoint. At a median follow-up of 36.5 months, the experimental arm yielded a superior DFS compared to chemotherapy (28.7 versus 18.0 months), with a hazard ratio (HR) of 0.60 (95% CI 0.42–0.87, $p = 0.0054$). The mature OS results were presented at the American Society of Clinical Oncology (ASCO) 2020 Meeting, with a median follow-up of 76.9 months. The median OS was not statistically different, at 75.5 months in the gefitinib arm and 79.2 months in the chemotherapy arm (HR 0.92, 95% CI 0.62–1.36). Furthermore, only 51.5% of patients in the chemotherapy arm were exposed to a TKI at progression [13].

The EVAN trial is a phase 2 randomized trial on a smaller cohort of 102 patients with fully-resected stage IIIA EGFR-mutant NSCLC [21]. Patients were randomized between adjuvant chemotherapy for four cycles and erlotinib for up to two years. The primary endpoint was two-year DFS, while OS was a secondary endpoint. At a median follow-up of 33 months, the two-year DFS was 81.4% in the erlotinib group and 44.6% in the chemotherapy arm (RR 1.823, 95% CI 1.194–2.784, $p = 0.0054$). While these results are promising, the sample size was small, and OS data are not yet available.

Table 1. Phase II/III studies with EGFR tyrosine kinase inhibitors (TKIs) in the adjuvant setting.

Study	Phase	Population	n	Arm(s)	Patients Receiving Adjuvant Chemotherapy (%)	Median DFS (mos)	2-Year DFS	3-Year DFS	Median OS (mos)
RADIANT [9]	III	IB-IIIA NSCLCs, EGFR-positive by IHC and/or FISH	623 vs. 250	Erlotinib for 2 years vs. placebo	50.6% vs. 57.1%	50.5 vs. 48.2 (HR 0.90)	75% vs. 54%	N.R.	Not reached vs. Not reached (HR 1.09)
BR19 [10]	III	IB-IIIA NSCLCs	251 vs. 252	Gefitinib for 2 years vs. placebo	17% vs. 17%	4.2 years vs. Not reached (HR 1.22)	N.R.	N.R.	5.1 years vs. Not reached (HR 1.24)
SELECT [11]	II	IA-IIIA EGFR-mutated NSCLC	100	Erlotinib for 2 years	N.R.	Not reached	88%	N.R.	Not reached
CTONG1104 ADJUVANT [12]	III	II-IIIA EGFR-mutated NSCLC	111 vs. 111	Gefitinib for 2 years vs. vinorelbine/cisplatin	0% vs. 100%	30.8 vs. 19.8 (HR 0.56)	N.R.	39.6% vs. 32.5%	75.5 vs. 62.8 (HR 0.92)
EVAN [20]	II	IIIA EGFR-mutated NSCLC	51 vs. 51	Erlotinib for 2 years vs. vinorelbine/cisplatin	0% vs. 100%	42.4 vs. 21.0 (HR 0.268)	81.4% vs. 44.6%	54.2% vs. 19.8%	Not reached vs. Not reached (HR 0.165)
ADAURA [21]	III	IB-IIIA EGFR-mutated NSCLC	339 vs. 343	Osimertinib for 3 years vs. placebo	55% vs. 56%	Not reached vs. 20.4 (HR 0.17) *	90% * vs. 44% *	80% * vs. 28% *	Not reached vs. Not reached (HR 0.40) *

* Intention-to-treat (ITT) population (stage II-IIIA NSCLC). Abbreviations: IHC, immunohistochemistry; FISH, fluorescent in situ hybridization; DFS, disease-free survival; OS, overall survival; mos, months; HR, hazard ratio; N.R., not reported.

Recently, the ADAURA trial was presented at the ASCO 2020 Meeting [22]. It evaluated the impact of adjuvant osimertinib, a third-generation EGFR TKI, compared to placebo in fully-resected stage IB to IIIA (TNM 7) EGFR-mutant NSCLC. Patients in both arms were eligible for chemotherapy, but the trial was not stratified based on whether it was administered. The experimental arm received osimertinib until progression or up to three years. The primary endpoint was once again DFS, among the stage II to IIIA patients, while DFS in the intention-to-treat population and OS were among the secondary endpoints. The results presented at the ASCO Meeting were from an off-protocol interim safety analysis after the data safety monitoring board asked to unblind the trial due to a very strong signal favoring the osimertinib arm. At this preliminary analysis, with a median follow-up of 22 months, the DFS for stage II to IIIA patients was not reached in the osimertinib arm and 20.4 months in the placebo arm, with an HR of 0.17 (95% CI 0.12–0.23, $p < 0.0001$). While the HR of DFS is certainly impressive, there is much controversy about whether these immature data should lead to a change in practice, as the real question is whether patients will live longer if treated earlier. Furthermore, at the ASCO 2020, the final analysis of the CTONG1104/ADJUVANT trial was presented, showing that treating EGFR positive patients with EGFR TKI delayed the relapse but did not translate into OS benefit [12]. Moreover, the ADAURA trial doesn't address the question about the importance and need for adjuvant chemotherapy in these patients, as the majority of them received it before being randomized in the trial.

Probably the largest effort to address the role of EGFR TKI in the adjuvant setting is the ALCHEMIST trial (NCT02194738). Patients with stage IB to IIIA after radical surgery will get molecular analysis, and patients whose tumor harbors an EGFR mutation will enter the EGFR mutation substudy. It is aiming to recruit 410 patients and will randomly assign patients to erlotinib for two years versus placebo. The primary endpoint is OS.

Neoadjuvant therapy has the potential to facilitate surgery by shrinking the tumor. A phase II single-arm study assessing the impact of 28 days of neoadjuvant gefitinib in stage I NSCLC found a 50% response rate among patients whose tumors harbored EGFR mutations. There was no safety signal for increased risk of surgery. Upon histologic analysis, there was more fibrotic scar tissue, lower cell proliferation, and residual tumor cells were concentrated in fibrous stroma with lymphocytic infiltration [23,24]. Three small phase II trials evaluated neoadjuvant erlotinib among patients with stage IIIA EGFR-mutant NSCLC. In spite of a response rate of 58%, the first failed to show a survival benefit compared to non-EGFR-mutated patients receiving chemotherapy [25]. The second was a single-arm trial and reported a 42% response rate, with 21% downstaging to T0-3 N0 M0. On a pathological level, 50% of patients had a partial response, while 50% had stable disease [26]. The third study compared neoadjuvant erlotinib among 15 patients whose tumors had EGFR mutations to chemotherapy in 16 patients without these alterations [27]. The authors report a trend towards better response rate, pathological response rates, and overall survival. Given the small number and heterogeneous prognosis inherent in oncogenic driven NSCLC, no conclusions can be drawn. One potential pitfall is disease flare after TKI interruption, and this will need to be evaluated in larger prospective trials [28]. The ongoing phase II EMERGING trial is comparing neoadjuvant erlotinib to cisplatin-gemcitabine in patients with stage IIIA EGFR-mutant NSCLC (NCT01407822). A neoadjuvant phase III trial of gefitinib versus carboplatin and vinorelbine among patients with stage II-IIIA EGFR-mutant NSCLC is planned (NCT03203590).

After the impressive early results of the adjuvant ADAURA trial, the neoADAURA (NCT04351555) trial will follow. It is a phase III trial that will compare neoadjuvant osimertinib, with or without chemotherapy, to chemotherapy alone in resectable NSCLC patients. The primary endpoint will be major pathological response rates, while OS and DFS will be among secondary endpoints. A phase II trial is ongoing (NCT03433469).

3. Anaplastic Lymphoma Kinase (ALK) Gene Fusions

ALK rearrangements are detected in 2 to 7% of NSCLC patients. Interestingly, this comprises under 5% of resected NSCLCs but up to 19% of stage IV cancers. This could possibly be due to the biology of ALK driven tumors, with rapid proliferation and spread. It is interesting to note that this contrasts with EGFR mutations, which are not stage dependent [29–31].

Given the infrequency of ALK translocations in localized NSCLC, it is no surprise that there are fewer clinical trials assessing ALK inhibitors in this setting. Nonetheless, two adjuvant phase III trials are ongoing. The aforementioned ALCHEMIST trial (NCT02194738) has an arm randomizing patients with stage IB to IIIA (TNM7) fully-resected ALK-driven NSCLC to observation versus crizotinib for up to 24 months after completing standard therapy, including chemotherapy and radiotherapy, where indicated. The primary endpoint is OS, and DFS is among secondary endpoints. The second, more recent, phase III multicenter randomized adjuvant trial, ALINA (NCT03456076), is comparing alectinib to standard of care in stage IB-IIIA (TNM 7) fully resected ALK-rearranged NSCLC. This trial excludes patients with N2 stage IIIA cancer who could be candidates for postoperative radiotherapy in some centers, as this could represent a confounding factor in a high-risk group. Alectinib is administered for up to 24 months, while the control arm receives four cycles of platinum-based chemotherapy. The primary endpoint is DFS, while OS is among secondary endpoints.

Finally, a multicenter phase II trial will soon begin recruiting patients with stage IB to IIIB resectable NSCLC with a variety of oncogenic drivers, assessing the effect of eight weeks of neoadjuvant therapy followed by the possibility of adjuvant treatment with the same drug (NCT04302025). One arm will assess alectinib in this setting. The NCT03088930 trial is evaluating neoadjuvant crizotinib in a similar design.

4. Others Oncogenic Drivers

EGFR and ALK are two of the multitude of currently identified and targetable oncogenic drivers in NSCLC. As a reminder, in lung adenocarcinoma, targetable alterations comprise roughly half of all diseases, and the list of actionable therapeutic targets is constantly growing [32].

Front-line therapy for metastatic disease has largely shifted or is gradually moving towards a TKI approach among patients whose tumors harbor these alterations. To name a few, BRAF V600E mutations are successfully targeted by combined BRAF/MEK inhibitors, mainly dabrafenib and trametinib, with high response rates and relatively low toxicity [33]. Similarly, ROS1 rearrangements can be targeted with drugs, including crizotinib, ceritinib, lorlatinib, entrectinib, or repotrectinib [26]. MET alterations, including exon 14 mutations and amplifications, can be treated with a variety of TKIs, among the most common, crizotinib, capmatinib, and tepotinib [34–37]. NTRK fusions are a recent addition to this list, with impressive responses to entrectinib and larotrectinib [38,39]. RET fusion-positive NSCLC also appears to benefit from treatment with selpercatinib or praseltinib [40,41] The very common KRAS G12C mutation, hitherto considered undruggable, is now being targeted by small molecules, with the potential to dramatically alter the therapeutic landscape [42]. HER2 targeting appears to be moving forward in strides, with the antibody-drug conjugate, trastuzumab–deruxtecan, yielding impressive preliminary results [43]. It is important to be aware of the efficacy of these drugs, even if only in early phase trials for some, to understand the rationale for testing them in the adjuvant or neoadjuvant setting.

The abovementioned ALK trial (NCT04302025) also includes arms for ROS1 and NTRK, each to be treated by entrectinib following the same treatment pattern as the ALK arm, as well as BRAF V600E mutations to be treated by vemurafenib and cobimetinib. The primary endpoint is major pathological response, defined as 10% or lower of viable tumor cells, while OS and DFS are among secondary objectives.

A small phase II trial is ongoing, assessing the efficacy of six weeks of neoadjuvant crizotinib in patients with MET, ROS1, or ALK alterations and resectable stage IA-IIIA NSCLC (NCT03088930). Objective response rate is the primary endpoint while OS and DFS are among secondary endpoints.

These trials are summarized in Table 2. To our knowledge, no other phase II to III trials are ongoing in this setting.

Table 2. Ongoing phase II–III neoadjuvant and adjuvant trials with tyrosine kinase inhibitors (TKIs).

Trial	Phase	Design	Population	Arm(s)	Primary Outcome	Clinical Trial Identification
ALCHEMIST	III	adjuvant	IB-IIIA NSCLCs, EGFR-mutated NSCLC	erlotinib for 2 years vs. placebo	OS	NCT02194738
ALCHEMIST	III	adjuvant	IB-IIIA NSCLCs, ALK-rearranged NSCLC	crizotinib for 2 years vs. placebo	OS	NCT02194738
ALINA	III	adjuvant	IB-IIIA NSCLCs, ALK-rearranged NSCLC	alectinib for 2 years vs. chemotherapy	DFS	NCT03456076
EMERGING	II	Neoadjuvant + adjuvant	IIIA EGFR-mutated NSCLCs	erlotinib for 6 weeks then 1 year post-op vs. cisplatin-gemcitabine	ORR	NCT01407822
NCT03203590	III	neoadjuvant	II-IIIA EGFR-mutated NSCLC	gefinitib for 8 weeks vs. carboplatin-vinorelbine	2 year DFS	NCT03203590
NeoADAURA	III	neoadjuvant	II-IIIA EGFR-mutated NSCLC	osimertinib +/− platinum-pemetrexed vs. platinum-pemetrexed	MPR	NCT04351555
NCT04302025	II	Neoadjuvant +/− adjuvant	IB-IIIB NSCLC with altered ALK, ROS1, NTRK or BRAF	8 weeks neoadjuvant +/− adjuvant with alectinib, entrectinib or vemurafenib+cobimetinib	MPR	NCT04302025
NCT03088930	II	neoadjuvant	IA-IIIA NSCLC with altered MET, ROS1 or ALK	crizotinib for 6 weeks	ORR	NCT03088930

Abbreviations: DFS, disease-free survival; OS, overall survival; MPR, major pathological response; ORR, objective response rate.

5. Limitations of TKIs in the Early Stage Setting

To date, no data have proven that TKIs can be curative in NSCLC, and patients 'compliance with TKI treatment could be lowered by the persistency of chronic adverse events due to long term use of TKIs in patients who are free from cancer [36,37]. In the metastatic setting, all TKIs eventually fail, through on- or off-target escape mechanisms. Similarly, treatment discontinuation can lead to a flare in tumor growth. The aim of TKI therapy in early-stage disease could comprise various options. Of course, the primordial question is whether treating microscopic residual disease could eliminate cells rather than simply suppress growth, thus increasing cure rates rather than just relapse-free survival. Only the overall survival of properly conducted trials will provide this answer. However, other goals could exist. In the neoadjuvant setting, for instance, the higher response rate of TKIs compared to chemotherapy could facilitate surgical management.

Under selective therapy-induced pressure, oncogenic-driven tumors can develop different resistant clones. While cell subpopulations develop a quiescent or dormant state of cell-cycle arrest when exposed to TKIs, some will acquire resistance alterations, whether through mutations or epigenetic changes [44].

The drug-tolerant cells are a reservoir for potential tumor growth and will lead to progression if they escape immune surveillance and proliferate. Unless we manage to reactivate quiescent cells selectively to target them, it is unlikely TKIs will be curative given this behavior [45].

Combinatory therapy targeting and inhibiting signal transduction and activator of transcription 3 (STAT3) and Src may potentially be more effective by reducing the level of lung cancer stem cells subpopulation

6. Future Directions and Challenges

As we await the results of ongoing and planned trials of TKIs in the localized NSCLC setting, the question of which patients to treat in the non-metastatic setting may emerge. Circulating tumor DNA (ctDNA) can detect minimal residual disease among patients with operated early-stage NSCLC [46]. Similarly, ctDNA has been used to identify acquired-resistance mechanisms to TKIs [41,42,47,48]. Such an approach could have potential implications for initiating therapy upon early disease detection rather than broadly among all operated patients. It could also be a tool to monitor patients on adjuvant therapy in order to detect early resistance or relapse. These questions and more will have to be revisited in light of results of neo(adjuvant) trials. Only time will provide answers about the best care for our patients.

At the ASCO 2020 Meeting, the ADAURA presentation may have provided a glimpse of the future for adjuvant treatment in oncogenic driven mutated NSCLC. The assumption that a TKI, proven to be very effective in the metastatic setting, could and will be even more effective in early stages has become prevalent among many oncologists. There may be significant implications if the follow-up data of the ADAURA trial and the readouts from the ALINA are very positive, albeit for surrogate endpoints for OS. Oncologists may be very tempted to emulate this early-stage TKI approach in less frequent mutations, such as NTRK or RET, arguing that large trials are not feasible given the rarity of these alterations. This will generate a debate on the appropriateness and validity of extrapolating results without a formal trial; however, it is undeniable that the oncology world has already, as a consequence of preliminary results, shifted toward wider molecular genomic sequencing in early-stage NSCLC.

Funding: This research received no external funding

Conflicts of Interest: A.F. reports receiving personal fees from Bristol-Myers Squibb, Roche Holdings AG, Pfizer, Astellas, and Merck Sharp & Dohme, outside of the present work. A.A. reports receiving personal fees from Bristol-Myers Squibb, AstraZeneca, Roche Holdings AG, Pfizer, Merck Sharp & Dohme, and Boehringer Ingelheim, outside of the present work. A.R. No conflict of interest. V.G. No conflict of interest. D.C. No conflict of interest. C.R. reports roles in speaker bureaus for MSD and AstraZeneca; advisory board roles with ARCHER, Inivata, and Merck Serono; consultant roles with Mylan and Oncompass; a supported research grant from the Lung Cancer Research Foundation/Pfizer; and research support from Guardant Health and Biomark

References

1. Siegel, R.; Ma, J.; Zou, Z.; Jemal, A. Cancer statistics, 2014. *CA Cancer J. Clin.* **2014**, *64*, 9–29. [CrossRef]
2. Herbst, R.S.; Morgensztern, D.; Boshoff, C. The biology and management of non-small cell lung cancer. *Nature* **2018**, *553*, 446–454. [CrossRef]
3. Molina, J.R.; Yang, P.; Cassivi, S.D.; Schild, S.E.; Adjei, A.A. Non-small cell lung cancer: Epidemiology, risk factors, treatment, and survivorship. *Mayo Clin. Proc.* **2008**, *83*, 584–594. [CrossRef]
4. Gloeckler Ries, L.A.; Reichman, M.E.; Lewis, D.R.; Hankey, B.F.; Edwards, B.K. Cancer survival and incidence from the Surveillance, Epidemiology, and End Results (SEER) program. *Oncologist* **2003**, *8*, 541–552. [CrossRef]
5. Pisters, K.M.W.; Chevalier, T.L. Adjuvant Chemotherapy in Completely Resected Non–Small-Cell Lung Cancer. *J. Clin. Oncol.* **2005**, *23*, 3270–3278. [CrossRef] [PubMed]
6. Group, N.M.-a.C.; Arriagada, R.; Auperin, A.; Burdett, S.; Higgins, J.P.; Johnson, D.H.; Le Chevalier, T.; Le Pechoux, C.; Parmar, M.K.B.; Pignon, J.P.; et al. Adjuvant chemotherapy, with or without postoperative radiotherapy, in operable non-small-cell lung cancer: Two meta-analyses of individual patient data. *Lancet* **2010**, *375*, 1267–1277. [CrossRef]
7. Pignon, J.P.; Tribodet, H.; Scagliotti, G.V.; Douillard, J.Y.; Shepherd, F.A.; Stephens, R.J.; Dunant, A.; Torri, V.; Rosell, R.; Seymour, L.; et al. Lung adjuvant cisplatin evaluation: A pooled analysis by the LACE Collaborative Group. *J. Clin. Oncol.* **2008**, *26*, 3552–3559. [CrossRef] [PubMed]
8. Preoperative chemotherapy for non-small-cell lung cancer: A systematic review and meta-analysis of individual participant data. *Lancet* **2014**, *383*, 1561–1571. [CrossRef]
9. Kelly, K.; Altorki, N.K.; Eberhardt, W.E.; O'Brien, M.E.; Spigel, D.R.; Crinò, L.; Tsai, C.M.; Kim, J.H.; Cho, E.K.; Hoffman, P.C.; et al. Adjuvant Erlotinib Versus Placebo in Patients With Stage IB-IIIA Non-Small-Cell Lung Cancer (RADIANT): A Randomized, Double-Blind, Phase III Trial. *J. Clin. Oncol.* **2015**, *33*, 4007–4014. [CrossRef]
10. Goss, G.D.; O'Callaghan, C.; Lorimer, I.; Tsao, M.-S.; Masters, G.A.; Jett, J.; Edelman, M.J.; Lilenbaum, R.; Choy, H.; Khuri, F.; et al. Gefitinib versus placebo in completely resected non-small-cell lung cancer: Results of the NCIC CTG BR19 study. *J. Clin. Oncol.* **2013**, *31*, 3320–3326. [CrossRef]
11. Pennell, N.A.; Neal, J.W.; Chaft, J.E.; Azzoli, C.G.; Jänne, P.A.; Govindan, R.; Evans, T.L.; Costa, D.B.; Wakelee, H.A.; Heist, R.S.; et al. SELECT: A Phase II Trial of Adjuvant Erlotinib in Patients With Resected Epidermal Growth Factor Receptor-Mutant Non-Small-Cell Lung Cancer. *J. Clin. Oncol.* **2019**, *37*, 97–104. [CrossRef] [PubMed]
12. Wu, Y.-L.; Zhong, W.; Wang, Q.; Mao, W.; Xu, S.-T.; Wu, L.; Chen, C.; Cheng, Y.; Xu, L.; Wang, J. CTONG1104: Adjuvant gefitinib versus chemotherapy for resected N1-N2 NSCLC with EGFR mutation—Final overall survival analysis of the randomized phase III trial 1 analysis of the randomized phase III trial. *J. Clin. Oncol.* **2020**, *38*, 9005. [CrossRef]
13. Lynch, T.J.; Bell, D.W.; Sordella, R.; Gurubhagavatula, S.; Okimoto, R.A.; Brannigan, B.W.; Harris, P.L.; Haserlat, S.M.; Supko, J.G.; Haluska, F.G. Activating mutations in the epidermal growth factor receptor underlying responsiveness of non–small-cell lung cancer to gefitinib. *N. Engl. J. Med.* **2004**, *350*, 2129–2139. [CrossRef]
14. Paez, J.G.; Jänne, P.A.; Lee, J.C.; Tracy, S.; Greulich, H.; Gabriel, S.; Herman, P.; Kaye, F.J.; Lindeman, N.; Boggon, T.J. EGFR mutations in lung cancer: Correlation with clinical response to gefitinib therapy. *Science* **2004**, *304*, 1497–1500. [CrossRef] [PubMed]
15. Ramalingam, S.; Gray, J.; Ohe, Y.; Cho, B.; Vansteenkiste, J.; Zhou, C.; Reungwetwattana, T.; Cheng, Y.; Chewaskulyong, B.; Shah, R. Osimertinib vs comparator EGFR-TKI as first-line treatment for EGFRm advanced NSCLC (FLAURA): Final overall survival analysis. *Ann. Oncol.* **2019**, *30*, v914–v915. [CrossRef]
16. Fukuoka, M.; Wu, Y.-L.; Thongprasert, S.; Sunpaweravong, P.; Leong, S.-S.; Sriuranpong, V.; Chao, T.-Y.; Nakagawa, K.; Chu, D.-T.; Saijo, N. Biomarker analyses and final overall survival results from a phase III, randomized, open-label, first-line study of gefitinib versus carboplatin/paclitaxel in clinically selected patients with advanced non–small-cell lung cancer in Asia (IPASS). *J. Clin. Oncol.* **2011**, *29*, 2866–2874. [CrossRef]

17. Rosell, R.; Carcereny, E.; Gervais, R.; Vergnenegre, A.; Massuti, B.; Felip, E.; Palmero, R.; Garcia-Gomez, R.; Pallares, C.; Sanchez, J.M. Erlotinib versus standard chemotherapy as first-line treatment for European patients with advanced EGFR mutation-positive non-small-cell lung cancer (EURTAC): A multicentre, open-label, randomised phase 3 trial. *Lancet Oncol.* **2012**, *13*, 239–246. [CrossRef]
18. Yang, J.C.-H.; Schuler, M.H.; Yamamoto, N.; O'Byrne, K.J.; Hirsh, V.; Mok, T.; Geater, S.L.; Orlov, S.V.; Tsai, C.-M.; Boyer, M.J.; et al. LUX-Lung 3: A randomized, open-label, phase III study of afatinib versus pemetrexed and cisplatin as first-line treatment for patients with advanced adenocarcinoma of the lung harboring EGFR-activating mutations. *J. Clin Oncol.* **2012**, *30*, LBA7500-LBA7500. [CrossRef]
19. Wu, Y.-L.; Zhou, C.; Hu, C.-P.; Feng, J.; Lu, S.; Huang, Y.; Li, W.; Hou, M.; Shi, J.H.; Lee, K.Y. Afatinib versus cisplatin plus gemcitabine for first-line treatment of Asian patients with advanced non-small-cell lung cancer harbouring EGFR mutations (LUX-Lung 6): An open-label, randomised phase 3 trial. *Lancet Oncol.* **2014**, *15*, 213–222. [CrossRef]
20. Chevallier, M.; Tsantoulis, P.; Addeo, A.; Friedlaender, A. Influence of Concurrent Mutations on Overall Survival in EGFR-mutated Non-small Cell Lung Cancer. *Cancer Genom.Proteom.* **2020**, *17*, 597–603. [CrossRef]
21. Yue, D.; Xu, S.; Wang, Q.; Li, X.; Shen, Y.; Zhao, H.; Chen, C.; Mao, W.; Liu, W.; Liu, J.; et al. Erlotinib versus vinorelbine plus cisplatin as adjuvant therapy in Chinese patients with stage IIIA EGFR mutation-positive non-small-cell lung cancer (EVAN): A randomised, open-label, phase 2 trial. *Lancet. Respir. Med.* **2018**, *6*, 863–873. [CrossRef]
22. Herbst, R.S.; Tsuboi, M.; John, T.; Grohé, C.; Majem, M.; Goldman, J.W.; Kim, S.-W.; Marmol, D.; Rukazenkov, Y.; Wu, Y.-L. Osimertinib as adjuvant therapy in patients (pts) with stage IB–IIIA EGFR mutation positive (EGFRm) NSCLC after complete tumor resection: ADAURA. *J. Clin. Oncol.* **2020**, *38*. [CrossRef]
23. Lara-Guerra, H.; Waddell, T.K.; Salvarrey, M.A.; Joshua, A.M.; Chung, C.T.; Paul, N.; Boerner, S.; Sakurada, A.; Ludkovski, O.; Ma, C.; et al. Phase II study of preoperative gefitinib in clinical stage I non-small-cell lung cancer. *J. Clin. Oncol.* **2009**, *27*, 6229–6236. [CrossRef] [PubMed]
24. Lara-Guerra, H.; Chung, C.T.; Schwock, J.; Pintilie, M.; Hwang, D.M.; Leighl, N.B.; Waddell, T.K.; Tsao, M.S. Histopathological and immunohistochemical features associated with clinical response to neoadjuvant gefitinib therapy in early stage non-small cell lung cancer. *Lung Cancer* **2012**, *76*, 235–241. [CrossRef] [PubMed]
25. Zhong, W.; Yang, X.; Yan, H.; Zhang, X.; Su, J.; Chen, Z.; Liao, R.; Nie, Q.; Dong, S.; Zhou, Q.; et al. Phase II study of biomarker-guided neoadjuvant treatment strategy for IIIA-N2 non-small cell lung cancer based on epidermal growth factor receptor mutation status. *J Hematol. Oncol.* **2015**, *8*, 54. [CrossRef]
26. Xiong, L.; Li, R.; Sun, J.; Lou, Y.; Zhang, W.; Bai, H.; Wang, H.; Shen, J.; Jing, B.; Shi, C.; et al. Erlotinib as Neoadjuvant Therapy in Stage IIIA (N2) EGFR Mutation-Positive Non-Small Cell Lung Cancer: A Prospective, Single-Arm, Phase II Study. *Oncologist* **2019**, *24*, e157–e164. [CrossRef]
27. Xiong, L.; Lou, Y.; Bai, H.; Li, R.; Xia, J.; Fang, W.; Zhang, J.; Han-Zhang, H.; Lizaso, A.; Li, B.; et al. Efficacy of erlotinib as neoadjuvant regimen in EGFR-mutant locally advanced non-small cell lung cancer patients. *J. Int. Med. Res.* **2019**. [CrossRef]
28. Chaft, J.E.; Oxnard, G.R.; Sima, C.S.; Kris, M.G.; Miller, V.A.; Riely, G.J. Disease flare after tyrosine kinase inhibitor discontinuation in patients with EGFR-mutant lung cancer and acquired resistance to erlotinib or gefitinib: Implications for clinical trial design. *Clin. Cancer Res.* **2011**, *17*, 6298–6303. [CrossRef]
29. Addeo, A.; Tabbò, F.; Robinson, T.; Buffoni, L.; Novello, S. Precision medicine in ALK rearranged NSCLC: A rapidly evolving scenario. *Crit. Rev. Oncol. Hematol.* **2018**, *122*, 150–156. [CrossRef]
30. Friedlaender, A.; Banna, G.; Patel, S.; Addeo, A. Diagnosis and Treatment of ALK Aberrations in Metastatic NSCLC. *Curr. Treat. Opt. Oncol.* **2019**, *20*, 79. [CrossRef]
31. Soda, M.; Choi, Y.L.; Enomoto, M.; Takada, S.; Yamashita, Y.; Ishikawa, S.; Fujiwara, S.; Watanabe, H.; Kurashina, K.; Hatanaka, H.; et al. Identification of the transforming EML4-ALK fusion gene in non-small-cell lung cancer. *Nature* **2007**, *448*, 561–566. [CrossRef] [PubMed]
32. Massó-Vallés, D.; Beaulieu, M.-E.; Soucek, L. MYC, MYCL, and MYCN as therapeutic targets in lung cancer. *Expert Opin. Ther. Targets* **2020**, *24*, 101–114. [CrossRef] [PubMed]
33. Frisone, D.; Friedlaender, A.; Malapelle, U.; Banna, G.; Addeo, A. A BRAF new world. *Crit. Rev. Oncol. Hematol.* **2020**, *152*, 103008. [CrossRef] [PubMed]

34. Shaw, A.T.; Solomon, B.J.; Chiari, R.; Riely, G.J.; Besse, B.; Soo, R.A.; Kao, S.; Lin, C.-C.; Bauer, T.M.; Clancy, J.S.; et al. Lorlatinib in advanced *ROS1*-positive non-small-cell lung cancer: A multicentre, open-label, single-arm, phase 1–2 trial. *Lancet Oncol.* **2019**, *20*, 1691–1701. [CrossRef]
35. Groen, H.J.; Akerley, W.L.; Souquet, P.J.; Laack, E.; Han, J.-Y.; Smit, E.F.; Mansfield, A.S.; Garon, E.B.; Wolf, J.; Tan, D.S.-W. Capmatinib in patients with METex14-mutated or high-level MET-amplified advanced non–small-cell lung cancer (NSCLC): Results from cohort 6 of the phase 2 GEOMETRY mono-1 study. *J. Clin. Oncol.* **2020**, *38*. [CrossRef]
36. Paik, P.K.; Felip, E.; Veillon, R.; Sakai, H.; Cortot, A.B.; Garassino, M.C.; Mazieres, J.; Viteri, S.; Senellart, H.; Van Meerbeeck, J.; et al. Tepotinib in Non–Small-Cell Lung Cancer with MET Exon 14 Skipping Mutations. *N. Engl. J. Med.* **2020**. [CrossRef]
37. Friedlaender, A.; Drilon, A.; Banna, G.L.; Peters, S.; Addeo, A. The METeoric rise of MET in lung cancer. *Cancer* **2020**.
38. Hong, D.S.; DuBois, S.G.; Kummar, S.; Farago, A.F.; Albert, C.M.; Rohrberg, K.S.; van Tilburg, C.M.; Nagasubramanian, R.; Berlin, J.D.; Federman, N. Larotrectinib in patients with TRK fusion-positive solid tumours: A pooled analysis of three phase 1/2 clinical trials. *Lancet Oncol.* **2020**, *21*, 531–540. [CrossRef]
39. Doebele, R.C.; Drilon, A.; Paz-Ares, L.; Siena, S.; Shaw, A.T.; Farago, A.F.; Blakely, C.M.; Seto, T.; Cho, B.C.; Tosi, D. Entrectinib in patients with advanced or metastatic NTRK fusion-positive solid tumours: Integrated analysis of three phase 1–2 trials. *Lancet Oncol.* **2020**, *21*, 271–282. [CrossRef]
40. Drilon, A.; Oxnard, G.R.; Tan, D.S.W.; Loong, H.H.F.; Johnson, M.; Gainor, J.; McCoach, C.E.; Gautschi, O.; Besse, B.; Cho, B.C.; et al. Efficacy of Selpercatinib in RET Fusion–Positive Non–Small-Cell Lung Cancer. *N. Engl. J. Med.* **2020**, *383*, 813–824. [CrossRef]
41. Lee, D.H.; Subbiah, V.; Gainor, J.F.; Taylor, M.H.; Zhu, V.W.; Doebele, R.C.; Lopes, G.; Baik, C.; Garralda, E.; Gadgeel, S.M.; et al. Treatment with pralsetinib (formerly BLU-667), a potent and selective RET inhibitor, provides rapid clearance of ctDNA in patients with RET-altered non-small cell lung cancer (NSCLC) and medullary thyroid cancer (MTC). *Ann. Oncol.* **2019**, *30*, ix122. [CrossRef]
42. Friedlaender, A.; Drilon, A.; Weiss, G.J.; Banna, G.L.; Addeo, A. KRAS as a druggable target in NSCLC: Rising like a phoenix after decades of development failures. *Cancer Treat. Rev.* **2020**, *85*, 101978. [CrossRef] [PubMed]
43. Smit, E.F.; Nakagawa, K.; Nagasaka, M.; Felip, E.; Goto, Y.; Li, B.T.; Pacheco, J.M.; Murakami, H.; Barlesi, F.; Saltos, A.N.; et al. Trastuzumab deruxtecan (T-DXd; DS-8201) in patients with HER2-mutated metastatic non-small cell lung cancer (NSCLC): Interim results of DESTINY-Lung01. *J. Clin. Oncol.* **2020**, *38*, 9504-9504. [CrossRef]
44. Vallette, F.M.; Olivier, C.; Lézot, F.; Oliver, L.; Cochonneau, D.; Lalier, L.; Cartron, P.F.; Heymann, D. Dormant, quiescent, tolerant and persister cells: Four synonyms for the same target in cancer. *Biochem. Pharmacol.* **2019**, *162*, 169–176. [CrossRef]
45. Recasens, A.; Munoz, L. Targeting Cancer Cell Dormancy. *Trends Pharmacol. Sci.* **2019**, *40*, 128–141. [CrossRef]
46. Chaudhuri, A.A.; Chabon, J.J.; Lovejoy, A.F.; Newman, A.M.; Stehr, H.; Azad, T.D.; Khodadoust, M.S.; Esfahani, M.S.; Liu, C.L.; Zhou, L.; et al. Early Detection of Molecular Residual Disease in Localized Lung Cancer by Circulating Tumor DNA Profiling. *Cancer Discov.* **2017**, *7*, 1394–1403. [CrossRef]
47. Dagogo-Jack, I.; Brannon, A.R.; Ferris, L.A.; Campbell, C.D.; Lin, J.J.; Schultz, K.R.; Ackil, J.; Stevens, S.; Dardaei, L.; Yoda, S.; et al. Tracking the Evolution of Resistance to ALK Tyrosine Kinase Inhibitors through Longitudinal Analysis of Circulating Tumor DNA. *JCO Precis Oncol.* **2018**, *2018*. [CrossRef]
48. Chabon, J.J.; Simmons, A.D.; Lovejoy, A.F.; Esfahani, M.S.; Newman, A.M.; Haringsma, H.J.; Kurtz, D.M.; Stehr, H.; Scherer, F.; Karlovich, C.A.; et al. Circulating tumour DNA profiling reveals heterogeneity of EGFR inhibitor resistance mechanisms in lung cancer patients. *Nat. Commun.* **2016**, *7*, 11815. [CrossRef]

International Journal of *Molecular Sciences*

MDPI

Review

Repurposing Tyrosine Kinase Inhibitors to Overcome Multidrug Resistance in Cancer: A Focus on Transporters and Lysosomal Sequestration

Maria Krchniakova [1,2], Jan Skoda [1,2], Jakub Neradil [1,2], Petr Chlapek [1,2] and Renata Veselska [1,2,*]

1 Laboratory of Tumor Biology, Department of Experimental Biology, Faculty of Science, Masaryk University, 61137 Brno, Czech Republic; maria.krchniakova@mail.muni.cz (M.K.); jan.skoda@sci.muni.cz (J.S.); jneradil@sci.muni.cz (J.N.); chlapek@sci.muni.cz (P.C.)

2 International Clinical Research Center, St. Anne's University Hospital, 65691 Brno, Czech Republic

* Correspondence: veselska@sci.muni.cz; Tel.: +420-549-49-7905

Received: 8 April 2020; Accepted: 27 April 2020; Published: 30 April 2020

Abstract: Tyrosine kinase inhibitors (TKIs) are being increasingly used to treat various malignancies. Although they were designed to target aberrant tyrosine kinases, they are also intimately linked with the mechanisms of multidrug resistance (MDR) in cancer cells. MDR-related solute carrier (SLC) and ATB-binding cassette (ABC) transporters are responsible for TKI uptake and efflux, respectively. However, the role of TKIs appears to be dual because they can act as substrates and/or inhibitors of these transporters. In addition, several TKIs have been identified to be sequestered into lysosomes either due to their physiochemical properties or via ABC transporters expressed on the lysosomal membrane. Since the development of MDR represents a great concern in anticancer treatment, it is important to elucidate the interactions of TKIs with MDR-related transporters as well as to improve the properties that would prevent TKIs from diffusing into lysosomes. These findings not only help to avoid MDR, but also help to define the possible impact of combining TKIs with other anticancer drugs, leading to more efficient therapy and fewer adverse effects in patients.

Keywords: tyrosine kinase inhibitor; multidrug resistance; cancer; ABC transporter; SLC transporter; lysosomal sequestration

1. Introduction

Tyrosine kinase inhibitors (TKIs) are low molecular weight (<800 Da) organic compounds that are able to penetrate the cell membrane and interact with targets inside the cell. They were developed to block the ATP-binding sites of protein tyrosine kinases, thereby inhibiting or attenuating the enzymatic activity of aberrant tyrosine kinases responsible for the malignant phenotype of cells. Such targeted therapy can be aimed at either cancer cells, by inhibiting their proliferation and affecting their susceptibility to apoptosis, or the tumor microenvironment, by affecting angiogenesis and the invasion or formation of metastases.

So far, a number of TKIs has been approved by FDA for clinical use (respective molecular targets are summarized in Supplementary Table S1), but many more are currently under investigation: a brief example of experimental TKIs is listed in Supplementary Table S2. Due to their convenient oral administration, TKIs are used not only in anticancer therapy, but also in treating diabetes, inflammation, severe bone disorders and arteriosclerosis [1–3].

However, even anticancer treatment using TKIs leads to the development of multidrug resistance (MDR), i.e., resistance to structurally and functionally different drugs [4]. Thus, the main focus of this review is to describe the noncanonical role of TKIs in selected MDR mechanisms, which involve membrane transporters and drug accumulation in lysosomes.

2. Effects of TKIs on Membrane Transporters

The ATP-binding cassette (ABC) and the solute carrier (SLC) membrane transporters are considered to be the most relevant transporters affecting the exposure to administered TKIs [4]. Both types are expressed ubiquitously throughout human tissue and can recognize and translocate various molecules across biological membranes, including TKIs. As such, they can affect the pharmacokinetic parameters of TKIs, such as drug absorption, distribution, metabolism, excretion and toxicity [4]. Therefore, these transporters expressed on cancer cells are considered a major determinant of MDR because increased efflux or decreased transporter-mediated influx can lead to inefficient intracellular drug concentrations and/or undesired drug interactions.

2.1. ABC Transporters

ABC transporters are transmembrane proteins that have been investigated in relation to their active drug efflux irrespective of the prevailing gradient, thus causing drug resistance. The most widely studied ABC transporters with respect to MDR include P-glycoprotein (Pgp, ABCB1), multidrug resistance protein 1 (MRP1, ABCC1) and breast cancer resistance protein (BCRP, ABCG2). Multiple mechanisms modulating the expression of ABC transporters have been proposed [5], including the loss of uL3 ribosomal protein, which has been recently associated with the upregulation of ABCB1 [6]. An overview of ABC transporters involved in MDR and interacting with TKIs is listed in Table 1.

Some TKIs are able to bind to the substrate-binding pocket of an ABC transporter (Table 1) [7–11], which leads to their efflux from cells and explains the reduced therapeutic efficacy and/or resistance acquired during the course of TKI therapy. ABCA3 protected leukemic stem cells from dasatinib, imatinib, and nilotinib, which target the BCR-ABL kinase [7]. Exposure to these TKIs led to a dose-dependent increase in ABCA3 transcription, supporting drug efflux, but when cells were cotreated with the COX2 inhibitor indomethacin, ABCA3 expression decreased, and the combination potentiated the antineoplastic efficacy of TKIs [7]. Similarly, gefitinib causes indirect induction of ABCG2 expression [12]. In fact, targeting EGFR with gefitinib results in its internalization, phosphorylation by Akt and translocation to the nucleus, where EGFR affects the *ABCG2* gene promoter enhancing its transcription [12].

In contrast, TKIs can also act as inhibitors of ABC transporters. Similarly to their interaction with protein tyrosine kinases, TKIs block the ATP-binding sites of membrane transporters, preventing the phosphorylation and inhibiting the efflux function of transporters [13–17]. Although cabozantinib affected the ATPase activity of the ABCG2 transporter, it also interacted with the transporter at the drug-substrate binding site, antagonizing the transporter by competitive inhibition [15]. TKIs usually inhibit ABC transporters directly and do not alter their expression or localization [13,16,17].

Interestingly, ponatinib treatment resulted in a decrease in ABCB1 and ABCG2 cell surface expression, and imatinib downregulated ABCG2 expression in BCR-ABL-positive cells [18,19]. However, these effects were most likely caused indirectly via inhibition of the Akt signaling that is downstream of the BCR-ABL axis that is inhibited by the TKIs [18,19].

When inhibiting ABC transporters, substrate drugs are no longer pumped outside of cells, and the cytotoxicity of substrate drugs in resistant cells overexpressing ABC transporters is significantly increased. In vitro studies demonstrated that TKI administration increased intracellular accumulation of rhodamine 123 or doxorubicin in multidrug-resistant cells overexpressing selected ABC transporters [20,21]. Treatment with TKIs inhibiting these transporters (Table 1) was able to enhance the cytotoxicity of substrate drugs, such as paclitaxel, docetaxel [14], vincristine, vinblastine [20,22], doxorubicin [20], etoposide [23], cytarabine [24], mitoxantrone and topotecan [15,19,25], while sensitivity to cisplatin, which is not a substrate for ABC transporters, was not significantly altered [26]. The inhibitory effect of TKIs (e.g., gefitinib or ibrutinib) was comparable to that of known inhibitors of ABC transporters [14,27]. Resensitizing multidrug-resistant cancer cells can also be achieved by combining a TKI with an ABC transporter substrate affinity together with a second TKI having an ABC transporter inhibitory activity. A low-dose treatment with the ABCB1 transporter substrate dasatinib, in combination with the ABCB1 inhibitor nilotinib, provided additive/synergistic effects in leukemic cells overexpressing ABCB1 [28]. Supporting these findings, in in vivo experiments in respective xenograft mouse models, TKIs combined with conventional chemotherapeutics showed a greater inhibitory effect on tumor growth than single drugs [20,29,30]. Furthermore, simultaneous inhibition of ABCB1 and ABCG2 by erlotinib at the mouse blood–brain barrier improved brain permeability and pazopanib accumulation [31].

Depending on their concentration and affinity for the transporter, a number of TKIs have been reported to interact with ABC transporters as both substrates and inhibitors (Figure 1A) [17,19,25,32,33]. At lower concentrations, TKIs usually possess substrate-like properties (Figure 1Ai), but they tend to act as ABC inhibitors at higher yet pharmacologically relevant concentrations (Figure 1Aii) [13,19]. Indeed, combining ponatinib with topotecan or mitoxantrone, substrates of both ABCB1 and ABCG2, resulted in antagonistic effects at lower ponatinib concentrations, whereas higher concentrations led to synergistic effects [19]. In addition, contradictory effects have also been attributed to pazopanib. While it was described as a substrate for both ABCB1 and ABCG2 in the canine kidney cell line MDCKII [31], another study reported that pazopanib was an ABCB1 inhibitor that inhibited dasatinib efflux from LLC-PK1 porcine kidney cells [34].

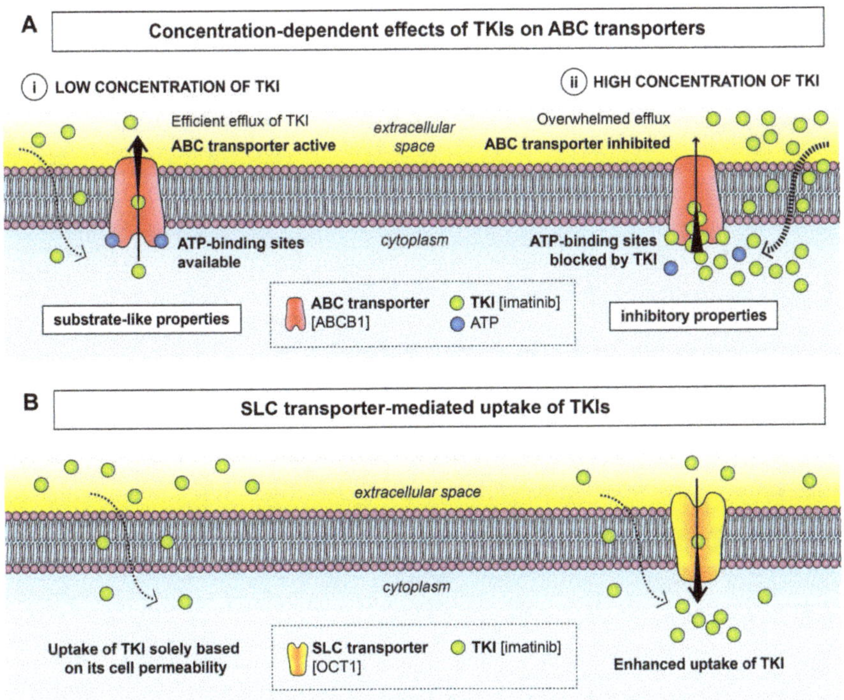

Figure 1. Transport of TKIs by ABC and SLC transporters. (**A**) At low concentrations (**i**), some TKIs exhibit substrate-like properties and are exported out of the cell by the respective ABC transporters. A high concentration of TKIs (**ii**) leads to blockage of the ATP-binding sites of ABC transporters, which results in inhibited efflux of the TKI. (**B**) Upregulated expression of SLC transporters can lead to enhanced uptake of some TKIs. Examples of TKIs and specific transporters are given in square brackets.

Table 1. Interactions of selected TKIs with ABC transporters.

ABC Transporter	Substrate	Inhibitor	Substrate/Inhibitor
ABCA3	dasatinib [7]; imatinib [7]; nilotinib [7]	–	–
ABCB1 (P-glycoprotein, MDR1)	brigatinib [9]; crizotinib [35]	cabozantinib [36]; canertinib * [31]; cediranib * [37]; ceritinib [38]; erlotinib [34]; gefitinib [14]; motesanib * [39]; neratinib [29]; osimertinib [40]; regorafenib [34]; saracatinib * [41]; sorafenib [34]; sunitinib [21]; vandetanib [42]; vatalanib * [43]	afatinib [44]; alectinib [33]; apatinib * [17]; bosutinib [45]; dasatinib [45]; ibrutinib [27]; imatinib [46]; lapatinib [47,48]; nilotinib [45]; nintedanib [22]; pazopanib [31,34]; ponatinib [19]
ABCC1 (MRP1)	–	cediranib * [37]; ibrutinib [49]; sunitinib [21]; vandetanib [42]	–
ABCC2 (MRP2)	sorafenib [50]	sunitinib [51]	–
ABCC3 (MRP3)	imatinib [52]; sorafenib [53]	–	–
ABCC4 (MRP4)	imatinib [8]	erlotinib [54]; gefitinib [54]; sorafenib [55]; sunitinib [51]	–
ABCC6 (MRP6)	dasatinib [10]; nilotinib [10]	–	–
ABCC10 (MRP7)	gefitinib [11]	erlotinib [16]; ibrutinib [27]; imatinib [56]; lapatinib [16]; linsitinib * [13]; masitinib * [30]; nilotinib [20]; ponatinib [57]; sorafenib [24]	–
ABCC11 (MRP8)	–	sorafenib [24]	–
ABCG2 (BRCP)	brigatinib [9]; gefitinib [58]	axitinib [34]; cabozantinib [15]; canertinib * [31]; ceritinib [38]; erlotinib [34]; icotinib * [59]; linsitinib * [13]; masitinib * [60]; osimertinib [40]; quizartinib * [61]; regorafenib [34]; sorafenib [24]; sunitinib [21]; tandutinib * [15]; vandetanib [42]; vatalanib * [43]	afatinib [32]; alectinib [33]; apatinib * [17]; bosutinib [45]; dasatinib [45]; imatinib [46]; lapatinib [47]; nilotinib [45]; pazopanib [31,34]; ponatinib [19]; telatinib * [25]

* experimental TKIs.

2.2. SLC Transporters

While ABC transporters harness energy from ATP hydrolysis and function as efflux transporters, SLC transporters are primarily involved in the uptake of small molecules into cells, including TKIs [62] (Figure 1B). Unlike the described MDR mediated by ABC transporters in a number of malignancies, knowledge about the interactions of SLC transporters with drugs used in anticancer treatment is limited. Table 2 contains an overview of TKIs known to interact with SLC transporters.

Table 2. Interactions of selected TKIs with SLC transporters.

SLC Transporter	Substrate	Inhibitor
OCT1 (SLC22A1)	imatinib [63,64] sorafenib [55]	crizotinib [51] erlotinib [65] gefitinib [65] nilotinib [66] sunitinib [65]
OCT2 (SLC22A2)	erlotinib [67]	crizotinib [68] gefitinib [65] nilotinib [65] saracatinib [69] sunitinib [65] vandetanib [68]
OCT3 (SLC22A3)	–	gefitinib [65] nilotinib [65] sunitinib [65]
OCTN2 (SLC22A5)	imatinib [8]	–
OAT3 (SLC22A8)	erlotinib [67]	–
OAT6 (SLC22A20)	sorafenib [70]	–
OATP1A2 (SLCO1A2)	imatinib [8]	–
OATP1B1 (SLCO1B1)	–	axitinib [71] lapatinib [51] nilotinib [71] pazopanib [71] sorafenib [71]
OATP1B3 (SLCO1B3)	imatinib [8]	–
OATP2B1 (SLCO2B1)	erlotinib [72]	–

The activity of imatinib was linked with the expression of organic cation transporter 1 (OCT1, SLC22A1), as it was found to be a substrate for this transporter in the CEM human leukemia cell line [73]. A positive correlation was found in patients with chronic myeloid leukemia (CML) in a phase II trial between survival and the functional activity of OCT1 that was assessed by measuring imatinib influx [63,64]. In addition, temperature-dependent uptake experiments demonstrated that the uptake of imatinib was an active process rather than a passive penetration of cell membranes [73]. Other transporters that might affect the oral absorption of imatinib and the liver access of imatinib include the uptake organic cation/carnitine transporter (OCTN2, SLC22A5) and the uptake organic anion-transporting polypeptides OATP1A2 (SLCO1A2) and OATP1B3 (SLCO1B3), for which imatinib is a substrate [8].

In contrast, the cellular uptake of nilotinib seems to be independent of OCT expression. This was observed in KCL22 human leukemia cell line overexpressing OCT1 [66] as well as in mononuclear cells from patients with CML [74]. In fact, nilotinib has been reported as a potential inhibitor of OCT1 [66], OCT2, OCT3 [65] and OATP1B1 [71].

The uptake of drugs into nontarget (nonneoplastic) cells by SLC transporters resulting in higher drug toxicity presents another obstacle in anticancer treatment. Organic anion transporter 6 (OAT6, SLC22A20) was found to regulate the entry of sorafenib into keratinocytes, contributing to sorafenib-induced skin toxicity [70].

3. Lysosomal Sequestration

Lysosomes contribute to MDR via a mechanism called lysosomal trapping. Compounds can be sequestered (trapped) in lysosomes based on their physiochemical properties: (i) basic pKa, an acid dissociation constant for the conjugated acid of the weak base, which affects the extent of lysosomal trapping, and (ii) logP, the partition coefficient between octanol and water, which regulates the kinetics of passive membrane permeability [75]. Accumulation in lysosomes is typical for lipophilic and amphiphilic compounds with lipophilic amines (logP > 1) and weak bases with ionizable amine groups (pKa > 6) [75]. Due to their hydrophobic character, these drugs are able to permeate the lipid membranes via passive diffusion. However, after entering the acidic environment of lysosomes, compounds become positively charged, which restricts their diffusion back into the cytoplasm and prevents them from reaching their cytoplasmic or nuclear targets [75] (Figure 2A). Furthermore, lysosomal sequestration is driven by the pH difference between the neutral cytosol (pH ~ 7.2) and the acidic lysosomal compartment (pH ~ 5) [76]. This process requires continuous acidification of the lysosomes by membrane-bound ATP-dependent lysosomal proton pumps of the vacuolar ATPase (V-ATPase) family. Agents that are sequestered in lysosomes are called lysosomotropic, and their accumulation within lysosomes is known as lysosomotropism [75].

Lysosomal sequestration has been recognized as another mechanism of resistance to TKIs [77], and TKIs known to be accumulated in lysosomes are summarized in Table 3. The ability of TKIs to be sequestered in lysosomes can be detected by fluorescence microscopy in the case of inhibitors that exhibit autofluorescence, such as sunitinib [23,77], lapatinib [78], imatinib [79,80] or nintedanib [81], and they colocalize with stained lysosomes. In the case of TKIs that are not autofluorescent (e.g., gefitinib or lapatinib), lysosomal sequestration can be demonstrated by their influence on the lysosomal accumulation of LysoTracker® Red [76].

Several TKIs do not harbor physiochemical properties of hydrophobic, weak base molecules but can be entrapped in the acidic milieu of lysosomes (Table 3) [82–84]. ABC transporters facilitate the active accumulation of drugs in lysosomes, as these pumps have been found on the membranes of intracellular compartments, including the Golgi apparatus and intracellular vesicles [85,86]. ABCA3 [87], ABCB1 [88], and ABCG2 [89] were demonstrated on lysosomal membranes, explaining the lysosomal sequestration of their respective substrate TKIs, including imatinib [87], sorafenib [83] and pazopanib [84].

Interestingly, the ABCB1-mediated resistance phenotype of leukemia cells was stronger when ABCB1 was expressed intracellularly than when it was expressed on the plasma membrane, indicating that the accumulation of drugs in lysosomes is most likely more effective than the efflux via membrane transporters [85]. Furthermore, stressors present in the tumor microenvironment (e.g., hypoxia, oxidants, or glucose starvation) were found to upregulate and relocalize ABCB1 to lysosomal membranes, resulting in increased drug resistance [88].

In many cases, resistance mediated by lysosomal sequestration is reversible. Removing sunitinib from tumor cell culture for several weeks resulted in normalization of cell lysosomal capacity and recovery of drug sensitivity [77]. Similarly to the in vitro data, patients with metastatic renal cell carcinoma developed resistance to sunitinib. However, it was transient after treatment interruption and subsequent rechallenge [90].

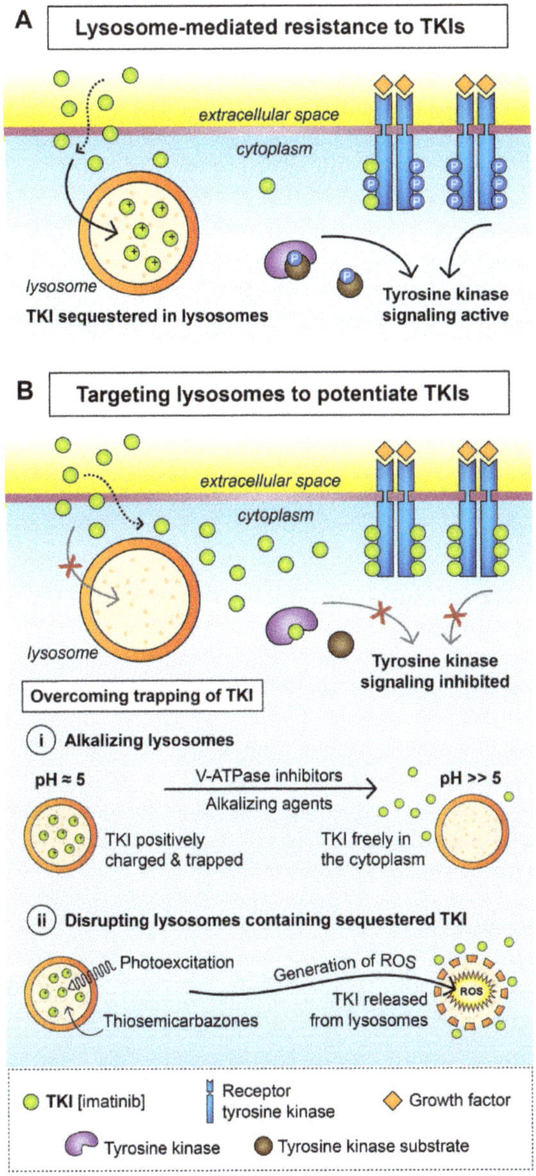

Figure 2. Lysosomes in resistance to TKIs. (**A**) Sequestration of TKIs into lysosomes provides a mechanism of resistance to TKIs. (**B**) Targeting lysosomes by alkalizing their milieu (**i**) or disrupting their integrity (**ii**) can potentiate the effects of TKI treatment.

Table 3. List of TKIs known to be sequestered into lysosomes.

TKI	pKa [1]	LogP [2]	Reference
dasatinib	8.49	3.82	[82]
gefitinib	6.85	3.75	[76]
imatinib	8.10	4.50	[79]
lapatinib	7.20	4.64	[76]
nilotinib	6.30	5.36	[80]
nintedanib	7.90	3.60	[81]
pazopanib	5.07	3.60	[84]
sorafenib	4.34	2.03	[83]
sunitinib	9.04	5.20	[77]

[1] acid dissociation constant for the conjugated acid of the weak base. [2] partition coefficient between octanol and water.

Overcoming Lysosomal Sequestration

There are several mechanisms that may reverse sequestration: either preventing the accumulation of TKIs in the lysosomes by alkalizing the lysosomal milieu or disrupting the lysosomal membrane leading to efflux of TKIs. Concomitant or sequential treatment with TKIs and drugs that interfere with lysosomal function could present an effective means of overcoming the MDR mediated by lysosomal trapping (Figure 2B).

Several alkalizing agents have been introduced to circumvent lysosomal trapping (Figure 2Bi). Bafilomycin A1 targets V-ATPase, an enzyme that acidifies lysosomes during biogenesis, and was reported to sensitize cells towards previously sequestered nintedanib [81]. Although it prevents lysosomal sequestration in vitro, efficient concentrations of bafilomycin A1 also exert cytotoxicity in normal cells, which hinders its use in clinical settings [77]. Chloroquine, originally established as an antimalarial agent, accumulates in lysosomes, increases lysosomal pH and triggers destabilization of the lysosomal membrane. Combined treatment using chloroquine and sunitinib resulted in enhanced inhibition of tumor growth in a xenograft mouse model [91]. Similarly, the chloroquine analogues hydroxychloroquine and Lys05 have been shown to target lysosome-mediated autophagy and have been tested with other anticancer therapies [92–94].

Interestingly, sunitinib itself is able to reduce the activity of acid sphingomyelinase that promotes lysosomal membrane stability, leading to destabilization of lysosomes and inducing nonapoptotic lysosome-dependent cell death [23].

Photodestruction of lysosomes with sequestered photoexcitable TKIs presents another approach for overcoming lysosomal trapping (Figure 2Bii). Exposing sequestered sunitinib to a specific wavelength in vitro resulted in the generation of reactive oxygen species (ROS) and almost immediate disruption of lysosomes, followed by the release of the drug into the cytoplasm [95]. Similar observations and markedly attenuated tumor growth were reported after sunitinib photoexcitation in a xenograft model [95].

However, phototherapy is limited due to superficial and local treatment options, and apart from chloroquine [91], not many effective drugs have been identified to accumulate in lysosomes and then disrupt the lysosomal membrane. Thiosemicarbazone iron chelators represent novel anticancer agents that are transported into the lysosomes via ABCB1 [96]. There, they create redox-active complexes with copper, and generated ROS permeate the lysosomal membrane (Figure 2Bii). Thiosemicarbazones were able to disrupt lysosomes and free sequestered doxorubicin in ABCB1-overexpressing cells [88,96]. Whether these agents can potentiate the effect of TKIs trapped in lysosomes is yet to be elucidated.

4. Clinical Trials Repurposing TKIs in Combinational Strategies

The ability of several TKIs to modulate ABC transporters was shown in cancer cell lines as well as in xenograft models and primary cells collected from patients [29,33,40,49]. TKIs inhibiting ABC transporters were able to reverse the MDR phenotype of cancer cells and enhance the effect of other anticancer drugs at the quite low, usually noncytotoxic concentrations achieved in patients [13,20,56]. This evidence underlines the potential clinical value of TKIs and provides a rationale for their repurposing in combinational strategies overcoming ABC transporter-mediated MDR. Table 4 lists examples of clinical trials combining TKIs with other anticancer drugs.

Table 4. Combinational strategies using TKIs in clinical trials.

Combination of Drugs		Malignancy	Reference
apatinib *	+ etoposide + irinotecan	ovarian cancer high-grade glioma	[97] [98]
cediranib *	+ carboplatin, paclitaxel + cisplatin, gemcitabine	cervical cancer biliary tract cancer	[99] [100]
crizotinib	+ methotrexate	NSCLC	[101]
erlotinib	+ cabozantinib + carboplatin + everolimus + gemcitabine + gemcitabine, oxaliplatin + topotecan	NSCLC ovarian carcinoma HNSCC pancreatic cancer pancreatic cancer solid tumors	[102] [103] [104] [105] [106] [107]
gefitinib	+ carboplatin, pemetrexed	NSCLC	[108]
lapatinib	+ capecitabine + paclitaxel	breast cancer breast cancer	[109] [110]
neratinib	+ capecitabine + paclitaxel	breast cancer breast cancer	[111] [112]
nilotinib	+ vincristine, daunorubucin	ALL	[113]
nintedanib	+ docetaxel	NSCLC	[114]
sorafenib	+ cytarabine, daunorubicin + doxorubicin + gemcitabine, cisplatin	AML hepatocellular carcinoma collecting duct carcinoma	[115] [116] [117]
sunitinib	+ capecitabine + docetaxel	breast cancer breast cancer, gastric cancer	[118] [119,120]
vandetanib	+ docetaxel + pemetrexed	urothelial cancer NSCLC	[121] [122]

* experimental TKIs; AML: acute myeloid leukemia; ALL: acute lymphoblastic leukemia; HNSCC: head and neck squamous cell carcinoma; NSCLC: non-small-cell lung carcinoma.

Promising efficacy and improved clinical outcomes were described when combining paclitaxel with neratinib [112] or lapatinib [110] in HER2-positive breast cancer patients. Favorable results were also observed in combinations of docetaxel with nintedanib in non-small-cell lung carcinoma patients [114] and with sunitinib in patients with gastric cancer [120]. Resistance to docetaxel and paclitaxel is often caused by ABCB1- and ABCC10-mediated efflux [26]. Hence, adding TKIs that inhibit these transporters (Table 1), e.g., applied lapatinib [16,48], neratinib [29], nintedanib [22], or sunitinib [21], could, in fact, decrease the efflux of chemotherapeutics and result in enhanced antitumor effects observed in the studies (Table 4).

Similar conclusions could be drawn from the trials that combined erlotinib and gemcitabine for the treatment of advanced pancreatic cancer [105,106]. Gemcitabine plus erlotinib showed additive efficacy compared to gemcitabine alone [105] and addition of oxaliplatin to this regimen resulted in higher response rate and improved progression-free survival [106]. In vitro studies revealed that the resistance to oxaliplatin develops after upregulation of ABCC1 and ABCC4 transporters [123]. Furthermore, a combined siRNA-mediated knockdown of *ABCC3*, *ABCC5*, and *ABCC10* significantly sensitized cells to gemcitabine [124]. As erlotinib was demonstrated as a potent inhibitor of multiple ABC transporters (Table 1), including those that efflux gemcitabine and oxaliplatin from cancer cells [16,54] and are known to cause resistance in pancreatic adenocarcinomas [125], these data possibly elucidate the increased efficacy of the combined treatment in the respective clinical trials [105,106].

These examples demonstrate that TKIs added to the treatment enhance the response by not only targeting aberrant tyrosine kinases in malignant cells but also by sensitizing resistant tumors to other anticancer agents. Although the clinical trials (Table 4) were focused on advanced, metastatic and/or recurrent malignancies with known resistance to therapy, not all drug combinations attained satisfactory outcomes in patients [103,107,115,116,118]. However, the mechanisms mediating MDR in tumors were usually not examined and most trials did not focus specifically on reversing the ABC transporter-mediated MDR. This urges the need for combination strategies that would select TKIs attentively with regard to multiple determinants, including the tumor type, its expression profile as well as presence of MDR-related factors, in order to tailor therapeutic regiments that may lead to overcoming resistance and improved clinical response.

Nanotechnology could present a valuable strategy in combining TKIs with conventional chemotherapeutics [126]. Polymeric nanoparticles allowed a co-delivery of erlotinib and doxorubicin on the same platform while facilitating a sequential release of the drugs, which resulted in the enhanced cytotoxic effect on breast cancer cells [127]. Therefore, nanomedicine offers a convenient multidrug delivery system where the first released drug (TKI, e.g., erlotinib) sensitizes the cancer cells to the second drug (conventional chemotherapeutic, e.g., doxorubicin), hence avoiding MDR development and making therapy more efficient [126,127].

5. Conclusions

A more personalized approach to therapy, such as targeted therapy using TKIs, has been increasingly used in treating various types of malignancies. Emerging evidence suggests that apart from identifying specific targets of TKIs, it is also important to evaluate other characteristics of tumor cells as well as the drug itself. The expression of uptake/efflux membrane transporters and the physiochemical qualities of TKIs affect the exposure of administered TKIs.

Furthermore, the dual effects of TKIs on membrane transporters allow them to not only exert anticancer effects but also act as chemosensitizers to reverse the transporter-mediated efflux of other anticancer drugs. For instance, high expression of specific membrane transporters could provide the perfect environment for the therapeutic application of the TKIs that are transported into the cancer cells by abundant SLC transporters but at the same time inhibit the drug efflux pumps. This allows for either sequential or simultaneous administration of TKIs with other cytotoxic agents, harboring great synergistic potential, improving the efficacy of therapy, avoiding or reversing drug resistance, and possibly reducing associated toxicity and adverse effects (Figure 3).

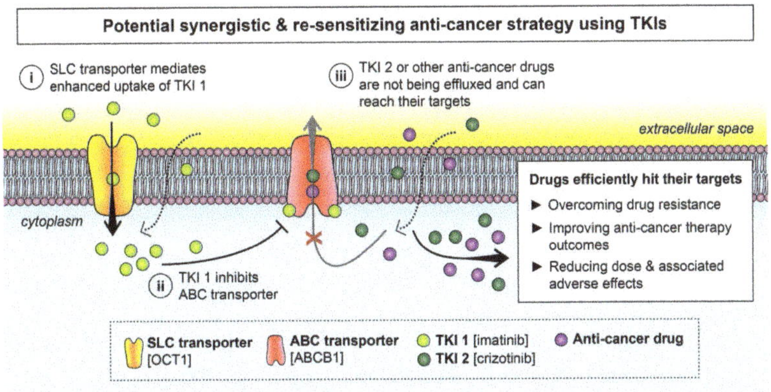

Figure 3. Schematic illustration of a potential anticancer strategy using TKIs that exploits the upregulated expression of SLC transporters to resensitize cells to anticancer drugs. In this scenario, high levels of certain SLC transporters (e.g., OCT1) are utilized to load a cancer cell with the first TKI (TKI 1; e.g., imatinib) (**i**). Apart from hitting its targets, TKI 1 also inhibits ABC transporters (e.g., ABCB1) (**ii**) so the second TKI (TKI 2; e.g., crizotinib) or other anti-cancer drugs are no longer effluxed from cancer cells (**iii**), which eventually results in synergistic effects of the drugs and improved treatment response.

Supplementary Materials: The following are available online at http://www.mdpi.com/1422-0067/21/9/3157/s1, Table S1: FDA-approved TKIs and their molecular targets, Table S2: An example of TKIs under investigation and their molecular targets.

Author Contributions: Conceptualization: M.K., J.S., R.V.; writing – original draft: M.K.; writing – review & editing: J.S., J.N., P.C., R.V; visualization: J.S. All authors have read and agreed to the published version of the manuscript.

Funding: This study was funded by project AZV MZCR 17-33104A, by project No. LQ1605 from the National Program of Sustainability II (MEYS CR) and by project Brno Ph.D. Talent 2017 from JCMM.

Conflicts of Interest: The authors declare no conflict of interest.

Abbreviations

ABC	ATP-binding cassette
ALL	acute lymphoblastic leukemia
AML	acute myeloid leukemia
BRCP	breast cancer resistance protein
CML	chronic myeloid leukemia
EGFR	epidermal growth factor receptor
HER2	human epidermal growth factor receptor 2
HNSCC	head and neck squamous cell carcinoma
LAMP2	lysosome-associated membrane protein 2
MDR	multidrug resistance
MRP	multidrug resistance protein
NSCLC	non-small-cell lung carcinoma
OAT	organic anion transporter
OATP	organic anion-transporting polypeptide
OCT	organic cation transporter
OCTN	organic cation/carnitine transporter
Pgp	P-glycoprotein
ROS	reactive oxygen species
SLC	solute carrier
TKI	tyrosine kinase inhibitor
V-ATPase	vacuolar ATPase

References

1. Louvet, C.; Szot, G.L.; Lang, J.; Lee, M.R.; Martinier, N.; Bollag, G.; Zhu, S.; Weiss, A.; Bluestone, J.A. Tyrosine kinase inhibitors reverse type 1 diabetes in nonobese diabetic mice. *Proc. Natl. Acad. Sci. USA* **2008**, *105*, 18895–18900. [CrossRef]
2. Weinblatt, M.E.; Kavanaugh, A.; Genovese, M.C.; Musser, T.K.; Grossbard, E.B.; Magilavy, D.B. An Oral Spleen Tyrosine Kinase (Syk) Inhibitor for Rheumatoid Arthritis. *N. Engl. J. Med.* **2010**, *363*, 1303–1312. [CrossRef]
3. Emami, H.; Vucic, E.; Subramanian, S.; Abdelbaky, A.; Fayad, Z.A.; Du, S.; Roth, E.; Ballantyne, C.M.; Mohler, E.R.; Farkouh, M.E.; et al. The effect of BMS-582949, a P38 mitogen-activated protein kinase (P38 MAPK) inhibitor on arterial inflammation: A multicenter FDG-PET trial. *Atherosclerosis* **2015**, *240*, 490–496. [CrossRef] [PubMed]
4. Wu, C.-P.; Hsieh, C.-H.; Wu, Y.-S. The Emergence of Drug Transporter-Mediated Multidrug Resistance to Cancer Chemotherapy. *Mol. Pharm.* **2011**, *8*, 1996–2011. [CrossRef] [PubMed]
5. Miller, D.S. Regulation of P-glycoprotein and other ABC drug transporters at the blood-brain barrier. *Trends Pharmacol. Sci.* **2010**, *31*, 246–254. [CrossRef] [PubMed]
6. Russo, A.; Saide, A.; Smaldone, S.; Faraonio, R.; Russo, G. Role of uL3 in multidrug resistance in p53-mutated lung cancer cells. *Int. J. Mol. Sci.* **2017**, *18*, 1–16. [CrossRef] [PubMed]
7. Hupfeld, T.; Chapuy, B.; Schrader, V.; Beutler, M.; Veltkamp, C.; Koch, R.; Cameron, S.; Aung, T.; Haase, D.; LaRosee, P.; et al. Tyrosinekinase inhibition facilitates cooperation of transcription factor SALL4 and ABC transporter A3 towards intrinsic CML cell drug resistance. *Br. J. Haematol.* **2013**, *161*, 204–213. [CrossRef] [PubMed]
8. Hu, S.; Franke, R.M.; Filipski, K.K.; Hu, C.; Orwick, S.J.; de Bruijn, E.A.; Burger, H.; Baker, S.D.; Sparreboom, A. Interaction of Imatinib with Human Organic Ion Carriers. *Clin. Cancer Res.* **2008**, *14*, 3034–3038. [CrossRef]
9. Li, W.; Sparidans, R.W.; Wang, Y.; Lebre, M.C.; Beijnen, J.H.; Schinkel, A.H. P-glycoprotein and breast cancer resistance protein restrict brigatinib brain accumulation and toxicity, and, alongside CYP3A, limit its oral availability. *Pharmacol. Res.* **2018**, *137*, 47–55. [CrossRef]
10. Eadie, L.N.; Dang, P.; Goyne, J.M.; Hughes, T.P.; White, D.L. ABCC6 plays a significant role in the transport of nilotinib and dasatinib, and contributes to TKI resistance in vitro, in both cell lines and primary patient mononuclear cells. *PLoS ONE* **2018**, *13*, e0192180. [CrossRef]
11. Zhao, H.; Huang, Y.; Shi, J.; Dai, Y.; Wu, L.; Zhou, H. ABCC10 plays a significant role in the transport of gefitinib and contributes to acquired resistance to gefitinib in NSCLC. *Front. Pharmacol.* **2018**, *9*, 1312. [CrossRef] [PubMed]
12. Huang, W.-C.; Chen, Y.-J.; Li, L.-Y.; Wei, Y.-L.; Hsu, S.-C.; Tsai, S.-L.; Chiu, P.-C.; Huang, W.-P.; Wang, Y.-N.; Chen, C.-H.; et al. Nuclear Translocation of Epidermal Growth Factor Receptor by Akt-dependent Phosphorylation Enhances Breast Cancer-resistant Protein Expression in Gefitinib-resistant Cells. *J. Biol. Chem.* **2011**, *286*, 20558–20568. [CrossRef] [PubMed]
13. Zhang, H.; Kathawala, R.J.; Wang, Y.-J.; Zhang, Y.-K.; Patel, A.; Shukla, S.; Robey, R.W.; Talele, T.T.; Ashby, C.R.; Ambudkar, S.V.; et al. Linsitinib (OSI-906) antagonizes ATP-binding cassette subfamily G member 2 and subfamily C member 10-mediated drug resistance. *Int. J. Biochem. Cell Biol.* **2014**, *51*, 111–119. [CrossRef] [PubMed]
14. Kitazaki, T.; Oka, M.; Nakamura, Y.; Tsurutani, J.; Doi, S.; Yasunaga, M.; Takemura, M.; Yabuuchi, H.; Soda, H.; Kohno, S. Gefitinib, an EGFR tyrosine kinase inhibitor, directly inhibits the function of P-glycoprotein in multidrug resistant cancer cells. *Lung Cancer* **2005**, *49*, 337–343. [CrossRef] [PubMed]
15. Zhang, G.-N.; Zhang, Y.-K.; Wang, Y.-J.; Barbuti, A.M.; Zhu, X.-J.; Yu, X.-Y.; Wen, A.-W.; Wurpel, J.N.D.; Chen, Z.-S. Modulating the function of ATP-binding cassette subfamily G member 2 (ABCG2) with inhibitor cabozantinib. *Pharmacol. Res.* **2017**, *119*, 89–98. [CrossRef]
16. Kuang, Y.-H.; Shen, T.; Chen, X.; Sodani, K.; Hopper-Borge, E.; Tiwari, A.K.; Lee, J.W.K.K.; Fu, L.-W.; Chen, Z.-S. Lapatinib and erlotinib are potent reversal agents for MRP7 (ABCC10)-mediated multidrug resistance. *Biochem. Pharmacol.* **2010**, *79*, 154–161. [CrossRef]
17. Mi, Y.-J.; Liang, Y.-J.; Huang, H.-B.; Zhao, H.-Y.; Wu, C.-P.; Wang, F.; Tao, L.-Y.; Zhang, C.-Z.; Dai, C.-L.; Tiwari, A.K.; et al. Apatinib (YN968D1) reverses multidrug resistance by inhibiting the efflux function of multiple ATP-binding cassette transporters. *Cancer Res.* **2010**, *70*, 7981–7991. [CrossRef]

18. Nakanishi, T.; Shiozawa, K.; Hassel, B.A.; Ross, D.D. Complex interaction of BCRP/ABCG2 and imatinib in BCR-ABL-expressing cells: BCRP-mediated resistance to imatinib is attenuated by imatinib-induced reduction of BCRP expression. *Blood* **2006**, *108*, 678–684. [CrossRef]
19. Sen, R.; Natarajan, K.; Bhullar, J.; Shukla, S.; Fang, H.-B.; Cai, L.; Chen, Z.-S.; Ambudkar, S.V.; Baer, M.R. The novel BCR-ABL and FLT3 inhibitor ponatinib is a potent inhibitor of the MDR-associated ATP-binding cassette transporter ABCG2. *Mol. Cancer Ther.* **2012**, *11*, 2033–2044. [CrossRef]
20. Tiwari, A.K.; Sodani, K.; Dai, C.-L.; Abuznait, A.H.; Singh, S.; Xiao, Z.-J.; Patel, A.; Talele, T.T.; Fu, L.; Kaddoumi, A.; et al. Nilotinib potentiates anticancer drug sensitivity in murine ABCB1-, ABCG2-, and ABCC10-multidrug resistance xenograft models. *Cancer Lett.* **2013**, *328*, 307–317. [CrossRef]
21. Shukla, S.; Robey, R.W.; Bates, S.E.; Ambudkar, S.V. Sunitinib (Sutent, SU11248), a small-molecule receptor tyrosine kinase inhibitor, blocks function of the ATP-binding cassette (ABC) transporters P-glycoprotein (ABCB1) and ABCG2. *Drug Metab. Dispos.* **2009**, *37*, 359–365. [CrossRef] [PubMed]
22. Englinger, B.; Lötsch, D.; Pirker, C.; Mohr, T.; van Schoonhoven, S.; Boidol, B.; Lardeau, C.-H.; Spitzwieser, M.; Szabó, P.; Heffeter, P.; et al. Acquired nintedanib resistance in FGFR1-driven small cell lung cancer: Role of endothelin-A receptor-activated ABCB1 expression. *Oncotarget* **2016**, *7*, 50161–50179. [CrossRef] [PubMed]
23. Ellegaard, A.-M.; Groth-Pedersen, L.; Oorschot, V.; Klumperman, J.; Kirkegaard, T.; Nylandsted, J.; Jaattela, M. Sunitinib and SU11652 Inhibit Acid Sphingomyelinase, Destabilize Lysosomes, and Inhibit Multidrug Resistance. *Mol. Cancer Ther.* **2013**, *12*, 2018–2020. [CrossRef] [PubMed]
24. Hu, S.; Chen, Z.; Franke, R.; Orwick, S.; Zhao, M.; Rudek, M.A.; Sparreboom, A.; Baker, S.D. Interaction of the Multikinase Inhibitors Sorafenib and Sunitinib with Solute Carriers and ATP-Binding Cassette Transporters. *Clin. Cancer Res.* **2009**, *15*, 6062–6069. [CrossRef] [PubMed]
25. Sodani, K.; Patel, A.; Anreddy, N.; Singh, S.; Yang, D.-H.; Kathawala, R.J.; Kumar, P.; Talele, T.T.; Chen, Z.-S. Telatinib reverses chemotherapeutic multidrug resistance mediated by ABCG2 efflux transporter in vitro and in vivo. *Biochem. Pharmacol.* **2014**, *89*, 52–61. [CrossRef] [PubMed]
26. Vispute, S.G.; Chen, J.-J.; Sun, Y.-L.; Sodani, K.S.; Singh, S.; Pan, Y.; Talele, T.; Ashby, C.R.; Chen, Z.-S. Vemurafenib (PLX4032, Zelboraf®), a BRAF Inhibitor, Modulates ABCB1-, ABCG2-, and ABCC10-Mediated Multidrug Resistance. *J. Can. Res. Updates* **2013**, *2*, 306–317.
27. Zhang, H.; Patel, A.; Wang, Y.-J.; Zhang, Y.-K.; Kathawala, R.J.; Qiu, L.-H.; Patel, B.A.; Huang, L.-H.; Shukla, S.; Yang, D.-H.; et al. The BTK Inhibitor Ibrutinib (PCI-32765) Overcomes Paclitaxel Resistance in ABCB1- and ABCC10-Overexpressing Cells and Tumors. *Mol. Cancer Ther.* **2017**, *16*, 1021–1030. [CrossRef]
28. Hiwase, D.K.; White, D.; Zrim, S.; Saunders, V.; Melo, J.V.; Hughes, T.P. Nilotinib-mediated inhibition of ABCB1 increases intracellular concentration of dasatinib in CML cells: Implications for combination TKI therapy. *Leukemia* **2010**, *24*, 658–660. [CrossRef]
29. Zhao, X.; Xie, J.; Chen, X.; Sim, H.M.; Zhang, X.; Liang, Y.; Singh, S.; Talele, T.T.; Sun, Y.; Ambudkar, S.V.; et al. Neratinib Reverses ATP-Binding Cassette B1-Mediated Chemotherapeutic Drug Resistance In Vitro, In Vivo, and Ex Vivo. *Mol. Pharmacol.* **2012**, *82*, 47–58. [CrossRef]
30. Kathawala, R.J.; Sodani, K.; Chen, K.; Patel, A.; Abuznait, A.H.; Anreddy, N.; Sun, Y.-L.; Kaddoumi, A.; Ashby, C.R.; Chen, Z.-S. Masitinib Antagonizes ATP-Binding Cassette Subfamily C Member 10-Mediated Paclitaxel Resistance: A Preclinical Study. *Mol. Cancer Ther.* **2014**, *13*, 714–723. [CrossRef]
31. Minocha, M.; Khurana, V.; Qin, B.; Pal, D.; Mitra, A.K. Enhanced brain accumulation of pazopanib by modulating P-gp and Bcrp1 mediated efflux with canertinib or erlotinib. *Int. J. Pharm.* **2012**, *436*, 127–134. [CrossRef] [PubMed]
32. van Hoppe, S.; Sparidans, R.W.; Wagenaar, E.; Beijnen, J.H.; Schinkel, A.H. Breast cancer resistance protein (BCRP/ABCG2) and P-glycoprotein (P-gp/ABCB1) transport afatinib and restrict its oral availability and brain accumulation. *Pharmacol. Res.* **2017**, *120*, 43–50. [CrossRef] [PubMed]
33. Yang, K.; Chen, Y.; To, K.K.W.; Wang, F.; Li, D.; Chen, L.; Fu, L. Alectinib (CH5424802) antagonizes ABCB1- and ABCG2-mediated multidrug resistance in vitro, in vivo and ex vivo. *Exp. Mol. Med.* **2017**, *49*, e303. [CrossRef] [PubMed]
34. D'Cunha, R.; Bae, S.; Murry, D.J.; An, G. TKI combination therapy: Strategy to enhance dasatinib uptake by inhibiting Pgp- and BCRP-mediated efflux. *Biopharm. Drug Dispos.* **2016**, *37*, 397–408. [CrossRef] [PubMed]

35. Chuan Tang, S.; Nguyen, L.N.; Sparidans, R.W.; Wagenaar, E.; Beijnen, J.H.; Schinkel, A.H. Increased oral availability and brain accumulation of the ALK inhibitor crizotinib by coadministration of the P-glycoprotein (ABCB1) and breast cancer resistance protein (ABCG2) inhibitor elacridar. *Int. J. Cancer* **2014**, *134*, 1484–1494. [CrossRef]
36. Xiang, Q.; Zhang, D.; Wang, J.; Zhang, H.; Zheng, Z.; Yu, D.; Li, Y.; Xu, J.; Chen, Y.; Shang, C. Cabozantinib reverses multidrug resistance of human hepatoma HepG2/adr cells by modulating the function of P-glycoprotein. *Liver Int.* **2015**, *35*, 1010–1023. [CrossRef]
37. Tao, L.; Liang, Y.; Wang, F.; Chen, L.; Yan, Y.; Dai, C.; Fu, L. Cediranib (recentin, AZD2171) reverses ABCB1- and ABCC1-mediated multidrug resistance by inhibition of their transport function. *Cancer Chemother. Pharmacol.* **2009**, *64*, 961–969. [CrossRef]
38. Hu, J.; Zhang, X.; Wang, F.; Wang, X.; Yang, K.; Xu, M.; To, K.K.W.; Li, Q.; Fu, L. Effect of ceritinib (LDK378) on enhancement of chemotherapeutic agents in ABCB1 and ABCG2 overexpressing cells in vitro and in vivo. *Oncotarget* **2015**, *6*, 44643–44659. [CrossRef]
39. Wang, Y.-J.; Kathawala, R.J.; Zhang, Y.-K.; Patel, A.; Kumar, P.; Shukla, S.; Fung, K.L.; Ambudkar, S.V.; Talele, T.T.; Chen, Z.-S. Motesanib (AMG706), a potent multikinase inhibitor, antagonizes multidrug resistance by inhibiting the efflux activity of the ABCB1. *Biochem. Pharmacol.* **2014**, *90*, 367–378. [CrossRef]
40. Chen, Z.; Chen, Y.; Xu, M.; Chen, L.; Zhang, X.; To, K.K.W.; Zhao, H.; Wang, F.; Xia, Z.; Chen, X.; et al. Osimertinib (AZD9291) Enhanced the Efficacy of Chemotherapeutic Agents in ABCB1- and ABCG2-Overexpressing Cells In Vitro, In Vivo, and Ex Vivo. *Mol. Cancer Ther.* **2016**, *15*, 1845–1858. [CrossRef]
41. Liu, K.-J.; He, J.-H.; Su, X.-D.; Sim, H.-M.; Xie, J.-D.; Chen, X.-G.; Wang, F.; Liang, Y.-J.; Singh, S.; Sodani, K.; et al. Saracatinib (AZD0530) is a potent modulator of ABCB1-mediated multidrug resistance in vitro and in vivo. *Int. J. Cancer* **2013**, *132*, 224–235. [CrossRef]
42. Zheng, L.; Wang, F.; Li, Y.; Zhang, X.; Chen, L.; Liang, Y.; Dai, C.; Yan, Y.; Tao, L.; Mi, Y.; et al. Vandetanib (Zactima, ZD6474) Antagonizes ABCC1- and ABCG2-Mediated Multidrug Resistance by Inhibition of Their Transport Function. *PLoS ONE* **2009**, *4*, e5172. [CrossRef] [PubMed]
43. To, K.K.W.; Poon, D.C.; Wei, Y.; Wang, F.; Lin, G.; Fu, L. Vatalanib sensitizes ABCB1 and ABCG2-overexpressing multidrug resistant colon cancer cells to chemotherapy under hypoxia. *Biochem. Pharmacol.* **2015**, *97*, 27–37. [CrossRef] [PubMed]
44. Zhang, Y.; Wang, C.; Duan, Y.; Huo, X.; Meng, Q.; Liu, Z.; Sun, H.; Ma, X.; Liu, K. Afatinib Decreases P-Glycoprotein Expression to Promote Adriamycin Toxicity of A549T Cells. *J. Cell. Biochem.* **2018**, *119*, 414–423. [CrossRef] [PubMed]
45. Hegedűs, C.; Özvegy-Laczka, C.; Apáti, Á.; Magócsi, M.; Német, K.; Őrfi, L.; Kéri, G.; Katona, M.; Takáts, Z.; Váradi, A.; et al. Interaction of nilotinib, dasatinib and bosutinib with ABCB1 and ABCG2: Implications for altered anti-cancer effects and pharmacological properties. *Br. J. Pharmacol.* **2009**, *158*, 1153–1164. [CrossRef]
46. Shukla, S.; Sauna, Z.E.; Ambudkar, S.V. Evidence for the interaction of imatinib at the transport-substrate site(s) of the multidrug-resistance-linked ABC drug transporters ABCB1 (P-glycoprotein) and ABCG2. *Leukemia* **2008**, *22*, 445–447. [CrossRef]
47. Dai, C.-L.; Tiwari, A.K.; Wu, C.-P.; Su, X.; Wang, S.-R.; Liu, D.; Ashby, C.R.; Huang, Y.; Robey, R.W.; Liang, Y.-J.; et al. Lapatinib (Tykerb, GW572016) Reverses Multidrug Resistance in Cancer Cells by Inhibiting the Activity of ATP-Binding Cassette Subfamily B Member 1 and G Member 2. *Cancer Res.* **2008**, *68*, 7905–7914. [CrossRef]
48. Radic-Sarikas, B.; Halasz, M.; Huber, K.V.M.; Winter, G.E.; Tsafou, K.P.; Papamarkou, T.; Brunak, S.; Kolch, W.; Superti-Furga, G. Lapatinib potentiates cytotoxicity of YM155 in neuroblastoma via inhibition of the ABCB1 efflux transporter. *Sci. Rep.* **2017**, *7*, 1–8. [CrossRef]
49. Zhang, H.; Patel, A.; Ma, S.L.; Li, X.J.; Zhang, Y.K.; Yang, P.Q.; Kathawala, R.J.; Wang, Y.J.; Anreddy, N.; Fu, L.W.; et al. In vitro, in vivo and ex vivo characterization of ibrutinib: A potent inhibitor of the efflux function of the transporter MRP1. *Br. J. Pharmacol.* **2014**, *171*, 5845–5857. [CrossRef]
50. Shibayama, Y.; Nakano, K.; Maeda, H.; Taguchi, M.; Ikeda, R.; Sugawara, M.; Iseki, K.; Takeda, Y.; Yamada, K. Multidrug resistance protein 2 implicates anticancer drug-resistance to sorafenib. *Biol. Pharm. Bull.* **2011**, *34*, 433–435. [CrossRef]
51. Gay, C.; Toulet, D.; Le Corre, P. Pharmacokinetic drug-drug interactions of tyrosine kinase inhibitors: A focus on cytochrome P450, transporters, and acid suppression therapy. *Hematol. Oncol.* **2017**, *35*, 259–280. [CrossRef] [PubMed]

52. Radich, J.P.; Dai, H.; Mao, M.; Oehler, V.; Schelter, J.; Druker, B.; Sawyers, C.; Shah, N.; Stock, W.; Willman, C.L.; et al. Gene expression changes associated with progression and response in chronic myeloid leukemia. *Proc. Natl. Acad. Sci. USA* **2006**, *103*, 2794–2799. [CrossRef] [PubMed]
53. Tomonari, T.; Takeishi, S.; Taniguchi, T.; Tanaka, T.; Tanaka, H.; Fujimoto, S.; Kimura, T.; Okamoto, K.; Miyamoto, H.; Muguruma, N.; et al. MRP3 as a novel resistance factor for sorafenib in hepatocellular carcinoma. *Oncotarget* **2016**, *7*, 7207–7215. [CrossRef] [PubMed]
54. Cheung, L.; Yu, D.M.T.; Neiron, Z.; Failes, T.W.; Arndt, G.M.; Fletcher, J.I. Identification of new MRP4 inhibitors from a library of FDA approved drugs using a high-throughput bioluminescence screen. *Biochem. Pharmacol.* **2015**, *93*, 380–388. [CrossRef]
55. Macias, R.I.R.; Sánchez-Martín, A.; Rodríguez-Macías, G.; Sánchez-Abarca, L.I.; Lozano, E.; Herraez, E.; Odero, M.D.; Díez-Martín, J.L.; Marin, J.J.G.; Briz, O. Role of drug transporters in the sensitivity of acute myeloid leukemia to sorafenib. *Oncotarget* **2018**, *9*, 28474–28485. [CrossRef]
56. Shen, T.; Kuang, Y.-H.; Ashby, C.R.; Lei, Y.; Chen, A.; Zhou, Y.; Chen, X.; Tiwari, A.K.; Hopper-Borge, E.; Ouyang, J.; et al. Imatinib and Nilotinib Reverse Multidrug Resistance in Cancer Cells by Inhibiting the Efflux Activity of the MRP7 (ABCC10). *PLoS ONE* **2009**, *4*, e7520. [CrossRef]
57. Sun, Y.-L.; Kumar, P.; Sodani, K.; Patel, A.; Pan, Y.; Baer, M.R.; Chen, Z.-S.; Jiang, W.-Q.; Pan, Y.; Pan, Y.; et al. Ponatinib enhances anticancer drug sensitivity in MRP7-overexpressing cells. *Oncol. Rep.* **2014**, *31*, 1605–1612. [CrossRef]
58. Chen, Y.-J.; Huang, W.-C.; Wei, Y.-L.; Hsu, S.-C.; Yuan, P.; Lin, H.Y.; Wistuba, I.I.; Lee, J.J.; Yen, C.-J.; Su, W.-C.; et al. Elevated BCRP/ABCG2 Expression Confers Acquired Resistance to Gefitinib in Wild-Type EGFR-Expressing Cells. *PLoS ONE* **2011**, *6*, e21428. [CrossRef]
59. Wang, D.-S.; Patel, A.; Shukla, S.; Zhang, Y.-K.; Wang, Y.-J.; Kathawala, R.J.; Robey, R.W.; Zhang, L.; Yang, D.-H.; Talele, T.T.; et al. Icotinib antagonizes ABCG2-mediated multidrug resistance, but not the pemetrexed resistance mediated by thymidylate synthase and ABCG2. *Oncotarget* **2014**, *5*, 4529–4542. [CrossRef]
60. Kathawala, R.J.; Chen, J.-J.; Zhang, Y.-K.; Wang, Y.-J.; Patel, A.; Wang, D.-S.; Talele, T.T.; Ashby, C.R.; Chen, Z.-S. Masitinib antagonizes ATP-binding cassette subfamily G member 2-mediated multidrug resistance. *Int. J. Oncol.* **2014**, *44*, 1634–1642. [CrossRef]
61. Li, J.; Kumar, P.; Anreddy, N.; Zhang, Y.-K.; Wang, Y.-J.; Chen, Y.; Talele, T.T.; Gupta, K.; Trombetta, L.D.; Chen, Z.-S. Quizartinib (AC220) reverses ABCG2-mediated multidrug resistance: In vitro and in vivo studies. *Oncotarget* **2017**, *8*, 93785–93799. [CrossRef] [PubMed]
62. Lin, L.; Yee, S.W.; Kim, R.B.; Giacomini, K.M. SLC transporters as therapeutic targets: Emerging opportunities. *Nat. Rev. Drug Discov.* **2015**, *14*, 543–560. [CrossRef] [PubMed]
63. Engler, J.R.; Frede, A.; Saunders, V.A.; Zannettino, A.C.W.; Hughes, T.P.; White, D.L. Chronic Myeloid Leukemia CD34 cells have reduced uptake of imatinib due to low OCT-1 Activity. *Leukemia* **2010**, *24*, 765–770. [CrossRef] [PubMed]
64. White, D.L.; Dang, P.; Engler, J.; Frede, A.; Zrim, S.; Osborn, M.; Saunders, V.A.; Manley, P.W.; Hughes, T.P. Functional activity of the OCT-1 protein is predictive of long-term outcome in patients with chronic-phase chronic myeloid leukemia treated with imatinib. *J. Clin. Oncol.* **2010**, *28*, 2761–2767. [CrossRef]
65. Minematsu, T.; Giacomini, K.M. Interactions of Tyrosine Kinase Inhibitors with Organic Cation Transporters and Multidrug and Toxic Compound Extrusion Proteins. *Mol. Cancer Ther.* **2011**, *10*, 531–539. [CrossRef]
66. Davies, A.; Jordanides, N.E.; Giannoudis, A.; Lucas, C.M.; Hatziieremia, S.; Harris, R.J.; Jørgensen, H.G.; Holyoake, T.L.; Pirmohamed, M.; Clark, R.E.; et al. Nilotinib concentration in cell lines and primary CD34+ chronic myeloid leukemia cells is not mediated by active uptake or efflux by major drug transporters. *Leukemia* **2009**, *23*, 1999–2006. [CrossRef]
67. Elmeliegy, M.A.; Carcaboso, A.M.; Tagen, M.; Bai, F.; Stewart, C.F. Role of ATP-Binding Cassette and Solute Carrier Transporters in Erlotinib CNS Penetration and Intracellular Accumulation. *Clin. Cancer Res.* **2011**, *17*, 89–99. [CrossRef]
68. Arakawa, H.; Omote, S.; Tamai, I. Inhibitory Effect of Crizotinib on Creatinine Uptake by Renal Secretory Transporter OCT2. *J. Pharm. Sci.* **2017**, *106*, 2899–2903. [CrossRef]
69. Morrow, C.J.; Ghattas, M.; Smith, C.; Bönisch, H.; Bryce, R.A.; Hickinson, D.M.; Green, T.P.; Dive, C. Src family kinase inhibitor Saracatinib (AZD0530) impairs oxaliplatin uptake in colorectal cancer cells and blocks organic cation transporters. *Cancer Res.* **2010**, *70*, 5931–5941. [CrossRef]

70. Zimmerman, E.I.; Gibson, A.A.; Hu, S.; Vasilyeva, A.; Orwick, S.J.; Du, G.; Mascara, G.P.; Ong, S.S.; Chen, T.; Vogel, P.; et al. Therapeutics, Targets, and Chemical Biology Multikinase Inhibitors Induce Cutaneous Toxicity through OAT6-Mediated Uptake and MAP3K7-Driven Cell Death. *Cancer Res.* **2016**, *76*, 117–126. [CrossRef]
71. Hu, S.; Mathijssen, R.H.J.; de Bruijn, P.; Baker, S.D.; Sparreboom, A. Inhibition of OATP1B1 by tyrosine kinase inhibitors: In vitro–in vivo correlations. *Br. J. Cancer* **2014**, *110*, 894–898. [CrossRef] [PubMed]
72. Bauer, M.; Matsuda, A.; Wulkersdorfer, B.; Philippe, C.; Traxl, A.; Özvegy-Laczka, C.; Stanek, J.; Nics, L.; Klebermass, E.-M.; Poschner, S.; et al. Influence of OATPs on Hepatic Disposition of Erlotinib Measured With Positron Emission Tomography. *Clin. Pharmacol. Ther.* **2018**, *104*, 139–147. [CrossRef] [PubMed]
73. Thomas, J.; Wang, L.; Clark, R.E.; Pirmohamed, M.; Reiffers, J.; Goldman, J.M.; Melo, J.V. Active transport of imatinib into and out of cells: Implications for drug resistance. *Blood* **2004**, *104*, 3739–3745. [CrossRef] [PubMed]
74. White, D.L.; Saunders, V.A.; Dang, P.; Engler, J.; Zannettino, A.C.W.; Cambareri, A.C.; Quinn, S.R.; Manley, P.W.; Hughes, T.P. OCT-1–mediated influx is a key determinant of the intracellular uptake of imatinib but not nilotinib (AMN107): Reduced OCT-1 activity is the cause of low in vitro sensitivity to imatinib. *Blood* **2006**, *108*, 697–704. [CrossRef] [PubMed]
75. De Duve, C.; De Barsy, T.; Poole, B.; Trouet, A.; Tulkens, P.; van Hoof, F. Lysosomotropic agents. *Biochem. Pharmacol.* **1974**, *23*, 2495–2531. [CrossRef]
76. Kazmi, F.; Hensley, T.; Pope, C.; Funk, R.S.; Loewen, G.J.; Buckley, D.B.; Parkinson, A. Lysosomal Sequestration (Trapping) of Lipophilic Amine (Cationic Amphiphilic) Drugs in Immortalized Human Hepatocytes (Fa2N-4 Cells). *Drug Metab. Dispos.* **2013**, *41*, 897–905. [CrossRef]
77. Gotink, K.J.; Broxterman, H.J.; Labots, M.; De Haas, R.R.; Dekker, H.; Honeywell, R.J.; Rudek, M.A.; Beerepoot, L.V.; Musters, R.J.; Jansen, G.; et al. Lysosomal sequestration of sunitinib: A novel mechanism of drug resistance. *Clin. Cancer Res.* **2011**, *17*, 7337–7346. [CrossRef]
78. Wilson, J.N.; Liu, W.; Brown, A.S.; Landgraf, R. Binding-induced, turn-on fluorescence of the EGFR/ERBB kinase inhibitor, lapatinib. *Org. Biomol. Chem.* **2015**, *13*, 5006–5011. [CrossRef]
79. Burger, H.; den Dekker, A.T.; Segeletz, S.; Boersma, A.W.M.; de Bruijn, P.; Debiec-Rychter, M.; Taguchi, T.; Sleijfer, S.; Sparreboom, A.; Mathijssen, R.H.J.; et al. Lysosomal Sequestration Determines Intracellular Imatinib Levels. *Mol. Pharmacol.* **2015**, *88*, 477–487. [CrossRef]
80. Fu, D.; Zhou, J.; Zhu, W.S.; Manley, P.W.; Wang, Y.K.; Hood, T.; Wylie, A.; Xie, X.S. Imaging the intracellular distribution of tyrosine kinase inhibitors in living cells with quantitative hyperspectral stimulated Raman scattering. *Nat. Chem.* **2014**, *6*, 614–622. [CrossRef]
81. Englinger, B.; Kallus, S.; Senkiv, J.; Heilos, D.; Gabler, L.; van Schoonhoven, S.; Terenzi, A.; Moser, P.; Pirker, C.; Timelthaler, G.; et al. Intrinsic fluorescence of the clinically approved multikinase inhibitor nintedanib reveals lysosomal sequestration as resistance mechanism in FGFR-driven lung cancer. *J. Exp. Clin. Cancer Res.* **2017**, *36*, 122. [CrossRef] [PubMed]
82. Nadanaciva, S.; Lu, S.; Gebhard, D.F.; Jessen, B.A.; Pennie, W.D.; Will, Y. A high content screening assay for identifying lysosomotropic compounds. *Toxicol. Vitr.* **2011**, *25*, 715–723. [CrossRef] [PubMed]
83. Colombo, F.; Trombetta, E.; Cetrangolo, P.; Maggioni, M.; Razini, P.; De Santis, F.; Torrente, Y.; Prati, D.; Torresani, E.; Porretti, L. Giant Lysosomes as a Chemotherapy Resistance Mechanism in Hepatocellular Carcinoma Cells. *PLoS ONE* **2014**, *9*, e114787. [CrossRef] [PubMed]
84. Gotink, K.J.; Rovithi, M.; de Haas, R.R.; Honeywell, R.J.; Dekker, H.; Poel, D.; Azijli, K.; Peters, G.J.; Broxterman, H.J.; Verheul, H.M.W. Cross-resistance to clinically used tyrosine kinase inhibitors sunitinib, sorafenib and pazopanib. *Cell. Oncol.* **2015**, *38*, 119–129. [CrossRef] [PubMed]
85. Ferrao, P.; Sincock, P.; Cole, S.; Ashman, L. Intracellular P-gp contributes to functional drug efflux and resistance in acute myeloid leukaemia. *Leuk. Res.* **2001**, *25*, 395–405. [CrossRef]
86. Molinari, A.; Calcabrini, A.; Meschini, S.; Stringaro, A.; Crateri, P.; Toccacieli, L.; Marra, M.; Colone, M.; Cianfriglia, M.; Arancia, G. Subcellular Detection and Localization of the Drug Transporter P-Glycoprotein in Cultured Tumor Cells. *Curr. Protein Pept. Sci.* **2002**, *3*, 653–670. [CrossRef]
87. Chapuy, B.; Panse, M.; Radunski, U.; Koch, R.; Wenzel, D.; Inagaki, N.; Haase, D.; Truemper, L.; Wulf, G.G. ABC transporter A3 facilitates lysosomal sequestration of imatinib and modulates susceptibility of chronic myeloid leukemia cell lines to this drug. *Haematologica* **2009**, *94*, 1528–1536. [CrossRef]

88. Al-Akra, L.; Bae, D.-H.; Sahni, S.; Huang, M.L.H.; Park, K.C.; Lane, D.J.R.; Jansson, P.J.; Richardson, D.R. Tumor stressors induce two mechanisms of intracellular P-glycoprotein-mediated resistance that are overcome by lysosomal-targeted thiosemicarbazones. *J. Biol. Chem.* **2018**, *293*, 3562–3587. [CrossRef]
89. Yamagishi, T.; Sahni, S.; Sharp, D.M.; Arvind, A.; Jansson, P.J.; Richardson, D.R. P-glycoprotein mediates drug resistance via a novel mechanism involving lysosomal sequestration. *J. Biol. Chem.* **2013**, *288*, 31761–31771. [CrossRef]
90. Zama, I.N.; Hutson, T.E.; Elson, P.; Cleary, J.M.; Choueiri, T.K.; Heng, D.Y.C.; Ramaiya, N.; Michaelson, M.D.; Garcia, J.A.; Knox, J.J.; et al. Sunitinib rechallenge in metastatic renal cell carcinoma patients. *Cancer* **2010**, *116*, 5400–5406. [CrossRef]
91. Gotink, K.J.; Broxterman, H.J.; Honeywell, R.J.; Dekker, H.; de Haas, R.R.; Miles, K.M.; Adelaiye, R.; Griffioen, A.W.; Peters, G.J.; Pili, R.; et al. Acquired tumor cell resistance to sunitinib causes resistance in a HT-29 human colon cancer xenograft mouse model without affecting sunitinib biodistribution or the tumor microvasculature. *Oncoscience* **2014**, *1*, 844–853. [CrossRef]
92. McAfee, Q.; Zhang, Z.; Samanta, A.; Levi, S.M.; Ma, X.-H.; Piao, S.; Lynch, J.P.; Uehara, T.; Sepulveda, A.R.; Davis, L.E.; et al. Autophagy inhibitor Lys05 has single-agent antitumor activity and reproduces the phenotype of a genetic autophagy deficiency. *Proc. Natl. Acad. Sci. USA* **2012**, *109*, 8253–8258. [CrossRef]
93. Rosenfeld, M.R.; Ye, X.; Supko, J.G.; Desideri, S.; Grossman, S.A.; Brem, S.; Mikkelson, T.; Wang, D.; Chang, Y.C.; Hu, J.; et al. A phase I/II trial of hydroxychloroquine in conjunction with radiation therapy and concurrent and adjuvant temozolomide in patients with newly diagnosed glioblastoma multiforme. *Autophagy* **2014**, *10*, 1359–1368. [CrossRef]
94. Rangwala, R.; Chang, Y.C.; Hu, J.; Algazy, K.M.; Evans, T.L.; Fecher, L.A.; Schuchter, L.M.; Torigian, D.A.; Panosian, J.T.; Troxel, A.B.; et al. Combined MTOR and autophagy inhibition: Phase I trial of hydroxychloroquine and temsirolimus in patients with advanced solid tumors and melanoma. *Autophagy* **2014**, *10*, 1391–1402. [CrossRef]
95. Nowak-Sliwinska, P.; Weiss, A.; van Beijnum, J.R.; Wong, T.J.; Kilarski, W.W.; Szewczyk, G.; Verheul, H.M.W.; Sarna, T.; van den Bergh, H.; Griffioen, A.W. Photoactivation of lysosomally sequestered sunitinib after angiostatic treatment causes vascular occlusion and enhances tumor growth inhibition. *Cell Death Dis.* **2015**, *6*, e1641. [CrossRef]
96. Jansson, P.J.; Yamagishi, T.; Arvind, A.; Seebacher, N.; Gutierrez, E.; Stacy, A.; Maleki, S.; Sharp, D.; Sahni, S.; Richardson, D.R. Di-2-pyridylketone 4,4-dimethyl-3-thiosemicarbazone (Dp44mT) overcomes multidrug resistance by a novel mechanism involving the hijacking of lysosomal P-glycoprotein (Pgp). *J. Biol. Chem.* **2015**, *290*, 9588–9603. [CrossRef]
97. Lan, C.-Y.; Wang, Y.; Xiong, Y.; Li, J.-D.; Shen, J.-X.; Li, Y.-F.; Zheng, M.; Zhang, Y.-N.; Feng, Y.-L.; Liu, Q.; et al. Apatinib combined with oral etoposide in patients with platinum-resistant or platinum-refractory ovarian cancer (AEROC): A phase 2, single-arm, prospective study. *Lancet Oncol.* **2018**, *19*, 1239–1246. [CrossRef]
98. Wang, L.; Liang, L.; Yang, T.; Qiao, Y.; Xia, Y.; Liu, L.; Li, C.; Lu, P.; Jiang, X. A pilot clinical study of apatinib plus irinotecan in patients with recurrent high-grade glioma: Clinical Trial/Experimental Study. *Medicine (Baltimore)* **2017**, *96*, e9053. [CrossRef]
99. Symonds, R.P.; Gourley, C.; Davidson, S.; Carty, K.; McCartney, E.; Rai, D.; Banerjee, S.; Jackson, D.; Lord, R.; McCormack, M.; et al. Cediranib combined with carboplatin and paclitaxel in patients with metastatic or recurrent cervical cancer (CIRCCa): A randomised, double-blind, placebo-controlled phase 2 trial. *Lancet Oncol.* **2015**, *16*, 1515–1524. [CrossRef]
100. Valle, J.W.; Wasan, H.; Lopes, A.; Backen, A.C.; Palmer, D.H.; Morris, K.; Duggan, M.; Cunningham, D.; Anthoney, D.A.; Corrie, P.; et al. Cediranib or placebo in combination with cisplatin and gemcitabine chemotherapy for patients with advanced biliary tract cancer (ABC-03): A randomised phase 2 trial. *Lancet Oncol.* **2015**, *16*, 967–978. [CrossRef]
101. Ahn, H.K.; Han, B.; Lee, S.J.; Lim, T.; Sun, J.-M.; Ahn, J.S.; Ahn, M.-J.; Park, K. ALK inhibitor crizotinib combined with intrathecal methotrexate treatment for non-small cell lung cancer with leptomeningeal carcinomatosis. *Lung Cancer* **2012**, *76*, 253–254. [CrossRef]

102. Neal, J.W.; Dahlberg, S.E.; Wakelee, H.A.; Aisner, S.C.; Bowden, M.; Huang, Y.; Carbone, D.P.; Gerstner, G.J.; Lerner, R.E.; Rubin, J.L.; et al. Erlotinib, cabozantinib, or erlotinib plus cabozantinib as second-line or third-line treatment of patients with EGFR wild-type advanced non-small-cell lung cancer (ECOG-ACRIN 1512): A randomised, controlled, open-label, multicentre, phase 2 trial. *Lancet Oncol.* **2016**, *17*, 1661–1671. [CrossRef]
103. Hirte, H.; Oza, A.; Swenerton, K.; Ellard, S.L.; Grimshaw, R.; Fisher, B.; Tsao, M.; Seymour, L. A phase II study of erlotinib (OSI-774) given in combination with carboplatin in patients with recurrent epithelial ovarian cancer (NCIC CTG IND.149). *Gynecol. Oncol.* **2010**, *118*, 308–312. [CrossRef]
104. Massarelli, E.; Lin, H.; Ginsberg, L.E.; Tran, H.T.; Lee, J.J.; Canales, J.R.; Williams, M.D.; Blumenschein, G.R.; Lu, C.; Heymach, J.V.; et al. Phase II trial of everolimus and erlotinib in patients with platinum-resistant recurrent and/or metastatic head and neck squamous cell carcinoma. *Ann. Oncol.* **2015**, *26*, 1476–1480. [CrossRef]
105. Yang, Z.Y.; Yuan, J.Q.; Di, M.Y.; Zheng, D.Y.; Chen, J.Z.; Ding, H.; Wu, X.Y.; Huang, Y.F.; Mao, C.; Tang, J.L. Gemcitabine Plus Erlotinib for Advanced Pancreatic Cancer: A Systematic Review with Meta-Analysis. *PLoS ONE* **2013**, *8*, e57528. [CrossRef]
106. Lim, S.H.; Yun, J.; Lee, M.-Y.; Kim, H.J.; Kim, K.H.; Kim, S.H.; Lee, S.-C.; Bae, S.B.; Kim, C.K.; Lee, N.; et al. A randomized phase II clinical trial of gemcitabine, oxaliplatin, erlotinib combination chemotherapy versus gemcitabine and erlotinib in previously untreated patients with locally advanced or metastatic pancreatic cancer. *J. Clin. Oncol.* **2018**, *36*, 344. [CrossRef]
107. Stewart, C.F.; Tagen, M.; Schwartzberg, L.S.; Blakely, L.J.; Tauer, K.W.; Smiley, L.M. Phase I dosage finding and pharmacokinetic study of intravenous topotecan and oral erlotinib in adults with refractory solid tumors. *Cancer Chemother. Pharmacol.* **2014**, *73*, 561–568. [CrossRef]
108. Hosomi, Y.; Morita, S.; Sugawara, S.; Kato, T.; Fukuhara, T.; Gemma, A.; Takahashi, K.; Fujita, Y.; Harada, T.; Minato, K.; et al. Gefitinib Alone Versus Gefitinib Plus Chemotherapy for Non-Small-Cell Lung Cancer With Mutated Epidermal Growth Factor Receptor: NEJ009 Study. *J. Clin. Oncol.* **2020**, *38*, 115–123. [CrossRef]
109. Cetin, B.; Benekli, M.; Turker, I.; Koral, L.; Ulas, A.; Dane, F.; Oksuzoglu, B.; Kaplan, M.A.; Koca, D.; Boruban, C.; et al. Lapatinib plus capecitabine for HER2-positive advanced breast cancer: A multicentre study of Anatolian Society of Medical Oncology (ASMO). *J. Chemother.* **2014**, *26*, 300–305. [CrossRef]
110. Di Leo, A.; Gomez, H.L.; Aziz, Z.; Zvirbule, Z.; Bines, J.; Arbushites, M.C.; Guerrera, S.F.; Koehler, M.; Oliva, C.; Stein, S.H.; et al. Phase III, double-blind, randomized study comparing lapatinib plus paclitaxel with placebo plus paclitaxel as first-line treatment for metastatic breast cancer. *J. Clin. Oncol.* **2008**, *26*, 5544–5552. [CrossRef]
111. Saura, C.; Garcia-saenz, J.A.; Xu, B.; Harb, W.; Moroose, R.; Pluard, T.; Cortés, J.; Kiger, C.; Germa, C.; Wang, K.; et al. Safety and Efficacy of Neratinib in Combination With Capecitabine in Patients With Metastatic Human Epidermal Growth Factor Receptor 2–Positive Breast Cancer. *J. Clin. Oncol.* **2014**, *32*, 1–9. [CrossRef] [PubMed]
112. Chow, L.; Xu, B.; Gupta, S.; Freyman, A.; Zhao, Y.; Abbas, R.; Van, M.V.; Bondarenko, I. Combination neratinib (HKI-272) and paclitaxel therapy in patients with HER2-positive metastatic breast cancer. *Br. J. Cancer* **2013**, *108*, 1985–1993. [CrossRef] [PubMed]
113. Kim, D.-Y.; Joo, Y.-D.; Lim, S.-N.; Kim, S.-D.; Lee, J.-H.; Lee, J.-H.; Kim, D.H.; Kim, K.; Jung, C.W.; Kim, I.; et al. Nilotinib combined with multiagent chemotherapy for newly diagnosed Philadelphia-positive acute lymphoblastic leukemia. *Blood* **2015**, *126*, 746–756. [CrossRef]
114. Reck, M.; Kaiser, R.; Mellemgaard, A.; Douillard, J.-Y.; Orlov, S.; Krzakowski, M.; von Pawel, J.; Gottfried, M.; Bondarenko, I.; Liao, M.; et al. Docetaxel plus nintedanib versus docetaxel plus placebo in patients with previously treated non-small-cell lung cancer (LUME-Lung 1): A phase 3, double-blind, randomised controlled trial. *Lancet Oncol.* **2014**, *15*, 143–155. [CrossRef]
115. Serve, H.; Brunnberg, U.; Ottmann, O.; Brandts, C.; Steffen, B.; Krug, U.; Wagner, R.; Müller-Tidow, C.; Berdel, W.E.; Cristina Sauerland, M.; et al. Sorafenib in combination with intensive chemotherapy in elderly patients with acute myeloid leukemia: Results from a randomized, placebo-controlled trial. *J. Clin. Oncol.* **2013**, *31*, 3110–3118. [CrossRef]
116. Abou-Alfa, G.K.; Shi, Q.; Knox, J.J.; Kaubisch, A.; Niedzwiecki, D.; Posey, J.; Tan, B.R.; Kavan, P.; Goel, R.; Lammers, P.E.; et al. Assessment of Treatment With Sorafenib Plus Doxorubicin vs Sorafenib Alone in Patients With Advanced Hepatocellular Carcinoma. *JAMA Oncol.* **2019**, *5*, 1582. [CrossRef]

117. Sheng, X.; Cao, D.; Yuan, J.; Zhou, F.; Wei, Q.; Xie, X.; Cui, C.; Chi, Z.; Si, L.; Li, S.; et al. Sorafenib in combination with gemcitabine plus cisplatin chemotherapy in metastatic renal collecting duct carcinoma: A prospective, multicentre, single-arm, phase 2 study. *Eur. J. Cancer* **2018**, *100*, 1–7. [CrossRef]
118. Crown, J.P.; Diéras, V.; Staroslawska, E.; Yardley, D.A.; Bachelot, T.; Davidson, N.; Wildiers, H.; Fasching, P.A.; Capitain, O.; Ramos, M.; et al. Phase III trial of sunitinib in combination with capecitabine versus capecitabine monotherapy for the treatment of patients with pretreated metastatic breast cancer. *J. Clin. Oncol.* **2013**, *31*, 2870–2878. [CrossRef]
119. Bergh, J.; Bondarenko, I.M.; Lichinitser, M.R.; Liljegren, A.; Greil, R.; Voytko, N.L.; Makhson, A.N.; Cortes, J.; Lortholary, A.; Bischoff, J.; et al. First-line treatment of advanced breast cancer with sunitinib in combination with docetaxel versus docetaxel alone: Results of a prospective, randomized phase III study. *J. Clin. Oncol.* **2012**, *30*, 921–929. [CrossRef]
120. Yi, J.H.; Lee, J.; Lee, J.; Park, S.H.; Park, J.O.; Yim, D.-S.; Park, Y.S.; Lim, H.Y.; Kang, W.K. Randomised phase II trial of docetaxel and sunitinib in patients with metastatic gastric cancer who were previously treated with fluoropyrimidine and platinum. *Br. J. Cancer* **2012**, *106*, 1469–1474. [CrossRef]
121. Choueiri, T.K.; Ross, R.W.; Jacobus, S.; Vaishampayan, U.; Yu, E.Y.; Quinn, D.I.; Hahn, N.M.; Hutson, T.E.; Sonpavde, G.; Morrissey, S.C.; et al. Double-blind, randomized trial of docetaxel plus vandetanib versus docetaxel plus placebo in platinum-pretreated metastatic urothelial cancer. *J. Clin. Oncol.* **2012**, *30*, 507–512. [CrossRef]
122. Yang, C.; Gottfried, M.; Chan, V.; Raats, J.; De Marinis, F.; Abratt, R.P.; Read, J.; Vansteenkiste, J.F. Vandetanib Plus Pemetrexed for the Second-Line Treatment of Advanced Non-Small-Cell Lung Cancer: A Randomized, Double-Blind Phase III Trial. *J. Clin. Oncol.* **2011**, *29*, 1067–1074.
123. Beretta, G.L.; Benedetti, V.; Cossa, G.; Assaraf, Y.G.A.; Bram, E.; Gatti, L.; Corna, E.; Carenini, N.; Colangelo, D.; Howell, S.B.; et al. Increased levels and defective glycosylation of MRPs in ovarian carcinoma cells resistant to oxaliplatin. *Biochem. Pharmacol.* **2010**, *79*, 1108–1117.
124. Rudin, D.; Li, L.; Niu, N.; Kalari, K.R.; Gilbert, J.A.; Ames, M.M.; Wang, L. Gemcitabine Cytotoxicity: Interaction of Efflux and Deamination. *J. Drug Metab. Toxicol.* **2011**, *2*, 1–10. [CrossRef]
125. Adamska, A.; Falasca, M. ATP-binding cassette transporters in progression and clinical outcome of pancreatic cancer: What is the way forward? *World J. Gastroenterol.* **2018**, *24*, 3222–3238. [CrossRef]
126. Kemp, J.A.; Shim, M.S.; Heo, C.Y.; Kwon, Y.J. "Combo" nanomedicine: Co-delivery of multi-modal therapeutics for efficient, targeted, and safe cancer therapy. *Adv. Drug Deliv. Rev.* **2016**, *98*, 3–18.
127. Zhou, Z.; Kennell, C.; Jafari, M.; Lee, J.Y.; Ruiz-Torres, S.J.; Waltz, S.E.; Lee, J.H. Sequential delivery of erlotinib and doxorubicin for enhanced triple negative Breast cancer treatment using polymeric nanoparticle. *Int. J. Pharm.* **2017**, *530*, 300–307. [CrossRef]

International Journal of
Molecular Sciences

MDPI

Review

Evolution of Urothelial Bladder Cancer in the Context of Molecular Classifications

Martina Minoli [1], Mirjam Kiener [1], George N. Thalmann [1,2], Marianna Kruithof-de Julio [1,2] and Roland Seiler [1,2,*]

1 Department of BioMedical Research, Urology Research Laboratory, University of Bern, 3008 Bern, Switzerland; martina.minoli@dbmr.unibe.ch (M.M.); mirjam.kiener@dbmr.unibe.ch (M.K.); George.Thalmann@insel.ch (G.N.T.); marianna.kruithofdejulio@dbmr.unibe.ch (M.K.-d.J.)

2 Department of Urology, Inselspital, Bern University Hospital, 3008 Bern, Switzerland

* Correspondence: roland.seiler@insel.ch or r_seiler@gmx.ch

Received: 15 July 2020; Accepted: 5 August 2020; Published: 7 August 2020

Abstract: Bladder cancer is a heterogeneous disease that is not depicted by current classification systems. It was originally classified into non-muscle invasive and muscle invasive. However, clinically and genetically variable tumors are summarized within both classes. A definition of three groups may better account for the divergence in prognosis and probably also choice of treatment. The first group represents mostly non-invasive tumors that reoccur but do not progress. Contrarily, the second group represent non-muscle invasive tumors that likely progress to the third group, the muscle invasive tumors. High throughput tumor profiling improved our understanding of the biology of bladder cancer. It allows the identification of molecular subtypes, at least three for non-muscle invasive bladder cancer (Class I, Class II and Class III) and six for muscle-invasive bladder cancer (luminal papillary, luminal non-specified, luminal unstable, stroma-rich, basal/squamous and neuroendocrine-like) with distinct clinical and molecular phenotypes. Molecular subtypes can be potentially used to predict the response to treatment (e.g., neoadjuvant chemotherapy and immune checkpoint inhibitors). Moreover, they may allow to characterize the evolution of bladder cancer through different pathways. However, to move towards precision medicine, the understanding of the biological meaning of these molecular subtypes and differences in the composition of cell subpopulations will be mandatory.

Keywords: bladder cancer; muscle invasive; non-muscle invasive; molecular subtypes; evolution; targeted therapy; classification

1. Introduction: The Clinical Problem of Bladder Cancer

Bladder cancer (BLCa) is one of the most common cancers of the genitourinary tract. It is the tenth most common cancer worldwide, accounting for 549,000 new cases and about 200,000 deaths per year [1]. The majority of BLCa are urothelial carcinoma, whereas squamous cell carcinoma, small cell carcinoma, adenocarcinoma and sarcoma are less frequent. In this review, BLCa refers to urothelial carcinoma and excludes the other rare forms. BLCa is around three times more common in men than in women and tobacco smoking is the main risk factor [2–5]. Family history increases the risk of BLCa and is mainly associated with early onset, however, only scarce data on genetically inherited BLCa is available [6–8]. BLCa is the costliest cancer to treat on a per-patient basis and causes large socio-economic burden [1,9–12]. Nevertheless, it is still relatively understudied [13].

BLCa has been classified into non-muscle invasive BLCa (NMIBC, 70–75% of patients) and muscle invasive BLCa (MIBC, 25–30% of patients), which are associated with distinct prognosis [14]. However, clinically and genetically very distinct tumors are summarized in NMIBC. Therefore, a definition of three different groups better accounts for the divergence in prognosis and treatment choice. The first group consists of NMIBC confined to the mucosa [14]. This group is divided into high- and low-grade pTa depending on cellular atypia and alterations in tumor architecture. Despite the distinct prognosis, the urothelial carcinoma in situ (CIS-stage pTis) is also part of the first group. CIS is defined as a high-grade, flat, non-invasive lesion confined to the mucosa with often vague borders and spreading across various areas of the bladder. The second group consists of NMIBC tumors that invade the submucosal layer, the lamina propria. These include stage pT1 and are of high-grade in most of the cases. Among NMIBCs, 70% are staged as pTa, 20% as pT1 and the remaining 10% as CIS lesions [15]. The third group consists of MIBC, either organ-confined, locally advanced and/or metastatic. MIBC tumors are staged from T2a to T4b depending on the extent of muscle invasion and are of high-grade in the majority of the cases.

The major clinical problem of NMIBC is the frequent short-term reoccurrence after the first resection (50–70%) [16,17]. Although most NMIBC have a low probability (10–20%) of progression and high survival rate (~90% 5-year overall survival) [16–18], lymph node metastases can be found in 5% of pT1 tumors, whereas 50% of CIS potentially progress locally and disseminate [19,20]. Around 10–20% of NMIBC-diagnosed patients progress to MIBC. Currently, these patients cannot be prospectively identified [17,21].

Contrarily, in MIBC, the rapid metastatic progression and the consequently high mortality of patients is of high clinical relevance [22]. The 5-year overall survival probability of 60% is dramatically reduced to less than 10% in case of early metastatic dissemination [23]. Currently, curative therapy options are missing and only a minority of patients with metastatic BLCa show long-term response to palliative treatments [24].

Physicians are often met with treatment failure and/or reoccurrence in all BLCa groups, indicating that the management of BLCa is complex and current classification systems do not depict the heterogeneity of this disease. Huge efforts are undertaken to understand the biology of BLCa. Tumor profiling on large-scale cohorts has dramatically improved our understanding of the biology of this disease and may allow us to identify molecular subtypes with a distinct clinical and molecular phenotype. The ultimate goal of these investigations should be to predict the likelihood of reoccurrence, response to treatment and progression early in patients' history to guide clinical decision making.

In this review, we will present the state-of-the-art, clinical guidelines and standard operating procedures of BLCa and elucidate their limitations. We will also discuss how molecular subtypes discovered from the profiling of tumor samples improved our understanding of BLCa and their potential clinical meaning and application.

2. Current BLCa Management

2.1. NMIBC Treatment Options and Follow-Up

NMIBCs are first diagnosed with cystoscopy and confirmed by subsequent transurethral resection of the bladder tumor (TURBT) [25]. The clinical tumor, node and metastases (TNM) stages are determined based on cross-sectional imaging and pathological evaluation of the TURBT [25]. Furthermore, the histological grade is defined according to the 2004/2006 World Health Organization (WHO) classification [26].

Based on these evaluations, the European Organization for Research and Treatment of Cancer (EORTC) risk tables allow stratifying patients into low-, intermediate- and high-risk for reoccurrence or progression [17]. However, this may not be appropriate given that they are defined by different prognostic factors and require different follow-up strategies [27]. Prior disease reoccurrence rate and the number of distinct tumors are the strongest predictors of BLCa reoccurrence, whereas the most significant prognostic factors for disease progression and disease-specific survival are tumor stage and grade, and presence of CIS [17]. Ideally, a better differentiation, including prognostic markers, among these two risks (progression and reoccurrence) should be defined. Due to the high recurrence rate, patients undergo life-long monitoring. Standard of care includes cystoscopy every 3–6 months in the first 1–2 years and every year thereafter for 5 years, or lifelong [28,29]. In particular, intermediate- and high-risk tumors might reoccur even after 10 years, therefore a life-long surveillance schedule is recommended [30]. Treatment and surveillance of NMIBC patients is not only associated with considerable costs but also with high morbidity due to the considerable side effects, such as pain, risk for infection and irritation/damage of the urothelium.

Low-risk NMIBC patients after TURBT receive intravesical instillations of chemotherapy (Mitomycin C, Epirubicin or Pirarubicin) within 24 h after the surgery [25]. This targets the released tumor cells after TURBT, and significantly reduces the three-month reoccurrence risk [28,31]. Intermediate-risk and high-risk patients receive a Bacillus Calmette-Guérin (BCG) vaccination, which is an attenuated strain of *Mycobacterium bovis*. This prolongs reoccurrence time decreasing disease progression [28,32–34]. BCG is not always well tolerated, and the risk of relapse varies between 30% and 40% [35].

2.2. MIBC Treatment Options and Follow-Up

Neoadjuvant cisplatin-based chemotherapy (NAC) prior to radical cystectomy is the golden-standard of care for all non-metastatic MIBC patients [36]. This treatment should be offered to all patients with non-metastatic BLCa, organ-confined (cT2N0M0) or locally advanced BLCa (cT3a-T4a, N0-NX, M0) [23]. The systemic treatment for NAC can be the combination of methotrexate, vinblastine, doxorubicin and cisplatin (MVAC) or the combination of cisplatin with gemcitabine (GC), which is used because of the lower toxicity but comparable efficacy [36,37]. Life expectancy is increased in 40% of NAC-treated patients [14,29,38], which can be attributed to the reduction of tumor size prior to surgery, the treatment of micrometastases and the prevention of metastasis formation [36]. While only 10% of MIBC patients present metastasis at diagnosis, 40–50% show metastatic formations after reoccurrence detected during follow-up after cystectomy. For these patients, an effective therapy is lacking and the treatment intent is palliative. First-line treatment is cisplatin-based chemotherapy [14,29,36], and second-line treatments for cisplatin-resistant and metastatic BLCa are the Food and Drug Administration(FDA)-approved immune checkpoint inhibitors (ICIs) which target PD-L1 (atezolizumab, avelumab and durvalumab) or PD-1 (nivolumab and pembrolizumab). However, only 20–25% of patients respond positively to treatment [14,39–41].

Novel targeted therapies have recently been approved for treatment of patients resistant to cisplatin-based chemotherapy and ICIs. Erdafitinib, an inhibitor of the tyrosine-kinase activity of fibroblast growth factor receptor (FGFR), is approved for patients with FGFR3 or FGFR2 alterations [42,43]. Finally, enfortumab vedotin, an antibody-drug conjugate that targets Nectin-4, has shown promising response rates [44].

Follow-up includes, in the three years after initial treatment, computed tomography (CT) scan of chest, abdomen and pelvis every 3–6 months, and reoccurrence is treated with radiotherapy.

Importantly, the survival rate of MIBC patients has remained unchanged for the last 30 years [5] and currently, the treatment is offered in "a one size fits all" manner and by trial and error. The response is documented during the follow-up and only a minority of patients show a positive treatment response (e.g., NAC 30–40%, ICI 20–25%). Therefore, selecting patients according to the likelihood of response might help to improve patients' outcomes and avoid overtreatment in likely non-responders. The identification of accurate and predictive biomarkers for non-responders vs. responders is one of the major challenges in BLCa research.

3. Lack of Predictive Biomarkers: First Cause of Treatment Failure

The main problem for NMIBC is the high recurrence rate and early identification of tumors with the potential to progress.

The risk of reoccurrence after installation of chemotherapy varies between low-, high- and intermediate-risk tumors but also between immediate and delayed treatments. For low-risk tumors, the 5-year reoccurrence risk does not change between immediate (43%) or delayed treatment (46%) [45]. In contrast, immediate treatment decreases the recurrence risk from 31% to 20% for the intermediate-risk tumors and from 35% to 28% for high-risk tumors [45]. Therefore, rapid selection of patients according to the likelihood of response is crucial to improve chemotherapy efficacy.

Although BCG therapy has been used for more than 30 years, its mechanism of action is still under investigation. About 40% of patients fail BCG therapy and tumor reoccurs within 2 years [14,46]. BCG failure refers to three different conditions: (1) BCG relapse, if the disease reoccurs after ≥6 months of disease-free survival, (2) BCG refractory, if the tumor persists after 6 months of treatment and (3) BCG intolerance, if the treatment has been discontinued due to toxic side effects of BCG [14,28,47]. Many factors can cause BCG failure, e.g., the presence of metastasis prior to therapy, inappropriate immune response or an adaptive immune resistance with overexpression of exhaustion markers (PD-1 and PD-L1) [48–50]. Although there is no standard of care for patients after BCG failure, radical cystectomy is considered the preferred treatment [25]. Alternative approaches, such as immune checkpoint inhibitors against PD-L1 or PD-1, have been taken into consideration as therapy in combination with BCG, given the role of exhaustion markers (PD-L1 and PD-1) in BCG resistance and relapse [51]. However, efforts to understand how to identify patients that will benefit from BCG, combination therapies or will have to undergo radical cystectomy are much needed [52].

Such predictive biomarkers are also missing for MIBC patients. The major clinical problem in MIBC is treatment failure due to inaccurate treatment selection, which is associated with the high mortality of MIBC patients. This is exemplified by a complete NAC response only in a minor fraction of patients. Several clinical trials have shown that NAC improves overall survival in only 30–40% of patients [38,53–57]. In NAC non-responder patients, the treatment delays the effective therapy, directly associated with adverse prognosis, and overtreats patients that will suffer from unnecessary side effects [58–60].

The mechanisms of resistance of MIBC to systemic treatment are largely unknown and molecular biomarkers for the selection of effective second-line treatments are currently lacking. A one treatment fits all approach for such a highly heterogeneous disease is certainly a limiting factor. Stratifying patients into "subtypes" based on the molecular landscape can be used to predict the response to treatment [61–63]. BLCa can be grouped into at least two molecular subtypes: basal/squamous-like and luminal subtypes, and some of these subtypes were associated with therapy response. McConkey and colleagues employed gene expression profiles of NAC-treated MIBC tumors to predict therapy response. They showed that basal tumors were associated with better survival in the context of NAC and that tumors expressing wild-type (WT) p53-associated gene expression (p53-like subtype) were associated with bone metastasis and chemo-resistance [61]. In a retrospective study, Seiler and colleagues investigated the association between molecular subtypes and the response to NAC. They proposed four molecular subtypes (Basal, Claudin-low, luminal and luminal-infiltrated subtypes) that can predict NAC response and confirmed that basal tumors were more chemo-sensitive and benefited the most from NAC compared to cystectomy only. In addition, they showed that irrespective of the treatment, the luminal subtype had better overall survival. On the contrary, the claudin-low subtype, defined as a subset of basal tumors with (1) a decreased expression of claudin-3 and claudin-4, (2) an enrichment for tumor-initiating cells (TIC) signatures and (3) an over-expression of epithelial-mesenchymal transition (EMT) transcription factors (e.g., SNAI1, TWIST1), had the worst overall survival and did not profit from NAC [62]. In the context of immunotherapy, ICIs are effective only in a subset of patients (20–25%) [39–41]. Mariathasan and colleagues performed transcriptomic analysis of a cohort of MIBC tumors treated with atezolizumab, an anti-PD-L1 drug. The responders were correlated with high neoantigen or tumor mutation burden, and with infiltration of $CD8^+$ T-effector cells. Non-responders were correlated with a gene expression signature of transforming growth factor β (TGF-β) signaling in fibroblast, especially in cases where $CD8^+$ T-effector cells were only present in the peritumoral stroma and not in the tumor parenchyma [63]. However, clinical trials testing these stratifications of patients into "subtypes" are needed.

In conclusion, treatment failure is frequent in both NMIBC and MIBC. This may indicate that the current classification systems do not depict the heterogeneity of BLCa and that treatment itself does not consider BLCa heterogeneity. Moreover, current pathological assessment is inaccurate and around 40% of tumors are clinically up-staged, which leads to an inadequate treatment choice [64–66]. Improvement of patient stratification to guide treatment options may be achieved using molecular classifications, and one way to further classify BLCa patients is according to molecular subtypes. Understanding of their biological role and therapy response is mandatory for clinical implementation.

4. Molecular Classification

4.1. NMIBC and MIBC: Two-Pathway Theory

Histological and cytogenetic analysis of BLCa tumors revealed that low-grade NMIBC tumors had few cytogenetics changes, while MIBC tumors were more genetically unstable [67]. Therefore, based on the fact that NMIBC and MIBC tumors present with different clinical behavior, histology and evolution, two distinct carcinogenic pathways, the papillary and the non-papillary pathway, have been proposed. The papillary pathway includes hyperplasia which will give rise to low-grade NMIBC; in the non-papillary pathway, instead, flat dysplasia and/or CIS are believed to be the precursors of MIBC tumors. However, high-grade NMIBCs seem to derive from the co-occurrence of hyperplasia and dysplasia, thus suggesting that the two-pathways intersect.

The papillary pathway is characterized by activation or overexpression of oncogenes leading to the genomic stable low-grade non-invasive papillary (Ta) tumor. FGFR3-activating mutations are present in almost all the Ta tumors but absent in CIS and less frequent in high-grade T1 NMIBC (30%) and MIBC (10%) [68–72]. Therefore, they are considered as driving alterations that induce the hyperplasia and tumorigenesis of the papillary pathway [68,70–74]. Activating FGFR3 mutations are associated with good prognosis and low recurrence rate. Mutations in the telomerase reverse transcriptase (TERT) gene promoter, an early genomic alteration associated with predisposition to develop BLCa, are present in the majority of tumors carrying FGFR3 mutations [75]. TERT re-activation may prevent senescence in FGFR3 mutant hyperproliferative tumors and protect genomic integrity. In normal conditions, FGFR3 regulates the activation of RTK/RAS/RAF/MAPK and PI3K/Akt/mTOR pathways. Therefore, in BLCa activation RTK/RAS/RAF/MAPK pathway was proposed as a likely mechanism that regulates cell growth in low-grade NMIBC [74,76]. Both activation of FGFR3 and RAS results in stimulation of the RTK/RAS/RAF/MAPK pathway, supporting the reason why in 10–15% of NMIBC, FGFR3 and HRAS/KRAS mutations are mutually exclusive [77,78]. Conversely, concerning the PI3K/Akt/mTOR pathway, activating mutations in PI3KCA are enriched in FGFR3 mutant tumors and present in 25% NMIBC and in 20% MIBC [68,79–81]. However, PIK3CA mutations are associated with low risk of progression [82]. Alterations in chromatin-modifying genes are frequent in all stages of BLCa, indicating that they can be early genomic alterations [68,70]. The lysine demethylase 6A (KDM6A) is mutated in 60% of low-grade Ta tumors and inhibits H3K27 methyltransferase EZH2 [70,79,83]. Given that EZH2 is often overexpressed and associated with MIBC, KDM6A activity might counteract tumor progression [84]. Moreover, truncating mutations in the STAG2 gene, which encodes a component of the cohesin complex, are highly frequent in low-grade Ta tumors and associated with low risk of progression and recurrence [82,85].

The most common copy number alteration (CNA) in Ta tumors is loss of heterozygosity (LOH) of 9q and 9p [79,82,86–88]. LOH 9 leads to deletions of several tumor suppressors genes, such as CDKN2A (9q), TSC1 (9p) and PTCH1 (9p) [86,87,89]. Loss of CDKN2A is present in more than 90% of FGFR3-mutant MIBC and homozygous deletions of CDKN2A are associated with a higher grade and higher risk of progression [70,82,90,91]. Moreover, CDKN2A gene encodes $p16^{INK4a}$ and $p14^{ARF}$, which induce cell cycle arrest via TP53 and RB1 signaling pathways [92]. In BLCa, CDKN2A deletion promotes proliferation and is mutually exclusive with TP53 and RB1 loss [68,92,93].

The genomic and transcriptomic landscape of low-grade Ta NMIBC indicates an important role for sustained proliferative signaling and is characterized by more clearly distinct mutations than MIBC. However, high-grade NMIBC shares certain characteristic genomic alterations with MIBC.

The non-papillary pathway gives rise to the majority of MIBCs. They share some characteristic mutations, CNA and chromosomal abnormalities resulting from defects in DNA replication/repair machinery genes and mutations in tumor suppressors genes [21]. Tumors with high genomic instability are associated with advanced stage and poor survival [88]. CIS might give rise to MIBC through the non-papillary pathway by the gradual accumulation of genomic abnormalities [88,94].

Mutations in tumor suppressor genes such as TP53, RB1 and PTEN are frequent in MIBC [21,95,96]. TP53 mutations are thought to have a central role in the non-papillary pathway and they are present in 50% of MIBC and 10–20% of high-grade T1 tumors [68–70,95,97]. MDM2, which negatively regulates p53, is overexpressed in 30% of MIBC and it is mutually exclusive with TP53 mutations in the majority of tumors [68]. In some tumors, the loss of function of p53 co-occurs with the loss of function of Rb1 [92,96,98–100]. The level of Rb1 is known to be inversely correlated to p16 (CDKN2A) level, and alterations in the Rb1/p16 tumor suppressor checkpoint pathway are associated with MIBC and increased risk of progression [95]. In some cases, Rb1 function is inhibited due to the overexpression of cyclin D1 (CCND1) or loss of CDKN2A, which are prevalent in high-grade tumors [68,70,97,101]. Therefore, despite loss of the RB1 gene in only 10% of MIBC, the Rb1 pathway is potentially dysregulated in a much higher fraction of tumors. The heterogeneous mechanisms resulting in a similar phenotype are part of the problem with treatment failure, resistance and lack of biomarkers.

Defects in DNA damage response genes are present in around 30% of MIBC and high-grade NMIBC, whereas they are absent in most of low-grade NMIBCs [102]. The most frequently altered gene is ERCC2, a component of the nucleotide excision repair machinery (NER). It is mutated in high-grade T1 tumors (17%) and MIBC (20%), but wild type (WT) in Ta [70,79]. Interestingly, mutations in NER pathway genes can occur in different types of cancer, but recurrent mutations are rare—ERCC2 in BLCa is an exception [103–105]. The Cancer Genome Atlas (TCGA) consortium was identified and described an ERCC2 signature and confirmed its association with smoking [68,92]. The function of ERCC2 in the urothelium is not known but given its association with smoking, it can be hypothesized that it protects against damage induced by carcinogenic metabolites accumulating in the urine.

Besides the induction of genetic instability, several genetic alterations associated with MIBC regulate cell growth and invasion. EGFR family genes such as EGFR and HER2 (ERBB2) are frequently overexpressed in CIS, MIBC and especially in metastatic BLCa [68,70,106]. Activation of EGFRs induce activation of RAS/MAPK and PI3K/AKT/mTOR pathways. Moreover, as transcription factors, EGFR family members can induce the expression of proliferation promoting oncogenes such as MYC and CCND1 or genes regulating migration/invasion such as COX2 and MMPs [107–109]. HER2 amplification is expressed higher in lymph node metastases than in primary tumors and both HER2 and EGFR overexpression are associated with a higher risk of recurrence and progression to invasive disease [69,110–113]. In contrast to HER2 and EGFR, ERBB3 overexpression has been linked to low-grade papillary NMIBC and good prognosis [69,110–112].

Alterations of the PI3K/AKT/mTOR pathway are involved in the progression of MIBC to metastatic disease [114,115]. PTEN, a negative regulator of PI3K, is downregulated in over 90% of MIBC and in 40% of NMIBC. MIBC frequently harbor mutations in PTEN, whereas Ta and T1 tumors retain WT PTEN [81,114,116]. Loss of PTEN is associated with aggressive MIBC and metastasis. In more than 40% of MIBC, loss of PTEN co-occurred with TP53 loss [117]. Loss of PTEN and TP53 is associated with poor survival [81,117], suggesting that the dual loss cooperates in tumor invasion and metastasis formation [117].

The genomic and transcriptomic characterization of BLCa revealed distinct molecular alterations in low-grade non-invasive papillary tumors and MIBC. However, these studies additionally discovered similarities between high-grade NMIBC and MIBC. As MIBCs, high-grade NMIBCs have more complex CNAs and they gradually accumulate genomic instability during progression compared to low-grade Ta tumors [71,79]. Mutations affecting DNA replication/repair machinery genes and tumor suppressors genes are shared among high-grade papillary tumors (Ta/T1) and MIBC developing via the non-papillary pathway. These alterations lead to the acquisition of the invasion potential in high-grade papillary tumors and can be the intersection between the two distinct carcinogenic pathways [69]. This indicates that the two pathways of BLCa development can intersect. Accordingly, 10–20% of MIBCs originate from the progression of high-grade NMIBCs.

4.2. Molecular Subtypes to Understand BLCa Biology

The two-pathway theory can be used to understand early diseases but, given the complexity and high heterogeneity of BLCa, it cannot explain the biology and progression of some tumors. Tumor profiling on large-scale cohorts, instead, allows us to identify molecular subtypes of BLCa with distinct clinical and molecular phenotype (Figure 1). This approach can improve our understanding of the biology of this complex disease. Multiple studies have classified BLCa into molecular subtypes based on transcriptomic data. The different subtypes and their clinical relevance are discussed in the following.

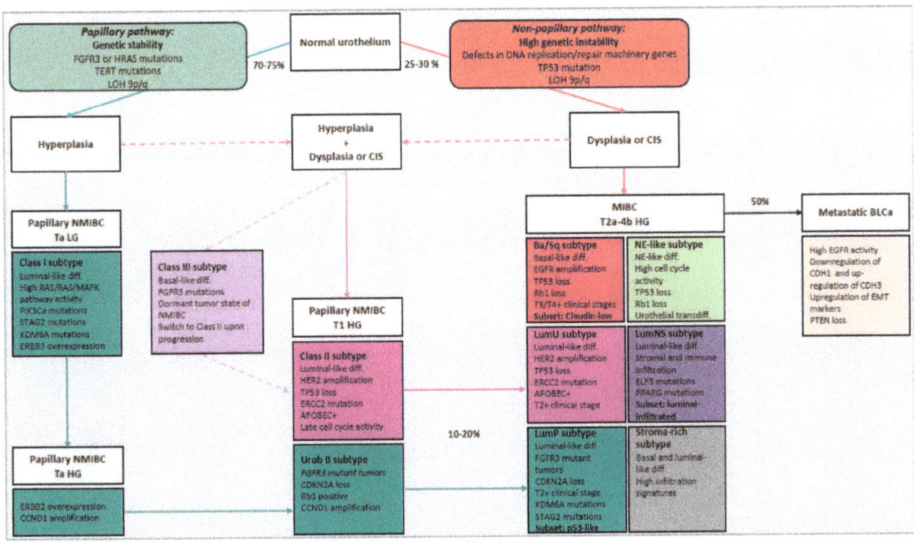

Figure 1. Simplified representation of the evolution of bladder cancer (BLCa) including the two distinct carcinogenic pathways (papillary and non-papillary pathways) and molecular subtypes of BLCa with their different characteristics. 75–80% of BLCa are non-muscle invasive bladder cancer (NMIBC) and 20–25% are muscle invasive bladder cancer (MIBC), of which 50% progress to metastatic BLCa. The papillary pathway includes hyperplasia which will give rise to Ta low-grade (LG) NMIBC and Ta/T1 high-grade (HG) NMIBC after the acquisition of mutations such as fibroblast growth factor receptor 3 (FGFR3) mutations or RAS mutations, loss of heterozygosity (LOH) 9p/q and telomerase reverse transcriptase (TERT) gene-promoter mutations. Class I/Urobasal A (Uro A) subtypes represent Ta LG NMIBC tumors and Uro B tumors were defined to be their progressed version. 10–20% of NMIBC potentially progress to MIBC and given the elevated expression of FGFR3 and the homozygous cyclin dependent kinase inhibitor 2A (CDKN2A) deletions in luminal papillary (LumP) tumors, they can originate from the progression of Class I/Uro A tumors through the papillary pathway. In the non-papillary pathway, instead, flat dysplasia and/or carcinoma in situ (CIS) are believed to be the precursors of MIBC tumors, after the gradual accumulation of genomic abnormalities such as loss of tumor suppressors TP53, defects in DNA replication/repair machinery genes (e.g., excision repair 2 (ERCC2)) and LOH 9q/p. Basal/Squamous (Ba/Sq) subtype represents the fraction of MIBC that originate

directly from the non-papillary pathway. Given their genomic instability, some HG NMIBC tumors seem to derive from the co-occurrence of hyperplasia and dysplasia shown by the dashed arrows. Class II subtypes represent T1 HG NMIBC tumors that originate from the co-occurrence of hyperplasia and dysplasia, and luminal unstable (LumU)/genomic unstable (GU) tumors are the progressed version that advanced through the non-papillary pathway. Luminal non-specified (LumNS) exhibits features of the LumP and LumU subtypes. Class III represents a dormant tumor state of NMIBC that seems to switch to Class II upon progression. However, the origin of Class III tumors is not fully clear. It is probably a fraction of HG tumors that originated from both papillary and non-papillary pathways given the basal phenotype and FGFR3 mutations. The neuroendocrine-like (NE-like) subtype expresses neuroendocrine differentiation markers and its origin is not fully clear. They maybe originate from transdifferentiation of urothelial cells upon treatment. The stroma-rich subtype, associated with high infiltration signatures, such as LumNS, represents a heterogeneous class of tumors that we need to investigate at a higher resolution. Blue arrow indicates reoccurrence of NMIBC. Dashed arrows are hypothetical. Abbreviation: PI3KCa: phosphatidylinositol-4,5-bisphosphate 3-binase catalytic subunit alpha, STAG2: stromal antigen 2, KDM6A: lysine demethylase 6A, ERBB2/3: Erb-B2 receptor tyrosine kinase 2/3, CCND1: cyclin D1, EGFR: epidermal growth factor receptor, HER2: human epidermal growth factor receptor 2, APOBEC: apoliprotein B mRNA-editing enzyme catalytic polypeptide-like, PPARG: peroxisome proliferator activated receptor gamma, ELF3: E74 Like ETS transcription factor 3, CDH1: cadherin 1, CDH3: cadherin 3, EMT: epithelial-to-mesenchymal transition, PTEN: phosphatase and tensin homolog, diff.: differentiation, inf: infiltration.

Hedegaard et al. identified three molecular subtypes of NMIBC (Class I, Class II and Class III) associated with distinct basal and luminal features and disease outcome (Table 1) [69]. Class I, with luminal-like features, associated with a good prognosis, is characterized by elevated expressions of early cell cycle regulators (CCDN1) and FGFR3 mutations. Given its features, Class I is comparable to the Urobasal A subtype (UroA) proposed by Sjödahl and colleagues who analyzed a mixed cohort of NMIBC and MIBC tumors [98]. Uro A was represented in the majority of the cases in low-grade non-invasive tumors. Class II, with luminal-like features, is characterized by elevated expressions of late cell cycle regulators and transcription factors associated with epithelial-to-mesenchymal transition (EMT) and ERBB2. Class II is comparable to the genomic unstable (GU) subtype from Sjödahl and colleagues. Class III, with basal-like features, associated with shorter overall survival, represents a dormant tumor state of NMIBC which shifted towards a basal phenotype with FGFR3 mutations. Only Class II and III tumors are associated with high stage and grade, high risk (according to EORTC), presence of CIS and progression to MIBC. Therefore, non-invasive tumors of Class II are associated with a high risk of progression through the non-papillary pathway (Figure 1) [69]. In contrast, class I tumors become muscle invasive via the papillary pathway, but this rarely happens. Interesting, it seems that Class III tumors switched to Class II upon progression. Therefore, given the fact that Class III tumors have a basal phenotype and FGFR3 mutations, it is not correct to associate the papillary pathway exclusively to luminal-like tumors. However, the origin of Class III tumors needs to be investigated.

It has been proven that NMIBC progression is also influenced by tumor differentiation status [118]. Therefore, NMIBC subtypes can represent three different development pathways of NMIBC that better differentiate between tumors with high and low risk of progression potential.

Currently, there is limited data on NMIBC molecular subtypes. Implementation of subclass stratification in NMIBC patients could facilitate biomarker discovery and ultimately assist in the identification of patients at higher risk of recurrent and/or progressive disease.

Concerning MIBC, to date, several groups have proposed comparable molecular classifications that have been associated to specific biological features, clinical outcomes and also responsiveness to therapy (Table 2) [62,68,91,92,98,118–123]. A consensus classification has been introduced to facilitate the implementation of molecular subtypes of MIBC in clinical trials and clinical practice [96]. The consensus classification is derived from the combination of the analysis of six previously published classifications of MIBC (Baylor [118], University of North Carolina (UNC) [121], Cartes d'Identité des Tumeurs (CIT)-Curie [91], MD Anderson Cancer Center (MDA) [120], Lund [123], TCGA [92]) with public transcriptome data of MIBC. This resulted in the description of six molecular subtypes for MIBC: luminal papillary (LumP), luminal non-specified (LumNS), luminal unstable (LumU), stroma-rich, basal/squamous (Ba/Sq) and neuroendocrine-like (NE-like) [96].

Table 1. Molecular subtypes of non-muscle invasive bladder cancer and their main features.

NMIBC Subtypes	Differentiation	Oncogenic Mechanism	Molecular Features	Stage/Grade	Associated Risk
Class I	Luminal	FGFR3 mutations	ERBB3 overexpression High RAS/RAF/MAPK pathway activity Early cell cycle regulators (CCDN1 amplification)	Low stage and grade	High risk of reoccurrence
Class II	Luminal	HER2 amplification	TP53/RB1 loss ERCC2 mutations Late cell cycle regulators EMT transcript factors APOBEC mutation signatures	High stage and grade	High risk of progression
Class III	Basal	FGFR3 mutations	Dormant tumor state of NMIBC that switch to Class II upon progression	High stage and grade	High risk of progression

Abbreviation: NMIBC: non-muscle invasive bladder cancer, EMT: epithelial-to-mesenchymal transition.

Table 2. Molecular subtypes of muscle invasive bladder cancer and their main features.

MIBC Subtypes	Differentiation	Oncogenic Mechanism	Molecular Features	Stage/Grade	Subsets
LumP	Luminal	FGFR3 mutations CDKN2A deletions	High RAS/RAF/MAPK pathway activity KDM6A mutations STAG2 mutations	T2+	p53-like: WT p53 gene expression signature
LumNS	Luminal	PPARG mutations	ELF3 mutations Elevated stromal and moderate immune infiltration	T2+	Luminal-infiltrated
LumU	Luminal	HER2 amplification Genomic instability	TP53 loss ERCC2 mutations High cell cycle activity APOBEC+ PPARG mutations	T2+	
Ba/Sq	Basal and squamous	EGFR mutations	TP53 and/or RB1 loss STAT3 regulon activation Elevated HIF1A activity	T3/T4	Claudin-low: Decreased expression of claudin-3 and claudin-4 Enrichment for TIC Overexpression of EMT transcript factors Stroma and immune infiltration High risk of metastatic disease
Stroma-rich	Basal and luminal		Elevated stroma and immune cells infiltration		
NE-like	Neuroendocrine	TP53 and/or RB1 loss	High cell cycle activity Urothelial transdifferentiation		

Abbreviation: MIBC: non-muscle invasive bladder cancer, LumP: luminal papillary, LumNS: luminal non-specified, LumU: luminal unstable, Ba/Sq: basal squamous, NE-like: neuroendocrine, STAT3: signal transducer and activator of transcription 3, HIF1A: hypoxia inducible factor 1 subunit alpha, TIC: tumor-initiating cells.

The Ba/Sq subtype is the most prevalent (35%) [96], characterized by a basal and squamous differentiation. Ba/Sq tumors express basal and stem cell markers such as KRT5/6, KRT14 and CD44, normally expressed in the basal or stem cell layer of the urothelium containing stem cell subpopulations [98,119]. It is strongly associated with STAT3 regulon activation, elevated HIF1A and downregulation of genes associated with urothelial differentiation such as FOXA1, GATA3 and PPARG [96]. Frequently mutated genes include the cell cycle regulator TP53 (61%), generally altered in advanced MIBC and with a central role in the non-papillary pathway [95,96]. TP53 is associated with a high risk of T2+ stage [124] and Ba/Sq subtype is over-represented in higher clinical stages (T3/T4) [96]. In 25% of Ba/Sq tumors, RB1 was mutated and TP53 and RB1 mutations co-occurred in 14% of Ba/Sq tumors, suggesting an interaction between the two cell cycle regulators. However, it was shown in mouse models that the dual loss of TP53 and RB1 is necessary but not sufficient to induce the invasion pathway in BLCa [125]. Therefore, given the prevalence of TP53 and RB1 mutations, Ba/Sq tumors originate from the non-papillary pathway, however other mutations sustain the invasion potential (Figure 1).

Concerning invasion and migration of bladder cancer cells, the interaction between cancer cells and the microenvironment is important. EMT is known to be an important process during tumor cell invasion. Downregulation of the cell adhesion molecule CDH1 (E-cadherin), and upregulation of its transcriptional repressors, such as ZEB1, ZEB2, SNAIL-1 and TWIST, in BLCa is associated with a poor prognosis and high risk of metastasis in BLCa [120,126]. Interestingly, patients with basal tumors at diagnosis present a more invasive and metastatic disease with a high EGFR activity and upregulation of EMT markers, such as ZEB1, ZEB2 and Vimentin [68,92,120,126]. It is known that EGFR activity can induce invasion by activating STAT3 which induces Twist gene expression [91,127]. The Ba/Sq subtype comprises a subset of Claudin-low tumors that is not included in the consensus classification but has been described in previous studies [62,122]. The Claudin-low subtype over-expresses transcription factors promoting EMT, resulting in a pronounced mesenchymal signature, and it is associated with poor survival irrespective of treatment [62,122]. Claudin-low tumors fundamentally differ from Ba/Sq tumors in their response to NAC, highlighting the importance of molecular subsets within subtypes of the consensus classification. The Claudin-low subtype is comparable to the Ba/Sq-infiltrate subtype proposed by Marzouka and colleagues which expresses CDH3 (P-cadherin) and EGFR but no ERBB2 or ERBB3 [123]. Similarly, Claudin-low tumors exhibit features of the Ba/Sq and stroma-rich subtype of the consensus classification. Both subtypes show higher immune infiltration compared to other subtypes [96]. In addition, the stroma-rich subtype exhibits an extraordinarily high contribution of stroma cells to the tumor mass. However, it is not yet fully understood whether the stroma-rich subtype classifies separately from Ba/Sq tumors due to the contamination with stroma cells or is biologically distinct from the Ba/Sq subtype. The stroma-rich subtype is better described later in the review. Given the differences in prognosis and response to therapy, functional comparisons between Claudin-low tumors and Ba/Sq tumors would help us to understand the most important pathways involved in the diversion of these two basal subtypes. This would facilitate the discovery of targeted treatments to block the rapid progression to metastasis formation of a subset of tumors sharing a basal phenotype.

Researchers aimed to identify transcription factors involved in basal expression signatures that determine the basal-like phenotype. They proposed ΔNp63 to be involved in the regulation of basal transcription programs. ΔNp63α, an isoform of p63 belonging to the p53 family, regulates epithelial development and stem cell biology and has also been implicated in basal breast cancer [120]. In BLCa, it was shown that ΔNp63α induces the expression of miR205, which inhibits EMT [128]. Downregulation of p63, as observed in MIBC, might therefore contribute to invasion [128]. However, high levels of ΔNp63α were associated with a lethal group of MIBC [129,130]. Both, STAT3 and EGFR activity seem to promote ΔNp63α expression [131]. The role of ΔNp63α in MIBC is not clear but it is probably involved in cell proliferation rather than in invasion. Moreover, p63 is highly expressed in the epithelial subset of BLCa and may therefore be involved in the biology of MIBC with epithelial and stemness phenotype [128]. Further functional investigations of the two major p63 isoforms (TA and ΔN) are needed to understand the p63–EMT relationship and to unravel the prognostic and therapeutic value of p63 in basal subtypes.

Luminal-like tumors (LumP, LumU and LumNS) presented a papillary morphology enriched in luminal markers such as low molecular weight KRT20 and uroplakins such as UPK1A and UPK2, normally expressed in terminally urothelial cells. Luminal MIBC exhibit gene expression signatures of PPARG, GATA3 and FOXA1 transcription factors involved in urothelial differentiation, and regulatory factors and receptors of the estrogen signaling (ESR2) [96,120]. Discrepancy in outcome of patients with luminal-like tumors has been reported [92,121,132]. However, mutations status and oncogenic mechanism were different between the three luminal subtypes, indicating that non-invasive luminal tumors can progress to invasion through different carcinogenic pathways and not only through the papillary pathway. The investigation of cellular mechanisms contributing to the differences between luminal subtypes would elucidate these distinct pathways of progression.

The LumP subtype was the most prevalent (24%), with the best overall survival compared to the other luminal subtypes. LumP tumors were strongly associated with high FGFR3 activity caused by gene fusion, mutation or amplification [96]. FGFR3-activating mutations and overexpression are predominantly associated with low-grade Ta tumors developing via the papillary pathway, which have a favorable outcome in most cases [95,133–135]. The LumP subtype is comparable to the Urothelial-like subtype of Sjödahl and colleagues formerly named "Urobasal" (Uro A and Uro B) [71,123]. Uro A is comparable to Class I NMIBC (luminal-like) [69]. The Uro B subtype was predominant in MIBC tumors and they are believed to be the progressed version of Uro A tumors. Comparable to UroB tumors, LumP tumors are believed to derive from Ta/T1 tumors. Given the elevated expression of FGFR3 and the homozygous CDKN2A deletions in 33% of LumP tumors, we can assume that LumP tumors represent the fraction of luminal FGFR3-mutant tumors that originated from Class I tumors and progressed through the papillary pathway (Figure 1) [121].

The LumP tumors are thought to have a low likelihood of response to cisplatin-based NAC in contrast to basal tumors [62]. Interestingly, Choi and colleagues described the existence of a portion of luminal tumors (p53-like subtype) which were associated with WT TP53 gene expression signature and were resistant to chemotherapy. However, it is known that p53 level increases in response to DNA damage to stop cell cycle and induce apoptosis [120]. Alternatively, WT p53 induced reversible senescence which impaired the apoptotic response after chemotherapy-induced DNA damage, as it was shown in mouse models for breast cancer [136]. The molecular basis of a positive or negative p53-signature in the context of NAC is not known in BLCa. In the analysis conducted by The Cancer Genome Atlas (TCGA) Consortium, it was shown that MDM2 amplification or overexpression was predicted to inactivate p53 in 76% of MIBC [68]. Investigation of mechanisms of chemotherapy resistance in this portion of luminal tumors is needed.

The LumU subtype instead was associated with worse overall survival compared to the other luminal subtypes. It is the most genomic unstable subtype among luminal subtypes and harbors the highest load of apoliprotein B mRNA-editing enzyme catalytic polypeptide-like (APOBEC)-induced mutations [96]. LumU had higher cell cycle activity than other subtypes and it was enriched in ERCC2 and TP53 mutations. The majority of NMIBC (85%) with mutated p53 are of high grade, indicating that p53 inactivation occurred in the majority of NMIBC that have the potential to invade [137]. The LumU subtype was associated with an overexpression of HER2, frequently observed in CIS, MIBC and metastatic BLCa [68,70,106]. The LumU subtype corresponds to the genomic unstable subtype, discovered by Sjödahl and colleagues. According to the fact that LumU tumors are enriched in mutations characteristic of the non-papillary pathway, we can assume that LumU/GU tumors originated from Class II NMIBCs through the non-papillary pathway (Figure 1) [69]. Given that LumU and LumP tumors progressed through distinct pathways, as shown by their distinct genomic properties, it is important to separate these tumors and characterize them distinctly.

The last luminal subtype proposed in the consensus classification was described as a non-specified luminal (LumNS) subtype and includes a minor fraction of tumors (8%). The LumNS subtype is characterized by an elevated stromal and moderate immune infiltration. It is enriched in mutations effecting regulatory factors such as ELF3 and PPARG. The LumNS is associated with the worst overall survival in the luminal group. It is comparable to the epithelial-infiltrated subtype of Sjödahl and colleagues, which combines features of the GU and Uro subtypes [98]. Accordingly, LumNS exhibits features of the LumP and LumU subtypes in the consensus classification.

An additional subtype, the stroma-rich subtype, is also highly infiltrated by stromal and immune cells. The stroma-rich subtype was associated with gene expression signatures of smooth muscle, endothelial cells, fibroblasts and myofibroblasts. Both LumNS and the stroma-rich subtypes were not well defined and may probably be the result of cluster analysis. Advanced MIBC tumors are composed of tumor cells and stromal and immune cells. This can confound global gene expression analysis and bias the results towards stromal gene signatures. To resolve this problem, Sjödahl and colleagues histologically analyzed MIBC tissue in addition to mRNA sequencing approaches [98,123]. They showed that the type and the level of infiltrating non-tumor cells can vary between tumors sharing the same tumor cell phenotype [98,123]. Therefore, LumNS and the stroma-rich subtypes represent a heterogeneous class of tumors that we need to investigate at a higher resolution and ideally, by functional assays, to understand their role and origin.

The neuroendocrine-like (NE-like) subtype is the rarest of the consensus classification and only found in 3% of MIBC. NE-like tumors are enriched with neuroendocrine differentiation markers including synaptophysin, chromogranin A (CGA) and neuron-specific enolase (NSE or CD56). Moreover, they are characterized by high cell cycle activity and the co-occurrence of TP53 and RB1 mutations or deletions [92,96,98]. Patients with tumors carrying alterations in both TP53 and RB1 have a high risk of BLCa progression and worse clinical outcome than patients with only one of these gene alterations [95]. The majority of NE-like tumors display partial or complete neuroendocrine histology. Given the large contribution of neuroendocrine histological variant in the NE-like subtype, it is not surprising that this subtype is associated with the worst prognosis. Neuroendocrine tumors comprise small cell carcinoma of the bladder (SCCB) and large cell carcinoma of the bladder (LCCB). SCCB is more prevalent than LCCB but is extremely rare, making up only 0.5% of BLCa, and it is usually diagnosed at late stages [138]. To date, no guidelines for the treatment of neuroendocrine tumors have been formulated. Treatment strategies commonly follow the therapy approaches for small cell carcinoma (SCC) of the lung, including surgery, chemotherapy and radiation [139]. However, this is clearly not sufficient to treat this most aggressive form of BLCa with a 5-year overall survival rate of only ~20% [138]. Very little is known about the biology and evolution of neuroendocrine tumors. However, recent evidence supports the idea that SCCB originates from urothelial cells. How urothelial cells transdifferentiate remains elusive [138,140]. Possibly, anti-cancer treatments can induce the transdifferentiation of urothelial cells into tumors with a neuroendocrine phenotype in a similar manner as it has been described for neuroendocrine transdifferentiation of prostate adenocarcinoma upon androgen-deprivation therapy [141]. A better understanding of the pathogenesis and molecular differences between the NE-like phenotype/SCCB and the other BLCa subtypes may serve to understand their different biological meaning and origin. However, NE-like tumors are rare, impeding large-scale genomic and transcriptomic analysis of patient datasets. In order to characterize this tumor subtype, it is mandatory to create tumor models reflecting the molecular features of NE-like tumors.

5. Molecular Classification to Guide Treatment Choice

Molecular subtypes could be used to stratify patients and may help to identify patients whom would benefit from conventional, targeted or combined therapies (Figure 2) [91,120].

Regarding conventional therapies, basal tumors, should be selected for NAC and not selected for radiotherapy given their hypoxic microenvironment (Figure 2) [62,126,142,143]. Radiotherapy, given the lack of predictive biomarkers, is underutilized and mainly used as palliative therapy. Recent findings suggest that tumors with an increased expression of genes associated with T-cell activation and the INFγ signaling pathway would benefit from radiotherapy [144]. However, clinical trials to analyze the impact of subtyping on response to radiotherapy are needed.

In the context of immunotherapy, no subtypes proposed in the consensus classification have a profile (infiltration of CD8$^+$ T cells, high IFNγ signaling and low TGFβ pathway) associated with a positive response to ICIs [63]. However, high infiltration of CD8$^+$ T cells has been associated with Claudin-low tumors [62,122] and luminal-infiltrate tumors [62]. Therefore, the authors hypothesize that these patients can benefit from checkpoint inhibitors (Figure 2) [39,40,145]. Interestingly, it was shown that NE-like tumors, given the low TGFβ pathway (low TGFβ1 and TGFβR1), and LumU, given the highest tumor mutation burden and tumor antigen burden, would benefit from ICIs (Figure 2) [146]. Stroma-rich tumors, instead, given the high activity of the TGFβ pathway, would probably be resistant to ICS [146]. More clinical trials to analyze the impact of subtyping on response to ICIs are needed.

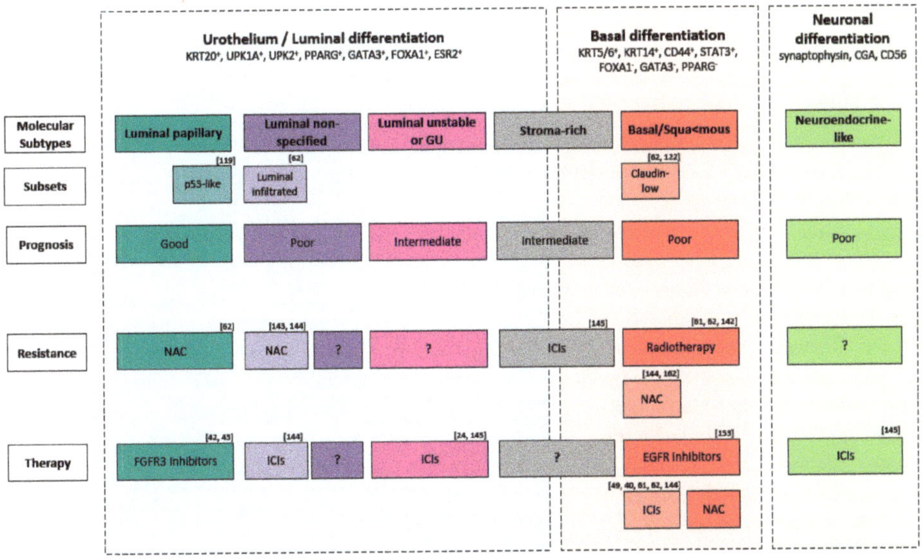

Figure 2. Molecular subtypes of muscle-invasive bladder cancer and their main expression markers, possible molecular subsets, prognosis, therapy resistance and therapy sensitive. Abbreviation: GU: Genomic Unstable, ICIs: immune checkpoint inhibitors, NAC: Neoadjuvant cisplatin-based chemotherapy.

There is evidence that a fraction of BCLa patients might respond to targeted therapy [97]. EGFR is frequently overexpressed in Ba/Sq tumors and exerts an oncogenic function. Therefore, selecting these tumors for EGFR-targeted therapies seems to be promising (Figure 2). Monoclonal antibodies or tyrosine kinase inhibitors (TKI) targeting EGFR are effective in different cancers like lung, head, neck and colorectal [147]. In contrast, targeting EGFR in BLCa showed conflicting results [148–153]. However, patient selection was not conducted according to EGFR status in some of these clinical trials, which may have concealed the response to EGFR-overexpressing tumors. Currently, afatinib, a second-generation of tyrosine-kinase inhibitor of ERBB2 and EGFR, is being tested in a clinical trial phase II for metastatic, chemotherapy refractory BLCa patients with HER2/ERBB3 mutations and HER2/EGFR amplification [154]. Notably, resistance to TKIs can occur by secondary mutations, activation of alternative pathways or anomalies in the downstream signaling transducers [155]. In the case of EGFR, it was shown in colon cancer that RAS mutations (KRAS and HRAS) can inhibit the response to EGFR-targeted therapy [91,156]. Therefore, the interaction between RAS mutation status and EGFR-targeted therapy should be further investigated. Combination targeted therapy should be considered in case of a Ras-mediated resistance to EGFR inhibitors [91].

In contrast to other solid cancers, targeting HER2 in MIBC did not show promising results and no HER-2 targeting drug has been approved yet [148,157]. Most likely, this is due to the fact that the majority of clinical trials evaluating HER-2 inhibitors in BLCa did not select patients based on HER-2 status. Those trials that did evaluate HER2 status mostly used immunohistochemistry (IHC) and/or fluorescence in situ hybridization (FISH). However, it was proven that patient selection based on DNA and RNA sequencing techniques better stratifies responders and non-responders [158,159]. LumU tumors overexpressing HER2 can be used to discover biomarkers to select patients that would probably benefit from HER-2-target therapy.

FGFR3 overexpression is believed to be the oncogenic mechanism in LumP tumors and can be considered a potential therapeutic target for these tumors (Figure 2) [96]. Erdafitinib, an inhibitor of the tyrosine kinase activity of FGFR, was recently approved as a treatment for patients with chemo-resistant locally advanced or metastatic BLCa carrying FGFR3 or FGFR2 alteration [42,43]. Given that LumP tumors would probably have a low likelihood of response to cisplatin-based NAC [62], they might be selected for FGFR3-target therapy.

6. Limitations and Perspectives of Molecular Classification

Current findings from the molecular investigation of BLCa cohorts showed that patients can be grouped into classes based on their molecular landscape. Of utmost importance, these molecular classes have been established based on investigations of mainly untreated tumor samples. However, the molecular classification has some limitations given the fact that BLCa is a very heterogeneous disease exhibiting vast histological and molecular intra- and inter-tumor heterogeneity was shown in some tumors [160,161]. Consequently, the existence of multiple subtypes within the same tumor has been demonstrated. Therefore, one molecular subtype cannot be representative for these heterogeneous tumors. Further investigations to study the impact of multiple subtypes on tumor heterogeneity and response to treatments are needed. There is current standardization and consensus on gene expression-based molecular subtyping [96]. However, further standardizations are needed for the implementation in the clinical setting. In addition, other analytical aspects such as tumor cell content, immune cell infiltration, selected genes, technique of expression profiling and selected antibodies in case of IHC need to be defined [162].

Different approaches can be used to overcome/complement the limitations of molecular subtyping. Intra-tumor heterogeneity is not entirely depicted by tumor bulk profiling but can be studied by single-cell profiling, spatial imaging techniques or imaging mass cytometry, such as cytometry by time-of-flight (CyToF). Inter-tumor heterogeneity and plasticity during tumor development, instead, can be analyzed by sequential analysis and repeated biopsies during patients' history. More molecular subtyping studies are needed to increase the number of patient sequencing datasets. This is crucial to identify more classes of BLCa, particularly rare subtypes, and investigate the evolution of heterogeneous tumors at higher resolution.

Before considering the use of molecular subtypes in a daily clinical practice, a better functional understanding is necessary. The behavior of molecular subtypes can be functionally studied in pre-clinical models from patient-derived material such as patient-derived organoids, ex vivo tissue cultures and xenografts. Patient-derived models that reflect the molecular phenotype of the different subtypes can be employed to investigate the biology of BLCa subtypes and evaluate differential drug responses in the context of different subtypes. Interestingly, there is the possibility to genetically engineer these models to study the relevance of specific mutations in a given subtype. Moreover, this technology can be used to better understand how the different subtypes evolve and progress through the two-pathway model.

Do we really need to go so far to treat each patient individually or is some sort of grouping necessary for clinical trials? There is an urgent need for clinical trials that investigate the impact of molecular classification.

7. Conclusions

To conclude, the management of BLCa is complex and current classification systems do not depict the heterogeneity of this disease. Promising findings from the molecular investigation of BLCa cohorts show that grouping patients into classes based on the molecular landscape can be used to understand the pathogenesis of this complex disease. Early data suggests that these investigations can also be used to predict the response to treatment and its implementation should be taken into consideration for the future management of this disease. Ideally, to have reliable results, we should complement mRNA with IHC analysis of protein for each patient's tumor. However, to move towards precision medicine, the understanding of the biological meaning of these molecular classes, the differences in the composition of cell subpopulations and clinical trials that investigate the impact of molecular classification will be mandatory.

Author Contributions: Conceptualization, M.M.; writing—original draft preparation, M.M.; writing—review and editing, M.K., R.S., M.K.-d.J., G.N.T.; supervision, R.S. and M.K.-d.J. All authors have read and agreed to the published version of the manuscript.

Funding: This research was funded by Swiss National Science Foundation, grant number 310030.

Acknowledgments: The authors would like to thank all members of the DBMR Urology group for critical discussions.

Conflicts of Interest: The authors declare no conflict of interest.

References

1. Bray, B.; Ferlay, J.; Soerjomataram, I.; Siegel, R.L.; Torre, L.A.; Jemal, A. Global Cancer Statistics 2018: GLOBOCAN Estimates of Incidence and Mortality Worldwide for 36 Cancers in 185 Countries. *CA Cancer J. Clin.* **2018**, *68*, 394–424. [CrossRef]
2. Letašiová, S.; Medveďová, A.; Šovčíková, A.; Dušinská, M.; Volkovová, K.; Mosoiu, C.; Bartonová, A. Bladder Cancer, a Review of the Environmental Risk Factors. *Environ. Health* **2012**, *11* (Suppl. 1), S11.
3. Burger, M.; Catto, J.W.F.; Dalbagni, G.; Grossman, H.B.; Herr, H.; Karakiewicz, P.; Kassouf, W.; Kiemeney, L.A.; la Vecchia, C.; Shariat, S.; et al. Epidemiology and Risk Factors of Urothelial Bladder Cancer. *Eur. Urol.* **2013**, *63*, 234–241. [CrossRef]
4. Ferlay, J.; Colombet, M.; Soerjomataram, I.; Dyba, T.; Randi, G.; Bettio, M.; Gavin, A.; Visser, O.; Bray, F. Cancer Incidence and Mortality Patterns in Europe: Estimates for 40 Countries and 25 Major Cancers in 2018. *Eur. J. Cancer* **2018**, *103*, 356–387. [CrossRef] [PubMed]
5. Berdik, C. Unlocking Bladder Cancer. *Nature* **2017**, *551*, S34–S35. [CrossRef] [PubMed]
6. Aben, K.K.H.; Witjes, J.A.; Schoenberg, M.P.; Kaa, C.H.d.; Verbeek, A.L.M.; Kiemeney, L.A.L.M. Familial Aggregation of Urothelial Cell Carcinoma. *Int. J. Cancer* **2002**, *98*, 274–278. [CrossRef] [PubMed]
7. Lin, J.; Spitz, M.R.; Dinney, C.P.; Etzel, C.J.; Grossman, H.B.; Wu, X. Bladder Cancer Risk as Modified by Family History and Smoking. *Cancer* **2006**, *107*, 705–711. [CrossRef]
8. Murta-Nascimento, C.; Silverman, D.T.; Kogevinas, M.; García-Closas, M.; Rothman, N.; Tardón, A.; García-Closas, R.; Serra, C.; Carrato, A.; Villanueva, C.; et al. Risk of Bladder Cancer Associated with Family History of Cancer: Do Low-Penetrance Polymorphisms Account for the Increase in Risk? *Cancer Epidemiol. Prev. Biomark.* **2007**, *16*, 1595–1600. [CrossRef]
9. Svatek, R.S.; Hollenbeck, B.K.; Holmäng, S.; Lee, R.; Kim, S.P.; Stenzl, A.; Lotan, Y. The Economics of Bladder Cancer: Costs and Considerations of Caring for This Disease. *Eur. Urol.* **2014**, *66*, 253–262. [CrossRef]
10. Hong, Y.M.; Loughlin, K.R. Economic Impact of Tumor Markers in Bladder Cancer Surveillance. *Urology* **2008**, *71*, 131–135. [CrossRef]
11. James, A.C.; Gore, J.L. The Costs of Non-Muscle Invasive Bladder Cancer. *Urol. Clin. N. Am.* **2013**, *40*, 261–269. [CrossRef] [PubMed]
12. Ploeg, M.; Aben, K.K.H.; Kiemeney, L.A. The Present and Future Burden of Urinary Bladder Cancer in the World. *World J. Urol.* **2009**, *27*, 289–293. [CrossRef] [PubMed]
13. Avritscher, E.B.C.; Cooksley, C.D.; Grossman, H.B.; Sabichi, A.L.; Hamblin, L.; Dinney, C.P.; Elting, L.S. Clinical Model of Lifetime Cost of Treating Bladder Cancer and Associated Complications. *Urology* **2006**, *68*, 549–553. [CrossRef] [PubMed]
14. Kamat, A.M.; Hahn, N.M.; Efstathiou, J.A.; Lerner, S.P.; Malmström, P.-U.; Choi, W.; Guo, C.C.; Lotan, Y.; Kassouf, W. Bladder Cancer. *Lancet* **2016**, *388*, 2796–2810. [CrossRef]
15. Rhijn, B.W.G.v.; Burger, M.; Lotan, Y.; Solsona, E.; Stief, C.G.; Sylvester, R.J.; Witjes, J.A.; Zlotta, A.R. Recurrence and Progression of Disease in Non–Muscle-Invasive Bladder Cancer: From Epidemiology to Treatment Strategy. *Eur. Urol.* **2009**, *56*, 430–442. [CrossRef] [PubMed]
16. Chamie, K.; Litwin, M.S.; Bassett, J.C.; Daskivich, T.J.; Lai, J.; Hanley, J.M.; Konety, B.R.; Saigal, C.S.; Urologic Diseases in America Project. Recurrence of High-Risk Bladder Cancer: A Population-Based Analysis. *Cancer* **2013**, *119*, 3219–3227. [CrossRef]
17. Sylvester, R.J.; van der Meijden, A.P.M.; Oosterlinck, W.; Witjes, J.A.; Bouffioux, C.; Denis, L.; Newling, D.W.W.; Kurth, K. Predicting Recurrence and Progression in Individual Patients with Stage Ta T1 Bladder Cancer Using EORTC Risk Tables: A Combined Analysis of 2596 Patients from Seven EORTC Trials. *Eur. Urol.* **2006**, *49*, 466–477. [CrossRef]

18. van den Bosch, S.; Alfred Witjes, J. Long-Term Cancer-Specific Survival in Patients with High-Risk, Non-Muscle-Invasive Bladder Cancer and Tumour Progression: A Systematic Review. *Eur. Urol.* **2011**, *60*, 493–500. [CrossRef]
19. van Rhijn, B.W.G.; van der Kwast, T.H.; Alkhateeb, S.S.; Fleshner, N.E.; van Leenders, G.J.L.H.; Bostrom, P.J.; van der Aa, M.N.M.; Kakiashvili, D.M.; Bangma, C.H.; Jewett, M.A.S.; et al. New and Highly Prognostic System to Discern T1 Bladder Cancer Substage. *Eur. Urol.* **2012**, *61*, 378–384. [CrossRef]
20. Tang, D.H.; Chang, S.S. Management of Carcinoma in Situ of the Bladder: Best Practice and Recent Developments. *Ther. Adv. Urol.* **2015**, *7*, 351–364. [CrossRef]
21. Knowles, M.A.; Hurst, C.D. Molecular Biology of Bladder Cancer: New Insights into Pathogenesis and Clinical Diversity. *Nat. Rev. Cancer* **2015**, *15*, 25–41. [CrossRef] [PubMed]
22. Shinagare, A.B.; Ramaiya, N.H.; Jagannathan, J.P.; Fennessy, F.M.; Taplin, M.; van den Abbeele, A.D. Metastatic Pattern of Bladder Cancer: Correlation With the Characteristics of the Primary Tumor. *Am. J. Roentgenol.* **2011**, *196*, 117–122. [CrossRef] [PubMed]
23. Hautmann, R.E.; Gschwend, J.E.; de Petriconi, R.C.; Kron, M.; Volkmer, B.G. Cystectomy for Transitional Cell Carcinoma of the Bladder: Results of a Surgery Only Series in the Neobladder Era. *J. Urol.* **2006**, *176*, 486–492. [CrossRef] [PubMed]
24. Kim, T.J.; Cho, K.S.; Koo, K.C. Current Status and Future Perspectives of Immunotherapy for Locally Advanced or Metastatic Urothelial Carcinoma: A Comprehensive Review. *Cancers* **2020**, *12*, 192. [CrossRef]
25. Babjuk, M.; Burger, M.; Compérat, E.M.; Gontero, P.; Mostafid, A.H.; Palou, J.; van Rhijn, B.W.G.; Rouprêt, M.; Shariat, S.F.; Sylvester, R.; et al. European Association of Urology Guidelines on Non-Muscle-Invasive Bladder Cancer (TaT1 and Carcinoma In Situ)—2019 Update. *Eur. Urol.* **2019**, *76*, 639–657. [CrossRef]
26. Soukup, V.; Čapoun, O.; Cohen, D.; Hernández, V.; Babjuk, M.; Burger, M.; Compérat, E.; Gontero, P.; Lam, T.; MacLennan, S.; et al. Prognostic Performance and Reproducibility of the 1973 and 2004/2016 World Health Organization Grading Classification Systems in Non–Muscle-Invasive Bladder Cancer: A European Association of Urology Non-Muscle Invasive Bladder Cancer Guidelines Panel Systematic Review. *Eur. Urol.* **2017**, *72*, 801–813.
27. Lipunova, N.; Wesselius, A.; Cheng, K.K.; van Schooten, F.J.; Cazier, J.-B.; Bryan, R.T.; Zeegers, M.P. Systematic Review: Genetic Associations for Prognostic Factors of Urinary Bladder Cancer. *Biomark. Cancer* **2019**, *11*. [CrossRef]
28. Babjuk, M.; Böhle, A.; Burger, M.; Capoun, O.; Cohen, D.; Compérat, E.M.; Hernández, V.; Kaasinen, E.; Palou, J.; Rouprêt, M.; et al. EAU Guidelines on Non–Muscle-Invasive Urothelial Carcinoma of the Bladder: Update 2016. *Eur. Urol.* **2017**, *71*, 447–461. [CrossRef]
29. Clark, P.E.; Agarwal, N.; Biagioli, M.C.; Eisenberger, M.A.; Greenberg, R.E.; Herr, H.W.; Inman, B.A.; Kuban, D.A.; Kuzel, T.M.; Lele, S.M.; et al. Bladder Cancer. *J. Natl. Compr. Canc. Netw.* **2013**, *11*, 446–475. [CrossRef]
30. Holmäng, S.; Ströck, V. Should Follow-up Cystoscopy in Bacillus Calmette-Guérin–Treated Patients Continue After Five Tumour-Free Years? *Eur. Urol.* **2012**, *61*, 503–507. [CrossRef]
31. Perlis, N.; Zlotta, A.R.; Beyene, J.; Finelli, A.; Fleshner, N.E.; Kulkarni, G.S. Immediate Post–Transurethral Resection of Bladder Tumor Intravesical Chemotherapy Prevents Non–Muscle-Invasive Bladder Cancer Recurrences: An Updated Meta-Analysis on 2548 Patients and Quality-of-Evidence Review. *Eur. Urol.* **2013**, *64*, 421–430. [CrossRef] [PubMed]
32. Arends, T.J.H.; Nativ, O.; Maffezzini, M.; de Cobelli, O.; Canepa, G.; Verweij, F.; Moskovitz, B.; van der Heijden, A.G.; Witjes, J.A. Results of a Randomised Controlled Trial Comparing Intravesical Chemohyperthermia with Mitomycin C Versus Bacillus Calmette-Guérin for Adjuvant Treatment of Patients with Intermediate- and High-Risk Non–Muscle-Invasive Bladder Cancer. *Eur. Urol.* **2016**, *69*, 1046–1052. [CrossRef] [PubMed]
33. Kamat, A.M.; Flaig, T.W.; Grossman, H.B.; Konety, B.; Lamm, D.; O'Donnell, M.A.; Uchio, E.; Efstathiou, J.A.; Taylor, J.A. Consensus Statement on Best Practice Management Regarding the Use of Intravesical Immunotherapy with BCG for Bladder Cancer. *Nat. Rev. Urol.* **2015**, *12*, 225–235. [CrossRef] [PubMed]
34. Sylvester, R.J.; Oosterlinck, W.; Holmang, S.; Sydes, M.R.; Birtle, A.; Gudjonsson, S.; de Nunzio, C.; Okamura, K.; Kaasinen, E.; Solsona, E.; et al. Systematic Review and Individual Patient Data Meta-Analysis of Randomized Trials Comparing a Single Immediate Instillation of Chemotherapy After Transurethral Resection with Transurethral Resection Alone in Patients with Stage PTa–PT1 Urothelial Carcinoma of the Bladder: Which Patients Benefit from the Instillation? *Eur. Urol.* **2016**, *69*, 231–244.

35. Sylvester, R.J.; van der Meijden, A.P.M.; Lamm, D.L. Intravesical Bacillus Calmette-Guerin Reduces the Risk of Progression in Patients with Superficial Bladder Cancer: A Meta-Analysis of the Published Results of Randomized Clinical Trials. *J. Urol.* **2002**, *168*, 1964–1970. [CrossRef]
36. Witjes, J.A.; Bruins, H.M.; Cathomas, R.; Compérat, E.M.; Cowan, N.C.; Gakis, G.; Hernández, V.; Linares Espinós, E.; Lorch, A.; Neuzillet, Y.; et al. European Association of Urology Guidelines on Muscle-Invasive and Metastatic Bladder Cancer: Summary of the 2020 Guidelines. *Eur. Urol.* **2020**. [CrossRef]
37. Sonpavde, G.; Watson, D.; Tourtellott, M.; Cowey, C.L.; Hellerstedt, B.; Hutson, T.E.; Zhan, F.; Vogelzang, N.J. Administration of Cisplatin-Based Chemotherapy for Advanced Urothelial Carcinoma in the Community. *Clin. Genitourin. Cancer* **2012**, *10*, 1–5. [CrossRef]
38. Zargar, H.; Espiritu, P.N.; Fairey, A.S.; Mertens, L.S.; Dinney, C.P.; Mir, M.C.; Krabbe, L.-M.; Cookson, M.S.; Jacobsen, N.-E.; Gandhi, N.; et al. Multicenter Assessment of Neoadjuvant Chemotherapy for Muscle-Invasive Bladder Cancer. *Eur. Urol.* **2015**, *67*, 241–249. [CrossRef]
39. Rosenberg, J.E.; Hoffman-Censits, J.; Powles, T.; van der Heijden, M.S.; Balar, A.V.; Necchi, A.; Dawson, N.; O'Donnell, P.H.; Balmanoukian, A.; Loriot, Y.; et al. Atezolizumab in Patients with Locally Advanced and Metastatic Urothelial Carcinoma Who Have Progressed Following Treatment with Platinum-Based Chemotherapy: A Single-Arm, Multicentre, Phase 2 Trial. *Lancet Lond. Engl.* **2016**, *387*, 1909–1920. [CrossRef]
40. Sharma, P.; Callahan, M.K.; Bono, P.; Kim, J.; Spiliopoulou, P.; Calvo, E.; Pillai, R.N.; Ott, P.A.; de Braud, F.; Morse, M.; et al. Nivolumab Monotherapy in Recurrent Metastatic Urothelial Carcinoma (CheckMate 032): A Multicentre, Open-Label, Two-Stage, Multi-Arm, Phase 1/2 Trial. *Lancet Oncol.* **2016**, *17*, 1590–1598. [CrossRef]
41. Hargadon, K.M.; Johnson, C.E.; Williams, C.J. Immune Checkpoint Blockade Therapy for Cancer: An Overview of FDA-Approved Immune Checkpoint Inhibitors. *Int. Immunopharmacol.* **2018**, *62*, 29–39. [CrossRef] [PubMed]
42. Montazeri, K.; Bellmunt, J. Erdafitinib for the Treatment of Metastatic Bladder Cancer. *Expert Rev. Clin. Pharmacol.* **2020**, *13*, 1–6. [CrossRef] [PubMed]
43. Loriot, Y.; Necchi, A.; Park, S.H.; Garcia-Donas, J.; Huddart, R.; Burgess, E.; Fleming, M.; Rezazadeh, A.; Mellado, B.; Varlamov, S.; et al. Erdafitinib in Locally Advanced or Metastatic Urothelial Carcinoma. *N. Engl. J. Med.* **2019**, *381*, 338–348. [CrossRef] [PubMed]
44. Rosenberg, J.E.; O'Donnell, P.H.; Balar, A.V.; McGregor, B.A.; Heath, E.I.; Yu, E.Y.; Galsky, M.D.; Hahn, N.M.; Gartner, E.M.; Pinelli, J.M.; et al. Pivotal Trial of Enfortumab Vedotin in Urothelial Carcinoma After Platinum and Anti-Programmed Death 1/Programmed Death Ligand 1 Therapy. *J. Clin. Oncol.* **2019**, *37*, 2592–2600. [CrossRef]
45. Bosschieter, J.; Nieuwenhuijzen, J.A.; van Ginkel, T.; Vis, A.N.; Witte, B.; Newling, D.; Beckers, G.M.A.; van Moorselaar, R.J.A. Value of an Immediate Intravesical Instillation of Mitomycin C in Patients with Non–Muscle-Invasive Bladder Cancer: A Prospective Multicentre Randomised Study in 2243 Patients. *Eur. Urol.* **2018**, *73*, 226–232. [CrossRef]
46. Steinberg, R.L.; Thomas, L.J.; O'Donnell, M.A. Bacillus Calmette-Guérin (BCG) Treatment Failures in Non-Muscle Invasive Bladder Cancer: What Truly Constitutes Unresponsive Disease. *Bladder Cancer* **2015**, *1*, 105–116. [CrossRef]
47. Lamm, D.; Persad, R.; Brausi, M.; Buckley, R.; Witjes, J.A.; Palou, J.; Böhle, A.; Kamat, A.M.; Colombel, M.; Soloway, M. Defining Progression in Nonmuscle Invasive Bladder Cancer: It Is Time for a New, Standard Definition. *J. Urol.* **2014**, *191*, 20–27. [CrossRef]
48. Power, C.A.; Wei, G.; Bretscher, P.A. Mycobacterial Dose Defines the Th1/Th2 Nature of the Immune Response Independently of Whether Immunization Is Administered by the Intravenous, Subcutaneous, or Intradermal Route. *Infect. Immun.* **1998**, *66*, 5743–5750. [CrossRef]
49. Zlotta, A.R.; Fleshner, N.E.; Jewett, M.A. The Management of BCG Failure in Non-Muscle-Invasive Bladder Cancer: An Update. *Can. Urol. Assoc. J.* **2009**, *3* (Suppl. 4), S199–S205. [CrossRef]
50. Kates, M.; Matoso, A.; Choi, W.; Baras, A.S.; Daniels, M.J.; Lombardo, K.; Brant, A.; Mikkilineni, N.; McConkey, D.J.; Kamat, A.M.; et al. Adaptive Immune Resistance to Intravesical BCG in Non-Muscle Invasive Bladder Cancer: Implications for Prospective BCG Unresponsive Trials. *Clin. Cancer Res.* **2019**. [CrossRef]

51. Pembrolizumab and BCG Solution in Treating Patients With Recurrent Non-Muscle-Invasive Bladder Cancer—Full Text View—ClinicalTrials.gov. Available online: https://clinicaltrials.gov/ct2/show/NCT02808143 (accessed on 10 July 2020).
52. Ślusarczyk, A.; Zapała, P.; Zapała, Ł.; Piecha, T.; Radziszewski, P. Prediction of BCG Responses in Non-Muscle-Invasive Bladder Cancer in the Era of Novel Immunotherapeutics. *Int. Urol. Nephrol.* **2019**, *51*, 1089–1099. [CrossRef] [PubMed]
53. International Collaboration of Trialists on Behalf of the Medical Research Council Advanced Bladder Cancer Working Party; EORTC Genito-Urinary Group; Australian Bladder Cancer Study Group; National Cancer Institute of Canada Clinical Trials Group; Finnbladder, Norwegian Bladder Cancer Study Group; Club Urologico Espanol de Tratamiento Oncologico (CUETO) Group. Neoadjuvant Cisplatin, Methotrexate, and Vinblastine Chemotherapy for Muscle-Invasive Bladder Cancer: A Randomised Controlled Trial. *Lancet* **1999**, *354*, 533–540. [CrossRef]
54. International Collaboration of Trialists. International Phase III Trial Assessing Neoadjuvant Cisplatin, Methotrexate, and Vinblastine Chemotherapy for Muscle-Invasive Bladder Cancer: Long-Term Results of the BA06 30894 Trial. *J. Clin. Oncol.* **2011**, *29*, 2171–2177. [CrossRef] [PubMed]
55. Plimack, E.R.; Hoffman-Censits, J.H.; Viterbo, R.; Trabulsi, E.J.; Ross, E.A.; Greenberg, R.E.; Chen DY, T.; Lallas, C.D.; Wong, Y.-N.; Lin, J.; et al. Accelerated Methotrexate, Vinblastine, Doxorubicin, and Cisplatin Is Safe, Effective, and Efficient Neoadjuvant Treatment for Muscle-Invasive Bladder Cancer: Results of a Multicenter Phase II Study With Molecular Correlates of Response and Toxicity. *J. Clin. Oncol.* **2014**, *32*, 1895–1901. [CrossRef] [PubMed]
56. Choueiri, T.K.; Jacobus, S.; Bellmunt, J.; Qu, A.; Appleman, L.J.; Tretter, C.; Bubley, G.J.; Stack, E.C.; Signoretti, S.; Walsh, M.; et al. Neoadjuvant Dose-Dense Methotrexate, Vinblastine, Doxorubicin, and Cisplatin With Pegfilgrastim Support in Muscle-Invasive Urothelial Cancer: Pathologic, Radiologic, and Biomarker Correlates. *J. Clin. Oncol.* **2014**, *32*, 1889–1894. [CrossRef]
57. Shah, J.B.; McConkey, D.J.; Dinney, C.P.N. New Strategies in Muscle-Invasive Bladder Cancer: On the Road to Personalized Medicine. *Clin. Cancer Res.* **2011**, *17*, 2608–2612. [CrossRef]
58. Fahmy, N.M.; Mahmud, S.; Aprikian, A.G. Delay in the Surgical Treatment of Bladder Cancer and Survival: Systematic Review of the Literature. *Eur. Urol.* **2006**, *50*, 1176–1182. [CrossRef]
59. Kassouf, W.; Agarwal, P.K.; Grossman, H.B.; Leibovici, D.; Munsell, M.F.; Siefker-Radtke, A.; Pisters, L.L.; Swanson, D.A.; Dinney, C.P.N.; Kamat, A.M. Outcome of Patients with Bladder Cancer with PN+ Disease after Preoperative Chemotherapy and Radical Cystectomy. *Urology* **2009**, *73*, 147–152. [CrossRef]
60. Dalbagni, G.; Vora, K.; Kaag, M.; Cronin, A.; Bochner, B.; Donat, S.M.; Herr, H.W. Clinical Outcome in a Contemporary Series of Restaged Patients with Clinical T1 Bladder Cancer. *Eur. Urol.* **2009**, *56*, 903–910. [CrossRef]
61. McConkey, D.J.; Choi, W.; Shen, Y.; Lee, I.-L.; Porten, S.; Matin, S.F.; Kamat, A.M.; Corn, P.; Millikan, R.E.; Dinney, C.; et al. A Prognostic Gene Expression Signature in the Molecular Classification of Chemotherapy-Naïve Urothelial Cancer Is Predictive of Clinical Outcomes from Neoadjuvant Chemotherapy: A Phase 2 Trial of Dose-Dense Methotrexate, Vinblastine, Doxorubicin, and Cisplatin with Bevacizumab in Urothelial Cancer. *Eur. Urol.* **2016**, *69*, 855–862.
62. Seiler, R.; Ashab, H.A.D.; Erho, N.; van Rhijn, B.W.G.; Winters, B.; Douglas, J.; van Kessel, K.E.; Fransen van de Putte, E.E.; Sommerlad, M.; Wang, N.Q.; et al. Impact of Molecular Subtypes in Muscle-Invasive Bladder Cancer on Predicting Response and Survival after Neoadjuvant Chemotherapy. *Eur. Urol.* **2017**, *72*, 544–554. [CrossRef]
63. Mariathasan, S.; Turley, S.J.; Nickles, D.; Castiglioni, A.; Yuen, K.; Wang, Y.; Kadel, E.E.; Koeppen, H.; Astarita, J.L.; Cubas, R.; et al. TGF-β Attenuates Tumour Response to PD-L1 Blockade by Contributing to Exclusion of T Cells. *Nature* **2018**, *554*, 544–548. [CrossRef] [PubMed]
64. Shariat, S.F.; Palapattu, G.S.; Karakiewicz, P.I.; Rogers, C.G.; Vazina, A.; Bastian, P.J.; Schoenberg, M.P.; Lerner, S.P.; Sagalowsky, A.I.; Lotan, Y. Discrepancy between Clinical and Pathologic Stage: Impact on Prognosis after Radical Cystectomy. *Eur. Urol.* **2007**, *51*, 137–151. [CrossRef] [PubMed]
65. Van Rhijn, B.W.G.; van der Kwast, T.H.; Kakiashvili, D.M.; Fleshner, N.E.; van der Aa, M.N.M.; Alkhateeb, S.; Bangma, C.H.; Jewett, M.A.S.; Zlotta, A.R. Pathological Stage Review Is Indicated in Primary PT1 Bladder Cancer. *BJU Int.* **2010**, *106*, 206–211. [CrossRef] [PubMed]

66. Svatek, R.S.; Shariat, S.F.; Novara, G.; Skinner, E.C.; Fradet, Y.; Bastian, P.J.; Kamat, A.M.; Kassouf, W.; Karakiewicz, P.I.; Fritsche, H.; et al. Discrepancy between Clinical and Pathological Stage: External Validation of the Impact on Prognosis in an International Radical Cystectomy Cohort: Stage discrepancy in urothelial carcinoma of the bladder. *BJU Int.* **2011**, *107*, 898–904. [CrossRef]
67. Jones, P.A.; Droller, M.J. Pathways of Development and Progression in Bladder Cancer: New Correlations between Clinical Observations and Molecular Mechanisms. *Semin. Urol.* **1993**, *11*, 177–192.
68. Cancer Genome Atlas Research Network. Comprehensive Molecular Characterization of Urothelial Bladder Carcinoma. *Nature* **2014**, *507*, 315–322. [CrossRef]
69. Hedegaard, J.; Lamy, P.; Nordentoft, I.; Algaba, F.; Høyer, S.; Ulhøi, B.P.; Vang, S.; Reinert, T.; Hermann, G.G.; Mogensen, K.; et al. Comprehensive Transcriptional Analysis of Early-Stage Urothelial Carcinoma. *Cancer Cell* **2016**, *30*, 27–42. [CrossRef]
70. Pietzak, E.J.; Bagrodia, A.; Cha, E.K.; Drill, E.N.; Iyer, G.; Isharwal, S.; Ostrovnaya, I.; Baez, P.; Li, Q.; Berger, M.F.; et al. Next-Generation Sequencing of Nonmuscle Invasive Bladder Cancer Reveals Potential Biomarkers and Rational Therapeutic Targets. *Eur. Urol.* **2017**, *72*, 952–959. [CrossRef]
71. Sjödahl, G.; Lauss, M.; Lövgren, K.; Chebil, G.; Gudjonsson, S.; Veerla, S.; Patschan, O.; Aine, M.; Fernö, M.; Ringnér, M.; et al. A Molecular Taxonomy for Urothelial Carcinoma. *Clin. Cancer Res. Off. J. Am. Assoc. Cancer Res.* **2012**, *18*, 3377–3386. [CrossRef]
72. van Rhijn, B.W.G.; Lurkin, I.; Radvanyi, F.; Kirkels, W.J.; van der Kwast, T.H.; Zwarthoff, E.C. The Fibroblast Growth Factor Receptor 3 (FGFR3) Mutation Is a Strong Indicator of Superficial Bladder Cancer with Low Recurrence Rate. *Cancer Res.* **2001**, *61*, 1265–1268. [PubMed]
73. Zieger, K.; Marcussen, N.; Borre, M.; Ørntoft, T.F.; Dyrskjøt, L. Consistent Genomic Alterations in Carcinoma in Situ of the Urinary Bladder Confirm the Presence of Two Major Pathways in Bladder Cancer Development. *Int. J. Cancer* **2009**, *125*, 2095–2103. [CrossRef]
74. Billerey, C.; Chopin, D.; Aubriot-Lorton, M.H.; Ricol, D.; Gil Diez de Medina, S.; Van Rhijn, B.; Bralet, M.P.; Lefrere-Belda, M.A.; Lahaye, J.B.; Abbou, C.C.; et al. Frequent FGFR3 Mutations in Papillary Non-Invasive Bladder (PTa) Tumors. *Am. J. Pathol.* **2001**, *158*, 1955–1959. [CrossRef]
75. Hosen, I.; Rachakonda, P.S.; Heidenreich, B.; de Verdier, P.J.; Ryk, C.; Steineck, G.; Hemminki, K.; Kumar, R. Mutations in TERT Promoter and FGFR3 and Telomere Length in Bladder Cancer. *Int. J. Cancer* **2015**, *137*, 1621–1629. [CrossRef] [PubMed]
76. Hussain, T.A.; Akhtar, M. Molecular Basis of Urinary Bladder Cancer. *Adv. Anat. Pathol.* **2013**, *20*, 53–60. [CrossRef] [PubMed]
77. Jebar, A.H.; Hurst, C.D.; Tomlinson, D.C.; Johnston, C.; Taylor, C.F.; Knowles, M.A. FGFR3 and Ras Gene Mutations Are Mutually Exclusive Genetic Events in Urothelial Cell Carcinoma. *Oncogene* **2005**, *24*, 5218–5225. [CrossRef]
78. Hurst, C.D.; Knowles, M.A. Mutational Landscape of Non-Muscle-Invasive Bladder Cancer. *Urol. Oncol.* **2018**. [CrossRef]
79. Hurst, C.D.; Alder, O.; Platt, F.M.; Droop, A.; Stead, L.F.; Burns, J.E.; Burghel, G.J.; Jain, S.; Klimczak, L.J.; Lindsay, H.; et al. Genomic Subtypes of Non-Invasive Bladder Cancer with Distinct Metabolic Profile, Clinical Outcome and Female Gender Bias in KDM6A Mutation Frequency. *Cancer Cell* **2017**, *32*, 701–715. [CrossRef]
80. López-Knowles, E.; Hernández, S.; Malats, N.; Kogevinas, M.; Lloreta, J.; Carrato, A.; Tardón, A.; Serra, C.; Real, F.X. PIK3CA Mutations Are an Early Genetic Alteration Associated with FGFR3 Mutations in Superficial Papillary Bladder Tumors. *Cancer Res.* **2006**, *66*, 7401–7404. [CrossRef]
81. Platt, F.M.; Hurst, C.D.; Taylor, C.F.; Gregory, W.M.; Harnden, P.; Knowles, M.A. Spectrum of Phosphatidylinositol 3-Kinase Pathway Gene Alterations in Bladder Cancer. *Clin. Cancer Res.* **2009**, *15*, 6008–6017. [CrossRef]
82. Meeks, J.J.; Carneiro, B.A.; Pai, S.G.; Oberlin, D.T.; Rademaker, A.; Fedorchak, K.; Balasubramanian, S.; Elvin, J.; Beaubier, N.; Giles, F.J. Genomic Characterization of High-Risk Non-Muscle Invasive Bladder Cancer. *Oncotarget* **2016**, *7*, 75176–75184. [CrossRef] [PubMed]
83. Martinez, V.G.; Munera-Maravilla, E.; Bernardini, A.; Rubio, C.; Suarez-Cabrera, C.; Segovia, C.; Lodewijk, I.; Dueñas, M.; Martínez-Fernández, M.; Paramio, J.M. Epigenetics of Bladder Cancer: Where Biomarkers and Therapeutic Targets Meet. *Front. Genet.* **2019**, *10*. [CrossRef] [PubMed]

84. Ramakrishnan, S.; Granger, V.; Rak, M.; Hu, Q.; Attwood, K.; Aquila, L.; Krishnan, N.; Osiecki, R.; Azabdaftari, G.; Guru, K.; et al. Inhibition of EZH2 Induces NK Cell-Mediated Differentiation and Death in Muscle-Invasive Bladder Cancer. *Cell Death Differ.* **2019**, *26*, 2100–2114. [CrossRef] [PubMed]
85. Solomon, D.A.; Kim, J.-S.; Bondaruk, J.; Shariat, S.F.; Wang, Z.-F.; Elkahloun, A.G.; Ozawa, T.; Gerard, J.; Zhuang, D.; Zhang, S.; et al. Frequent Truncating Mutations of STAG2 in Bladder Cancer. *Nat. Genet.* **2013**, *45*, 1428–1430. [CrossRef]
86. Aboulkassim, T.O.; LaRue, H.; Lemieux, P.; Rousseau, F.; Fradet, Y. Alteration of the PATCHED Locus in Superficial Bladder Cancer. *Oncogene* **2003**, *22*, 2967–2971. [CrossRef]
87. Knowles, M.A.; Habuchi, T.; Kennedy, W.; Cuthbert-Heavens, D. Mutation Spectrum of the 9q34 Tuberous Sclerosis Gene TSC1 in Transitional Cell Carcinoma of the Bladder. *Cancer Res.* **2003**, *63*, 7652–7656.
88. Blaveri, E.; Simko, J.P.; Korkola, J.E.; Brewer, J.L.; Baehner, F.; Mehta, K.; DeVries, S.; Koppie, T.; Pejavar, S.; Carroll, P.; et al. Bladder Cancer Outcome and Subtype Classification by Gene Expression. *Clin. Cancer Res.* **2005**, *11*, 4044–4055. [CrossRef]
89. Williamson, M.P.; Elder, P.A.; Shaw, M.E.; Devlin, J.; Knowles, M.A. P16 (CDKN2) Is a Major Deletion Target at 9p21 in Bladder Cancer. *Hum. Mol. Genet.* **1995**, *4*, 1569–1577. [CrossRef]
90. Bartoletti, R.; Cai, T.; Nesi, G.; Roberta Girardi, L.; Baroni, G.; Dal Canto, M. Loss of P16 Expression and Chromosome 9p21 LOH in Predicting Outcome of Patients Affected by Superficial Bladder Cancer. *J. Surg. Res.* **2007**, *143*, 422–427. [CrossRef]
91. Rebouissou, S.; Bernard-Pierrot, I.; de Reyniès, A.; Lepage, M.-L.; Krucker, C.; Chapeaublanc, E.; Hérault, A.; Kamoun, A.; Caillault, A.; Letouzé, E.; et al. EGFR as a Potential Therapeutic Target for a Subset of Muscle-Invasive Bladder Cancers Presenting a Basal-like Phenotype. *Sci. Transl. Med.* **2014**, *6*, 244ra91. [CrossRef]
92. Robertson, A.G.; Kim, J.; Al-Ahmadie, H.; Bellmunt, J.; Guo, G.; Cherniack, A.D.; Hinoue, T.; Laird, P.W.; Hoadley, K.A.; Akbani, R.; et al. Comprehensive Molecular Characterization of Muscle-Invasive Bladder Cancer. *Cell* **2017**, *171*, 540–556.e25. [CrossRef] [PubMed]
93. Baud, E.; Catilina, P.; Bignon, Y.J. P16 Involvement in Primary Bladder Tumors: Analysis of Deletions and Mutations. *Int. J. Oncol.* **1999**, *14*, 441–446. [CrossRef] [PubMed]
94. Morrison, C.D.; Liu, P.; Woloszynska-Read, A.; Zhang, J.; Luo, W.; Qin, M.; Bshara, W.; Conroy, J.M.; Sabatini, L.; Vedell, P.; et al. Whole-Genome Sequencing Identifies Genomic Heterogeneity at a Nucleotide and Chromosomal Level in Bladder Cancer. *Proc. Natl. Acad. Sci. USA* **2014**, *111*, E672–E681. [CrossRef] [PubMed]
95. Shariat, S.F.; Tokunaga, H.; Zhou, J.; Kim, J.; Ayala, G.E.; Benedict, W.F.; Lerner, S.P. P53, P21, PRB, and P16 Expression Predict Clinical Outcome in Cystectomy With Bladder Cancer. *J. Clin. Oncol.* **2004**, *22*, 1014–1024. [CrossRef] [PubMed]
96. Kamoun, A.; de Reyniès, A.; Allory, Y.; Sjödahl, G.; Robertson, A.G.; Seiler, R.; Hoadley, K.A.; Groeneveld, C.S.; Al-Ahmadie, H.; Choi, W.; et al. A Consensus Molecular Classification of Muscle-Invasive Bladder Cancer. *Eur. Urol.* **2019**. [CrossRef]
97. Ross, J.S.; Wang, K.; Al-Rohil, R.N.; Nazeer, T.; Sheehan, C.E.; Otto, G.A.; He, J.; Palmer, G.; Yelensky, R.; Lipson, D.; et al. Advanced Urothelial Carcinoma: Next-Generation Sequencing Reveals Diverse Genomic Alterations and Targets of Therapy. *Mod. Pathol. Off. J. U. S. Can. Acad. Pathol. Inc.* **2014**, *27*, 271–280. [CrossRef]
98. Sjödahl, G.; Eriksson, P.; Liedberg, F.; Höglund, M. Molecular Classification of Urothelial Carcinoma: Global MRNA Classification versus Tumour-Cell Phenotype Classification. *J. Pathol.* **2017**, *242*, 113–125. [CrossRef]
99. Cordon-Cardo, C.; Zhang, Z.-F.; Dalbagni, G.; Drobnjak, M.; Charytonowicz, E.; Hu, S.-X.; Xu, H.-J.; Reuter, V.E.; Benedict, W.F. Cooperative Effects of P53 and PRB Alterations in Primary Superficial Bladder Tumors. *Cancer Res.* **1997**, *57*, 1217–1221.
100. Grossman, H.B.; Liebert, M.; Antelo, M.; Dinney, C.P.; Hu, S.X.; Palmer, J.L.; Benedict, W.F. P53 and RB Expression Predict Progression in T1 Bladder Cancer. *Clin. Cancer Res.* **1998**, *4*, 829–834.
101. Sherr, C.J.; McCormick, F. The RB and P53 Pathways in Cancer. *Cancer Cell* **2002**, *2*, 103–112. [CrossRef]
102. Esteller, M. Epigenetics in Cancer. *N. Engl. J. Med.* **2008**, *358*, 1148–1159. [CrossRef] [PubMed]
103. Kim, J.; Mouw, K.W.; Polak, P.; Braunstein, L.Z.; Kamburov, A.; Kwiatkowski, D.J.; Rosenberg, J.E.; Van Allen, E.M.; D'Andrea, A.; Getz, G. Somatic ERCC2 Mutations Are Associated with a Distinct Genomic Signature in Urothelial Tumors. *Nat. Genet.* **2016**, *48*, 600–606. [CrossRef] [PubMed]

104. Van Allen, E.M.; Mouw, K.W.; Kim, P.; Iyer, G.; Wagle, N.; Al-Ahmadie, H.; Zhu, C.; Ostrovnaya, I.; Kryukov, G.V.; O'Connor, K.W.; et al. Somatic ERCC2 Mutations Correlate with Cisplatin Sensitivity in Muscle-Invasive Urothelial Carcinoma. *Cancer Discov.* **2014**, *4*, 1140–1153. [CrossRef] [PubMed]
105. Guo, G.; Sun, X.; Chen, C.; Wu, S.; Huang, P.; Li, Z.; Dean, M.; Huang, Y.; Jia, W.; Zhou, Q.; et al. Whole-Genome and Whole-Exome Sequencing of Bladder Cancer Identifies Frequent Alterations in Genes Involved in Sister Chromatid Cohesion and Segregation. *Nat. Genet.* **2013**, *45*, 1459–1463. [CrossRef]
106. Carlsson, J.; Wester, K.; De La Torre, M.; Malmström, P.-U.; Gårdmark, T. EGFR-Expression in Primary Urinary Bladder Cancer and Corresponding Metastases and the Relation to HER2-Expression. On the Possibility to Target These Receptors with Radionuclides. *Radiol. Oncol.* **2015**, *49*, 50–58. [CrossRef]
107. Gildea, J.J.; Harding, M.A.; Seraj, M.J.; Gulding, K.M.; Theodorescu, D. The Role of Ral A in Epidermal Growth Factor Receptor-Regulated Cell Motility. *Cancer Res.* **2002**, *62*, 982–985.
108. Seshacharyulu, P.; Ponnusamy, M.P.; Haridas, D.; Jain, M.; Ganti, A.p.a.r.K.; Batra, S.K. Targeting the EGFR Signaling Pathway in Cancer Therapy. *Expert Opin. Ther. Targets* **2012**, *16*, 15–31. [CrossRef]
109. Theodorescu, D.; Laderoute, K.R.; Gulding, K.M. Epidermal Growth Factor Receptor-Regulated Human Bladder Cancer Motility Is in Part a Phosphatidylinositol 3-Kinase-Mediated Process. *Cell Growth Differ.* **1998**, *9*, 919.
110. Memon, A.A.; Sorensen, B.S.; Melgard, P.; Fokdal, L.; Thykjaer, T.; Nexo, E. Expression of HER3, HER4 and Their Ligand Heregulin-4 Is Associated with Better Survival in Bladder Cancer Patients. *Br. J. Cancer* **2004**, *91*, 2034–2041. [CrossRef]
111. Coogan, C.L.; Estrada, C.R.; Kapur, S.; Bloom, K.J. HER-2/Neu Protein Overexpression and Gene Amplification in Human Transitional Cell Carcinoma of the Bladder. *Urology* **2004**, *63*, 786–790. [CrossRef]
112. Maida, F.D.; Mari, A.; Gesolfo, C.S.; Cangemi, A.; Allegro, R.; Sforza, S.; Cocci, A.; Tellini, R.; Masieri, L.; Russo, A.; et al. Epidermal Growth Factor Receptor (EGFR) Cell Expression During Adjuvant Treatment After Transurethral Resection for Non-Muscle-Invasive Bladder Cancer: A New Potential Tool to Identify Patients at Higher Risk of Disease Progression. *Clin. Genitourin. Cancer* **2019**, *17*, e751–e758. [CrossRef] [PubMed]
113. Fleischmann, A.; Rotzer, D.; Seiler, R.; Studer, U.E.; Thalmann, G.N. Her2 Amplification Is Significantly More Frequent in Lymph Node Metastases from Urothelial Bladder Cancer than in the Primary Tumours. *Eur. Urol.* **2011**, *60*, 350–357. [CrossRef] [PubMed]
114. Garcia, J.A.; Danielpour, D. MTOR Inhibition as a Therapeutic Strategy in the Management of Urologic Malignancies. *Mol. Cancer Ther.* **2008**, *7*, 1347–1354. [CrossRef] [PubMed]
115. Gildea, J.J.; Herlevsen, M.; Harding, M.A.; Gulding, K.M.; Moskaluk, C.A.; Frierson, H.F.; Theodorescu, D. PTEN Can Inhibit in Vitro Organotypic and in Vivo Orthotopic Invasion of Human Bladder Cancer Cells Even in the Absence of Its Lipid Phosphatase Activity. *Oncogene* **2004**, *23*, 6788–6797. [CrossRef] [PubMed]
116. Aveyard, J.S.; Skilleter, A.; Habuchi, T.; Knowles, M.A. Somatic Mutation of PTEN in Bladder Carcinoma. *Br. J. Cancer* **1999**, *80*, 904–908. [CrossRef]
117. Puzio-Kuter, A.M.; Castillo-Martin, M.; Kinkade, C.W.; Wang, X.; Shen, T.H.; Matos, T.; Shen, M.M.; Cordon-Cardo, C.; Abate-Shen, C. Inactivation of P53 and Pten Promotes Invasive Bladder Cancer. *Genes Dev.* **2009**, *23*, 675–680. [CrossRef] [PubMed]
118. Mo, Q.; Nikolos, F.; Chen, F.; Tramel, Z.; Lee, Y.-C.; Hayashi, K.; Xiao, J.; Shen, J.; Chan, K.S. Prognostic Power of a Tumor Differentiation Gene Signature for Bladder Urothelial Carcinomas. *JNCI J. Natl. Cancer Inst.* **2018**, *110*, 448–459. [CrossRef]
119. Choi, W.; Czerniak, B.; Ochoa, A.; Su, X.; Siefker-Radtke, A.; Dinney, C.; McConkey, D.J. Intrinsic Basal and Luminal Subtypes of Muscle-Invasive Bladder Cancer. *Nat. Rev. Urol.* **2014**, *11*, 400–410. [CrossRef]
120. Choi, W.; Porten, S.; Kim, S.; Willis, D.; Plimack, E.R.; Hoffman-Censits, J.; Roth, B.; Cheng, T.; Tran, M.; Lee, I.-L.; et al. Identification of Distinct Basal and Luminal Subtypes of Muscle-Invasive Bladder Cancer with Different Sensitivities to Frontline Chemotherapy. *Cancer Cell* **2014**, *25*, 152–165. [CrossRef]
121. Damrauer, J.S.; Hoadley, K.A.; Chism, D.D.; Fan, C.; Tiganelli, C.J.; Wobker, S.E.; Yeh, J.J.; Milowsky, M.I.; Iyer, G.; Parker, J.S.; et al. Intrinsic Subtypes of High-Grade Bladder Cancer Reflect the Hallmarks of Breast Cancer Biology. *Proc. Natl. Acad. Sci. USA* **2014**, *111*, 3110–3115. [CrossRef]
122. Kardos, J.; Chai, S.; Mose, L.E.; Selitsky, S.R.; Krishnan, B.; Saito, R.; Iglesia, M.D.; Milowsky, M.I.; Parker, J.S.; Kim, W.Y.; et al. Claudin-Low Bladder Tumors Are Immune Infiltrated and Actively Immune Suppressed. *JCI Insight 1* **2016**, *1*, e85902. [CrossRef] [PubMed]

123. Marzouka, N.; Eriksson, P.; Rovira, C.; Liedberg, F.; Sjödahl, G.; Höglund, M. A Validation and Extended Description of the Lund Taxonomy for Urothelial Carcinoma Using the TCGA Cohort. *Sci. Rep.* **2018**, *8*, 1–12. [CrossRef] [PubMed]
124. Nag, S.; Qin, J.; Srivenugopal, K.S.; Wang, M.; Zhang, R. The MDM2-P53 Pathway Revisited. *J. Biomed. Res.* **2013**, *27*, 254–271. [PubMed]
125. He, F.; Mo, L.; Zheng, X.-Y.; Hu, C.; Lepor, H.; Lee, E.Y.-H.P.; Sun, T.-T.; Wu, X.-R. Deficiency of PRb Family Proteins and P53 in Invasive Urothelial Tumorigenesis. *Cancer Res.* **2009**, *69*, 9413–9421. [CrossRef] [PubMed]
126. McConkey, D.J.; Choi, W.; Marquis, L.; Martin, F.; Williams, M.B.; Shah, J.; Svatek, R.; Das, A.; Adam, L.; Kamat, A.; et al. Role of Epithelial-to-Mesenchymal Transition (EMT) in Drug Sensitivity and Metastasis in Bladder Cancer. *Cancer Metastasis Rev.* **2009**, *28*, 335–344. [CrossRef] [PubMed]
127. Lo, H.-W.; Hsu, S.-C.; Xia, W.; Cao, X.; Shih, J.-Y.; Wei, Y.; Abbruzzese, J.L.; Hortobagyi, G.N.; Hung, M.-C. Epidermal Growth Factor Receptor Cooperates with Signal Transducer and Activator of Transcription 3 to Induce Epithelial-Mesenchymal Transition in Cancer Cells via Up-Regulation of TWIST Gene Expression. *Cancer Res.* **2007**, *67*, 9066–9076. [CrossRef]
128. Kurzrock, E.A.; Lieu, D.K.; deGraffenried, L.A.; Chan, C.W.; Isseroff, R.R. Label-Retaining Cells of the Bladder: Candidate Urothelial Stem Cells. *Am. J. Physiol.-Ren. Physiol.* **2008**, *294*, F1415–F1421. [CrossRef]
129. Choi, W.; Shah, J.B.; Tran, M.; Svatek, R.; Marquis, L.; Lee, I.-L.; Yu, D.; Adam, L.; Wen, S.; Shen, Y.; et al. P63 Expression Defines a Lethal Subset of Muscle-Invasive Bladder Cancers. *PLoS ONE* **2012**, *7*. [CrossRef]
130. Karni-Schmidt, O.; Castillo-Martin, M.; HuaiShen, T.; Gladoun, N.; Domingo-Domenech, J.; Sanchez-Carbayo, M.; Li, Y.; Lowe, S.; Prives, C.; Cordon-Cardo, C. Distinct Expression Profiles of P63 Variants during Urothelial Development and Bladder Cancer Progression. *Am. J. Pathol.* **2011**, *178*, 1350–1360. [CrossRef]
131. Chu, W.-K.; Dai, P.-M.; Li, H.-L.; Chen, J.-K. Transcriptional Activity of the ΔNp63 Promoter Is Regulated by STAT3. *J. Biol. Chem.* **2008**, *283*, 7328–7337. [CrossRef]
132. Kollberg, P.; Chebil, G.; Eriksson, P.; Sjödahl, G.; Liedberg, F. Molecular Subtypes Applied to a Population-Based Modern Cystectomy Series Do Not Predict Cancer-Specific Survival. *Urol. Oncol. Semin. Orig. Investig.* **2019**, *37*, 791–799. [CrossRef] [PubMed]
133. van Rhijn BW, G.; van der Kwast, T.H.; Vis, A.N.; Kirkels, W.J.; Boevé, E.R.; Jöbsis, A.C.; Zwarthoff, E.C. FGFR3 and P53 Characterize Alternative Genetic Pathways in the Pathogenesis of Urothelial Cell Carcinoma. *Cancer Res.* **2004**, *64*, 1911–1914. [CrossRef] [PubMed]
134. Tomlinson, D.C.; Baldo, O.; Harnden, P.; Knowles, M.A. FGFR3 Protein Expression and Its Relationship to Mutation Status and Prognostic Variables in Bladder Cancer. *J. Pathol.* **2007**, *213*, 91–98. [CrossRef] [PubMed]
135. van Rhijn Bas, W.G.; van der Kwast Theo, H.; Liu, L.; Fleshner Neil, E.; Bostrom Peter, J.; Vis André, N.; Alkhateeb Sultan, S.; Bangma Chris, H.; Jewett Michael, A.S.; Zwarthoff Ellen, C.; et al. The FGFR3 Mutation Is Related to Favorable PT1 Bladder Cancer. *J. Urol.* **2012**, *187*, 310–314. [CrossRef] [PubMed]
136. Jackson, J.G.; Pant, V.; Li, Q.; Chang, L.L.; Quintás-Cardama, A.; Garza, D.; Tavana, O.; Yang, P.; Manshouri, T.; Li, Y.; et al. P53 Mediated Senescence Impairs the Apoptotic Response to Chemotherapy and Clinical Outcome in Breast Cancer. *Cancer Cell* **2012**, *21*, 793–806. [CrossRef]
137. López-Knowles, E.; Hernández, S.; Kogevinas, M.; Lloreta, J.; Amorós, A.; Tardón, A.; Carrato, A.; Kishore, S.; Serra, C.; Malats, N.; et al. The P53 Pathway and Outcome among Patients with T1G3 Bladder Tumors. *Clin. Cancer Res.* **2006**, *12*, 6029–6036. [CrossRef]
138. Wang, L.; Smith, B.A.; Balanis, N.G.; Tsai, B.L.; Nguyen, K.; Cheng, M.W.; Obusan, M.B.; Esedebe, F.N.; Patel, S.J.; Zhang, H.; et al. A Genetically Defined Disease Model Reveals That Urothelial Cells Can Initiate Divergent Bladder Cancer Phenotypes. *Proc. Natl. Acad. Sci. USA* **2020**, *117*, 563–572. [CrossRef]
139. Ghervan, L.; Zaharie, A.; Ene, B.; Elec, F.I. Small-Cell Carcinoma of the Urinary Bladder: Where Do We Stand? *Clujul Med.* **2017**, *90*, 13–17. [CrossRef]
140. Chang, M.T.; Penson, A.; Desai, N.B.; Socci, N.D.; Shen, R.; Seshan, V.E.; Kundra, R.; Abeshouse, A.; Viale, A.; Cha, E.K.; et al. Small Cell Carcinomas of the Bladder and Lung Are Characterized by a Convergent but Distinct Pathogenesis. *Clin. Cancer Res. Off. J. Am. Assoc. Cancer Res.* **2018**, *24*, 1965–1973. [CrossRef]
141. Nadal, R.; Schweizer, M.; Kryvenko, O.N.; Epstein, J.I.; Eisenberger, M.A. Small Cell Carcinoma of the Prostate. *Nat. Rev. Urol.* **2014**, *11*, 213–219. [CrossRef]

142. Seiler, R.; Gibb, E.A.; Wang, N.Q.; Oo, H.Z.; Lam, H.-M.; van Kessel, K.E.; Voskuilen, C.S.; Winters, B.; Erho, N.; Takhar, M.M.; et al. Divergent Biological Response to Neoadjuvant Chemotherapy in Muscle-Invasive Bladder Cancer. *Clin. Cancer Res. Off. J. Am. Assoc. Cancer Res.* **2019**, *25*, 5082–5093. [CrossRef] [PubMed]
143. Yang, L.; Taylor, J.; Eustace, A.; Irlam, J.J.; Denley, H.; Hoskin, P.J.; Alsner, J.; Buffa, F.M.; Harris, A.L.; Choudhury, A. Gene Signature for Selecting Benefit from Hypoxia Modification of Radiotherapy for High-Risk Bladder Cancer Patients. *Clin. Cancer Res.* **2017**, *23*, 4761–4768. [CrossRef] [PubMed]
144. Efstathiou, J.A.; Mouw, K.W.; Gibb, E.A.; Liu, Y.; Wu, C.-L.; Drumm, M.R.; da Costa, J.B.; du Plessis, M.; Wang, N.Q.; Davicioni, E.; et al. Impact of Immune and Stromal Infiltration on Outcomes Following Bladder-Sparing Trimodality Therapy for Muscle-Invasive Bladder Cancer. *Eur. Urol.* **2019**, *76*, 59–68. [CrossRef] [PubMed]
145. Necchi, A.; Raggi, D.; Gallina, A.; Ross, J.S.; Farè, E.; Giannatempo, P.; Marandino, L.; Colecchia, M.; Lucianò, R.; Bianchi, M.; et al. Impact of Molecular Subtyping and Immune Infiltration on Pathological Response and Outcome Following Neoadjuvant Pembrolizumab in Muscle-Invasive Bladder Cancer. *Eur. Urol.* **2020**, *77*, 701–710. [CrossRef] [PubMed]
146. Kim, J.; Kwiatkowski, D.; McConkey, D.J.; Meeks, J.J.; Freeman, S.S.; Bellmunt, J.; Getz, G.; Lerner, S.P. The Cancer Genome Atlas Expression Subtypes Stratify Response to Checkpoint Inhibition in Advanced Urothelial Cancer and Identify a Subset of Patients with High Survival Probability. *Eur. Urol.* **2019**, *75*, 961–964. [CrossRef]
147. Dutta, P.R.; Maity, A. Cellular Responses to EGFR Inhibitors and Their Relevance to Cancer Therapy. *Cancer Lett.* **2007**, *254*, 165–177. [CrossRef]
148. Powles, T.; Huddart, R.A.; Elliott, T.; Sarker, S.-J.; Ackerman, C.; Jones, R.; Hussain, S.; Crabb, S.; Jagdev, S.; Chester, J.; et al. Phase III, Double-Blind, Randomized Trial That Compared Maintenance Lapatinib Versus Placebo After First-Line Chemotherapy in Patients With Human Epidermal Growth Factor Receptor 1/2–Positive Metastatic Bladder Cancer. *J. Clin. Oncol.* **2016**, *35*, 48–55. [CrossRef]
149. Hussain, M.; Daignault, S.; Agarwal, N.; Grivas, P.D.; Siefker-Radtke, A.O.; Puzanov, I.; MacVicar, G.R.; Levine, E.G.; Srinivas, S.; Twardowski, P.; et al. Randomized Phase 2 Trial of Gemcitabine/Cisplatin with or without Cetuximab in Patients with Advanced Urothelial Carcinoma. *Cancer* **2014**, *120*, 2684–2693. [CrossRef]
150. Stephenson, J.J.; Gregory, C.; Burris, H.; Larson, T.; Verma, U.; Cohn, A.; Crawford, J.; Cohen, R.B.; Martin, J.; Lum, P.; et al. An Open-Label Clinical Trial Evaluating Safety and Pharmacokinetics of Two Dosing Schedules of Panitumumab in Patients with Solid Tumors. *Clin. Colorectal Cancer* **2009**, *8*, 29–37. [CrossRef]
151. Wong, Y.-N.; Litwin, S.; Vaughn, D.; Cohen, S.; Plimack, E.R.; Lee, J.; Song, W.; Dabrow, M.; Brody, M.; Tuttle, H.; et al. Phase II Trial of Cetuximab With or Without Paclitaxel in Patients With Advanced Urothelial Tract Carcinoma. *J. Clin. Oncol.* **2012**, *30*, 3545–3551. [CrossRef]
152. Pruthi, R.S.; Nielsen, M.; Heathcote, S.; Wallen, E.M.; Rathmell, W.K.; Godley, P.; Whang, Y.; Fielding, J.; Schultz, H.; Grigson, G.; et al. A Phase II Trial of Neoadjuvant Erlotinib in Patients with Muscle-Invasive Bladder Cancer Undergoing Radical Cystectomy: Clinical and Pathological Results. *BJU Int.* **2010**, *106*, 349–354. [CrossRef] [PubMed]
153. Petrylak, D.P.; Tangen, C.M.; Van Veldhuizen Jr, P.J.; Goodwin, J.W.; Twardowski, P.W.; Atkins, J.N.; Kakhil, S.R.; Lange, M.K.; Mansukhani, M.; Crawford, E.D. Results of the Southwest Oncology Group Phase II Evaluation (Study S0031) of ZD1839 for Advanced Transitional Cell Carcinoma of the Urothelium. *BJU Int.* **2010**, *105*, 317–321. [CrossRef] [PubMed]
154. Boehringer Ingelheim. *LUX-Bladder 1: Phase II Open Label Single Arm Exploratory Trial of Oral Afatinib Monotherapy Following Platinum Failure for Patients with Advanced/Metastatic Urothelial Tract Carcinoma with Genetic Alterations in ERBB Receptors.*; Boehringer Ingelheim: Ingelheim am Rhein, Germany, 2019; Clinical trial registration NCT02780687; clinicaltrials.gov.
155. Huang, L.; Fu, L. Mechanisms of Resistance to EGFR Tyrosine Kinase Inhibitors. *Acta Pharm. Sin. B* **2015**, *5*, 390–401. [CrossRef] [PubMed]
156. Douillard, J.-Y.; Oliner, K.S.; Siena, S.; Tabernero, J.; Burkes, R.; Barugel, M.; Humblet, Y.; Bodoky, G.; Cunningham, D.; Jassem, J.; et al. Panitumumab–FOLFOX4 Treatment and RAS Mutations in Colorectal Cancer. Available online: https://www.nejm.org/doi/10.1056/NEJMoa1305275?url_ver=Z39.88-2003&rfr_id=ori%3Arid%3Acrossref.org&rfr_dat=cr_pub%3Dwww.ncbi.nlm.nih.gov (accessed on 16 April 2020).

157. Oudard, S.; Culine, S.; Vano, Y.; Goldwasser, F.; Théodore, C.; Nguyen, T.; Voog, E.; Banu, E.; Vieillefond, A.; Priou, F.; et al. Multicentre Randomised Phase II Trial of Gemcitabine+platinum, with or without Trastuzumab, in Advanced or Metastatic Urothelial Carcinoma Overexpressing Her2. *Eur. J. Cancer* **2015**, *51*, 45–54. [CrossRef] [PubMed]
158. Kiss, B.; Wyatt, A.W.; Douglas, J.; Skuginna, V.; Mo, F.; Anderson, S.; Rotzer, D.; Fleischmann, A.; Genitsch, V.; Hayashi, T.; et al. Her2 Alterations in Muscle-Invasive Bladder Cancer: Patient Selection beyond Protein Expression for Targeted Therapy. *Sci. Rep.* **2017**, *7*. [CrossRef] [PubMed]
159. Koshkin, V.S.; O'Donnell, P.; Yu, E.Y.; Grivas, P. Systematic Review: Targeting HER2 in Bladder Cancer. *Bladder Cancer* **2019**, *5*, 1–12. [CrossRef]
160. Thomsen MB, H.; Nordentoft, I.; Lamy, P.; Vang, S.; Reinert, L.; Mapendano, C.K.; Høyer, S.; Ørntoft, T.F.; Jensen, J.B.; Dyrskjøt, L. Comprehensive Multiregional Analysis of Molecular Heterogeneity in Bladder Cancer. *Sci. Rep.* **2017**, *7*. [CrossRef]
161. Genitsch, V.; Kollár, A.; Vandekerkhove, G.; Blarer, J.; Furrer, M.; Annala, M.; Herberts, C.; Pycha, A.; de Jong, J.J.; Liu, Y.; et al. Morphologic and Genomic Characterization of Urothelial to Sarcomatoid Transition in Muscle-Invasive Bladder Cancer†. *Urol. Oncol. Semin. Orig. Investig.* **2019**, *37*, 826–836. [CrossRef]
162. Sjödahl, G.; Jackson, C.L.; Bartlett, J.M.; Siemens, D.R.; Berman, D.M. Molecular Profiling in Muscle-Invasive Bladder Cancer: More than the Sum of Its Parts. *J. Pathol.* **2019**, *247*, 563–573. [CrossRef]

International Journal of *Molecular Sciences*

MDPI

Article

A Peptidic Thymidylate-Synthase Inhibitor Loaded on Pegylated Liposomes Enhances the Antitumour Effect of Chemotherapy Drugs in Human Ovarian Cancer Cells

Gaetano Marverti [1,*], Gaia Gozzi [1], Eleonora Maretti [2], Angela Lauriola [1,†], Leda Severi [2], Francesca Sacchetti [2], Lorena Losi [2], Salvatore Pacifico [3], Stefania Ferrari [2], Glauco Ponterini [2], Eliana Leo [2], Maria Paola Costi [2,*] and Domenico D'Arca [1,*]

[1] Department of Biomedical, Metabolic and Neural Sciences, Via G. Campi 287, University of Modena and Reggio Emilia, 41125 Modena, Italy; g.gozzi@holostem.com (G.G.); angela.lauriola@univr.it (A.L.)

[2] Department of Life Sciences, Via G. Campi 213/d, University of Modena and Reggio Emilia, 41125 Modena, Italy; eleonora.maretti@unimore.it (E.M.); leda.severi@gmail.com (L.S.); francesca.sacchetti@alice.it (F.S.); lorena.losi@unimore.it (L.L.); sferrari591@gmail.com (S.F.); glauco.ponterini@unimore.it (G.P.); elianagrazia.leo@unimore.it (E.L.)

[3] Department of Chemical and Pharmaceutical Sciences, via Fossato di Mortara 17–19, University of Ferrara, 44100 Ferrara, Italy; salvatore.pacifico@unife.it

* Correspondence: gaetano.marverti@unimore.it (G.M.); mariapaola.costi@unimore.it (M.P.C.); domenico.darca@unimore.it (D.D.)

† Present address: Department of Biotechnology, University of Verona, 37134 Verona, Italy.

Received: 30 May 2020; Accepted: 20 June 2020; Published: 23 June 2020

Abstract: There is currently no effective long-term treatment for ovarian cancer (OC) resistant to poly-chemotherapy regimens based on platinum drugs. Preclinical and clinical studies have demonstrated a strong association between development of Pt-drug resistance and increased thymidylate synthase (hTS) expression, and the consequent cross-resistance to the hTS inhibitors 5-fluorouracil (5-FU) and raltitrexed (RTX). In the present work, we propose a new tool to combat drug resistance. We propose to treat OC cell lines, both Pt-sensitive and -resistant, with dual combinations of one of the four chemotherapeutic agents that are widely used in the clinic, and the new peptide, hTS inhibitor, [D-Gln^4]LR. This binds hTS allosterically and, unlike classical inhibitors that bind at the catalytic pocket, causes cell growth inhibition without inducing hTS overexpression. The dual drug combinations showed schedule-dependent synergistic antiproliferative and apoptotic effects. We observed that the simultaneous treatment or 24h pre-treatment of OC cells with the peptide followed by either agent produced synergistic effects even in resistant cells. Similar synergistic or antagonistic effects were obtained by delivering the peptide into OC cells either by means of a commercial delivery system (SAINT-PhD) or by pH sensitive PEGylated liposomes. Relative to non-PEGylated liposomes, the latter had been previously characterized and found to allow macrophage escape, thus increasing their chance to reach the tumour tissue. The transition from the SAINT-PhD delivery system to the engineered liposomes represents an advancement towards a more drug-like delivery system and a further step towards the use of peptides for in vivo studies. Overall, the results suggest that the association of standard drugs, such as cDDP and/or 5-FU and/or RTX, with the novel peptidic TS inhibitor encapsulated into PEGylated pH-sensitive liposomes can represent a promising strategy for fighting resistance to cDDP and anti-hTS drugs.

Keywords: human thymidylate synthase peptidic-inhibitors; pH-sensitive PEGylated liposomes; ovarian cancer; drug-resistance; raltitrexed; 5-fluorouracil

---

1. Introduction

Ovarian cancer is the fifth cause of mortality among women of all ages in the Western world. Although higher response rates have been achieved using the poly-chemotherapy regimens, mainly based on the combination of cyclophosphamide or paclitaxel and a platinum compound, such as cisplatin (cDDP) or carboplatin [1], many patients still die during cancer relapse and progression due to the intrinsic or acquired resistance to chemotherapy as well as the genetic flexibility of the cancer cell's genome, resulting in multiple and often compensatory survival and proliferative signals, limiting the activity of anti-cancer strategies and compelling the continuous search for new drug combinations.

Cisplatin is an alkylating agent that interacts with the DNA double strain structure and causes formation of adducts that prevent transcription and cellular replication and induces apoptotic cell death. In the multifactorial process of the acquired resistance to cDDP and its derivatives, the phenotypes show overexpression of DNA-repair enzymes and enzymes necessary for the synthesis of thymidine, such as thymidylate synthase (TS) and dihydrofolate reductase (DHFR), since they cause a rapid turnover of DNA [2]. As a consequence, cells resistant to cDDP are also cross-resistant to the traditional antifolates such as 5-fluorouracil (5-FU) and methotrexate (MTX) [3].

It is well known that tumours express higher TS levels than normal tissues and that the ovarian tissue is among those with the highest levels of TS expression, a feature that correlates with poorer patients' overall survival [4]. 5-Fluorouracil and other antifolates, such as raltitrexed (RTX), inhibit TS, but they may cause TS upregulation in cDDP-resistant cells. The active metabolite of 5-FU, 5-fluorodeoxyuridine monophosphate (5-FdUMP), inhibits the dTMP synthesis by forming a stable ternary complex with TS and the methyl donor 5,10-methylene tetrahydrofolate (5,10-CH2THF) [5,6]. On the other hand, RTX is a TS cofactor analogue that inhibits TS with response rates similar to those of 5-FU and, differently from the pyrimidine analogues, is not incorporated into DNA. It has been licensed in many countries for the treatment of metastatic colorectal cancer [7,8]. Preclinical and clinical studies have demonstrated a strong association between development of resistance to both 5-FU and RTX and increased TS expression [9–11].

Exposure of cancer cells to 5-FU or other antifolate TS inhibitors acutely upregulates TS, probably due to the inhibition of a negative-feedback mechanism in which the TS protein binds its own mRNA and inhibits its translation [7,12,13]. In addition, TS behaves as an oncogene [14] by interfering with the expression of proteins involved in the regulation of proliferative and survival pathways such as c-myc and p53 by binding their mRNAs [15,16]. However, expression of wild-type p53 has been shown to be required for 5-FU- and RTX-induced antitumor effects [17,18]. Therefore, new strategies helpful to overcoming these TS-focused molecular mechanisms of chemo-resistance are required.

As one such strategy to obtain tumour cell death without inducing TS overexpression, we designed oligopeptides able to modulate the equilibrium between the active TS conformation and the inactive one. The novel mechanism of action involves binding at the monomer-monomer interface of the enzyme, rather than at the catalytic pocket, with stabilization of the inactive conformation and decrease of the abundance of the active form of the protein. Among these, the LR (LSCQLYQR) peptide inhibited the growth of ovarian carcinomas (OCs) following transfection by means of a commercial peptide delivery system, SAINT-PhD, while it left the TS cellular levels essentially unchanged [19,20]. More recently, among the peptides developed by modifying the LR lead to improve TS inhibition and the anticancer effect, the D-glutamine-modified peptide at position 4 ([D-Gln^4]LR) displayed the best growth inhibition of both cDDP-sensitive and -resistant OC cell lines [21]; it was more active than LR and 5-FU and affected the TS/DHFR expression pattern similarly to LR.

Proteomic studies have shown that, in human OC cell lines, these peptides modulate the expression of a panel of six proteins, including TS and other important folate-related enzymes such as phosphoribosylglycinamide formyltransferase (GART) and serine hydroxymethyltransferase (SHMT1) [22]. The modulation of these proteins was markedly different from that induced by pemetrexed (PMX), another TS inhibitor and a 5,10-CH2THF analogue, which, again, unlike the peptide, binds at the TS active site.

The possibility of combining conventional cytotoxic drugs with new agents that specifically interfere with key pathways controlling cancer cell survival and proliferation is considered an interesting and promising therapeutic approach. Indeed, if the cellular targets for these new agents and/or their mechanism of action are different from those of conventional cytotoxic drugs, and if the effects of the drugs combined on key pathways are concordant, their combination may act on the sensitivity of cancer cells synergistically.

Recently, the hydrophilic peptide LR has been delivered into OC cells by exploiting such biocompatible and efficient tools as solid lipid nanoparticles (SLNs) [23] and pH sensitive liposomes [24]. The ability to inhibit the cDDP-resistant OC cell growth demonstrated the efficacy of these nanosystems in the internalization of the peptide into cells while preserving its activity. Enhanced effects are expected when employing PEGylated liposomes. In fact, PEGylation represents a very useful method for achieving long circulation time for liposomes, hence allowing them more time for targeting tumours via the enhanced permeability and retention (EPR) effect [25]. The PEGylated pH-sensitive liposomes were characterized and compared favourably with non-PEGylated ones regarding surface hydrophilicity, stability in serum, interaction with macrophages, drug encapsulation efficiency and drug release [26]. Also, the well-characterized intracellular mechanism of action of the [D-Gln^4]LR peptide delivered using the SAINT-PhD delivery system was unaltered when the latter was replaced by these liposomes. While the commercial delivery system proved useful in the early cellular studies, the PEGylated liposomes strategy is deemed necessary to implement further in vivo studies of the [D-Gln^4]LR peptide.

In the present study, we test the efficacy of combinations of RTX, cDDP and 5-FU with the ([D-Gln^4]LR octapeptide, delivered into OC cells by either the SAINT-PhD delivery system or by PEGylated pH-sensitive liposomes. More specifically, we investigate their cytotoxicities and abilities to perturbate the cell cycle phase distribution. The microtubule polymer stabilizer paclitaxel was also included in the described combination list, because the drug is largely used in ovarian cancer chemotherapy and shows a different mechanism of action. To this aim, three different tumour-cell treatment schedules were tested. Special attention was focused on the synergistic effects exhibited by the combined cell treatments and the role of sequentially thereon.

2. Results

2.1. Sequence-Dependent Synergistic Antiproliferative Effects of [DGln^4]LR/Drug Combinations

In cancer cells treated with 5-FU or RTX, alterations in TS expression were associated with drug resistance response [7,12–14]. Despite the fact that these compounds and peptides share the same target (TS), they have different molecular mechanisms of action and, thus, are not a priori reciprocally competitive. Therefore, we investigated the potential cooperative antitumor effect of the [DGln^4]LR peptide in combination with these drugs, as well as with cDDP and paclitaxel, two other chemotherapeutic agents used for the treatment of OC patients. We used cDDP-sensitive A2780 and 2008 cell lines and their cDDP-resistant counterparts, A2780/CP and C13*, to evaluate the combined effect on cell proliferation of [DGln^4]LR with either 5-FU, RTX, cDDP or paclitaxel by the synergism Quotient (SQ) analysis. Among the different cell lines examined the IC_{50} values for the drugs alone varied depending on the resistance phenotype (Table S1).

The [D-Gln^4]LR peptide was delivered by the SAINT-PhD system to A2780 and A2780/CP cells either concurrently or sequentially with the four drugs, each administered at concentrations lower than the IC_{50} value (IC30) obtained with the drug alone on the tested cell line. As shown in the diagrams in Figure 1, and synthesized by the heat map in Figure 4, most of the results of the concurrent administration (sequence I) demonstrate additive or moderately supra-additive effects, as indicated by SQ values around or slightly higher than 1, even in the cDDP-resistant A2780/CP cell line. On the other hand, administration of the peptide followed by the drug (sequence II) caused an overall increase in cell killing. This was evident with both cell lines in the treatment with the lower doses of all four drugs, while, at the higher doses, only 5-FU showed an improved performance with sequence II over the concurrent administration.

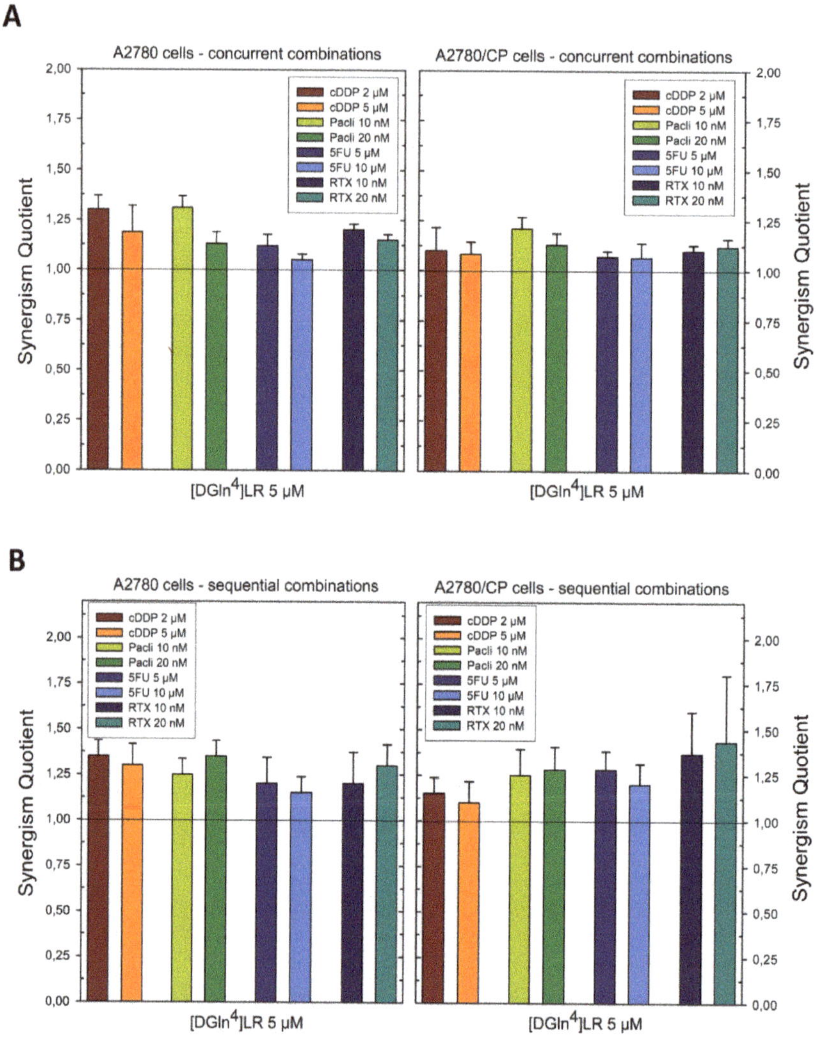

Figure 1. Effects of scheduled combinations of the [D-Gln4]LR peptide with cDDP, paclitaxel, 5FU and RTX on the SQ values in A2780 and A2780/CP cell lines. (**A**) concurrent combinations for 72 h. (**B**) Sequential combinations, as described in Section 4. The bars represent the mean of duplicate cell counts on three separate experiments and indicate the result of the inhibition of drug combination divided by the sum of the inhibition of a single drug to obtain the values of SQ. Error bars, SD.

With the second pair of cell lines, 2008 and C13*, concurrent administration was additive for cDDP, 5-FU, and paclitaxel at the lower dose, supra-additive at the higher dose of paclitaxel and for RTX at both doses (Figure 2 and Figure 4). Sequential exposure to [D-Gln^4]LR and then to the drugs did not enhance cell killing with the cDDP-sensitive 2008 cells with respect to the concurrent administration. On the contrary, with the resistant C13* cells both 5-FU and RTX increased their cytotoxicities when administered after [D-Gln^4]LR; in particular, RTX exhibited a synergistic efficacy at the lower dose. A general enhancement of cytotoxicity, with SQ values higher than unity, was obtained with both combined treatments of IGROV-1 cells, at almost all drug doses employed (Figures 3 and 4). Advantages of the sequential treatment according to schedule II over the concurrent treatments were demonstrated by supra-additive or synergistic SQ values obtained with 5-FU and RTX using this schedule, in particular at the lower concentrations tested. Interestingly, this scheduled combination with RTX 10 nM reached a synergistic SQ value of about 1.7. On the other hand, when drug treatment preceded exposure to the peptide (sequence III) the anti-proliferative results were antagonistic with all drugs and cell lines tested (Figure 4). As summarized in Figure 4, the IGROV-1 and A2780 cell lines show similar overall response profiles to the tested combinations. In fact, almost all the combinations, except 5-FU at low dose administered after [D-Gln^4]LR peptide, showed synergistic activity. On the other hand, all the combinations in which the [D-Gln^4]LR peptide was administered after the drugs showed antagonism or, in a few cases, additive activities. Regarding the tested combinations: (i) RTX-high dose in concurrent administration with the [D-Gln^4]LR peptide has a synergistic activity towards all the tested cell lines; (ii) low-dose RTX and high-dose 5-FU administered after the [D-Gln^4]LR peptide show synergistic activity towards all cell lines, except the 2008 cells; (iii) high-dose RTX and paclitaxel in the concurrent administration and high-dose paclitaxel administered after the [D-Gln^4]LR peptide showed synergistic activity towards all cell lines, except C13*.

Taken together these results show that scheduling may be crucial for potentiating the antitumor effect of [D-Gln^4]LR. In particular, its combination with 5-FU and RTX, i.e., drugs that target folate cycle enzymes, but also with cDDP and paclitaxel, shows enhanced effectiveness when the drugs are administered after the peptide.

2.2. *[DGln^4]LR Combination with Chemotherapy Drugs Cause Great Perturbation of Cell Cycle and Promotes Apoptosis*

Looking for a possible mechanism underlying the antiproliferative activity and correlating with the additive or synergistic effects of the combination between [DGln^4]LR with the chemotherapy drugs cDDP, 5-FU and RTX, a perturbation of the cell distribution in the different phases of the cell cycle was tested in cytofluorimetric experiments (DNA content analysis) on the 2008 and C13* cell lines. The percentage of cells in the different phases of the cell cycle is reported in Figures 5 and 6, Table S2 and Figures S1–S3.

Figure 5 and Table S2 report the results obtained with 2008 and C13* cell lines treated with 5 μM peptide and two pairs of 5-FU concentrations (5–10 μM and 10–20 μM), in sensitive and resistant cells, respectively. After 72 h, untreated cells of both lines showed a normal diploid distribution presenting fast proliferation characteristics. 5-FU, both alone and in combination, had a very scant effect in perturbing the distribution of the C13*-resistant cells in the different phases of cell cycle, even if a synergistic but modest accumulation of cells in sub-G1 phase was observed with 20 μM 5-FU, a finding indicative of apoptotic cell death (hypodiploid cells).

On the contrary, 2008 cells were more sensitive to both concentrations of 5-FU (5–10 μM) showing a reduction of cell accumulation in the G0/G1 phase, coupled with an increase of cell accumulation in both S and sub-G1 phases. The increase of sub-G1 phases (apoptotic cells) was almost doubled by combination with the peptide, reaching approximately 20% of synergistic cell killing (SQ = 1.38-1.33).

The other two chemotherapy drugs, cDDP and RTX caused a great decrease of sensitive cell accumulation in the G0/G1 phase that paralleled with a remarkable increase of percentage of the sub-G1 phase cell population (apoptotic cells), The combination of the peptide with each chemotherapy drug produced a synergistic accumulation of hypodiploid 2008 cells in comparison to single drug treatment (SQ = 1.1-1.2). This effect was also accompanied by a decrease of cell accumulation in all phases of cell cycle (Figure 6, Figures S1 and S2, Table S3). Notably, treatment with each drug alone deranged the cell-cycle phase distribution even in the resistant C13* cells, in which the percentage of the sub-G1 phase population was synergistically increased by combining the peptide with cDDP (SQ = 1.7) and with RTX (SQ = 1.13). RTX alone was also very effective as an apoptosis-inducing agent, particularly in sensitive cells.

Interestingly, an additive and a synergistic accumulation of cell population in the sub-G1 phase was also observed in the sensitive and resistant cells, respectively, already after 48 hr treatment with combinations of the peptide with either cDDP or RTX (Table S3 and Figure S3).

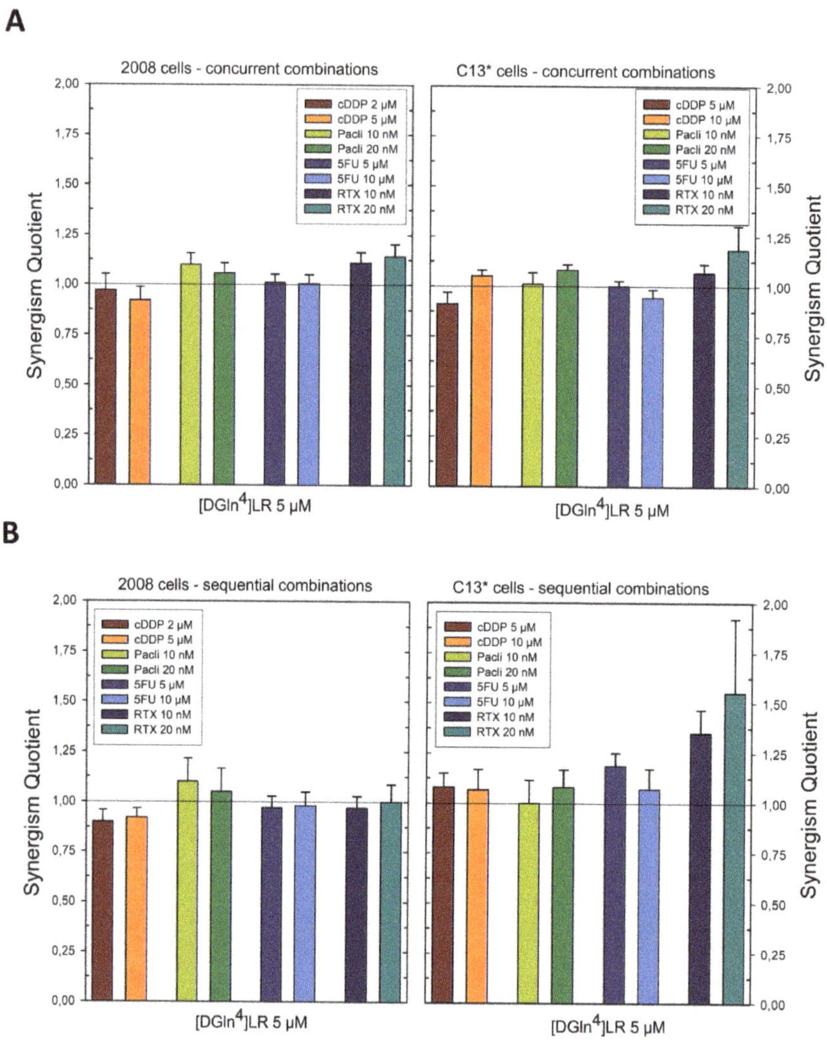

Figure 2. Effects of scheduled combinations of [DGln4]LR peptide with cDDP, paclitaxel, 5FU and RTX on the SQ values in 2008 and C13* cell lines. (**A**) concurrent combinations for 72 h. (**B**) Sequential combinations as described in Section 4. The bars represent the mean of duplicate cell counts on three separate experiments and indicate the results of the inhibition of drug combinations divided by the sum of the inhibition of a single drug to obtain the values of SQ. Error bars, SD.

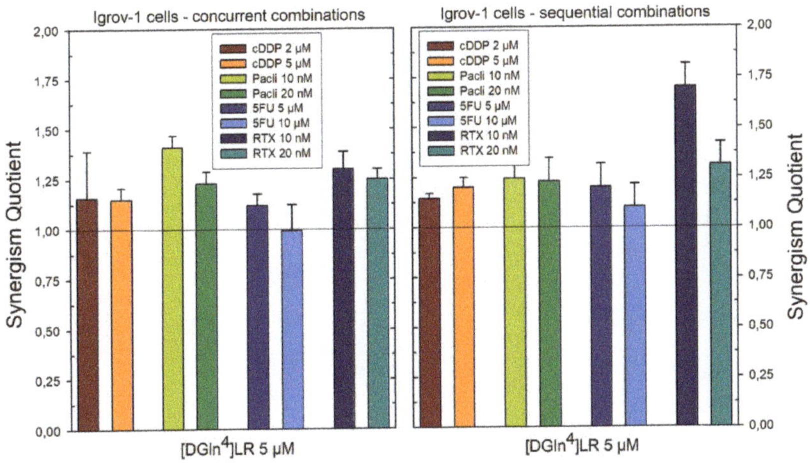

Figure 3. Effects of scheduled combinations of [DGln4]LR peptide with cDDP, paclitaxel, 5FU and RTX on the SQ values in IGROV-1 cell line. (Left panel) Concurrent combinations for 72 hr. (Right panel) Sequential combinations as described in Section 4. The bars represent the mean of duplicate cell counts on three separate experiments and indicate the results of the inhibition of drug combinations divided by the sum of the inhibition of a single drug to obtain the values of SQ. Error bars, SD.

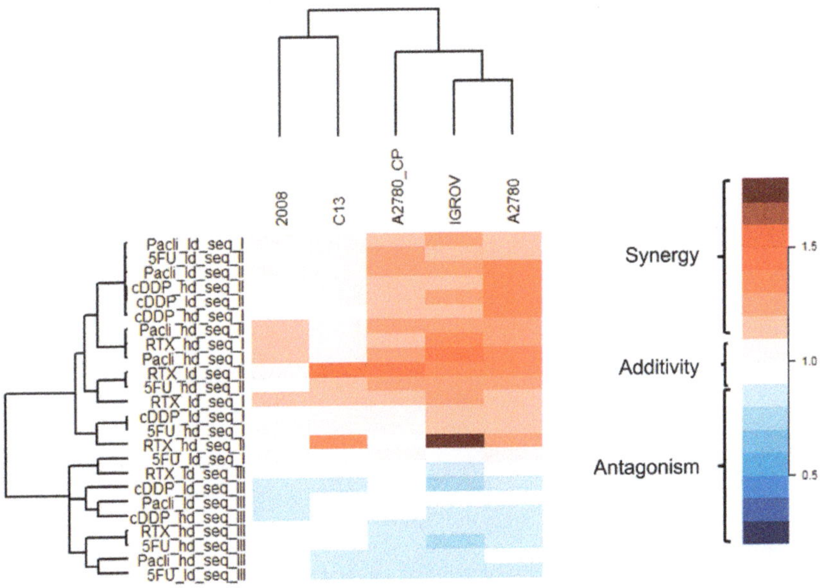

Figure 4. Heatmap of the synergism quotient values of the tested combinations (rows) against the different cell lines (columns). Colour code: red, SQ > 1; blue, SQ < 1. The reported dendrogram was built based on the dissimilarity matrix using Euclidean distances and the complete linkage method. ld: lower dose; hd: higher dose; Pacli: paclitaxel, Seq I, II, III: combination sequences I, II and III as described in the text.

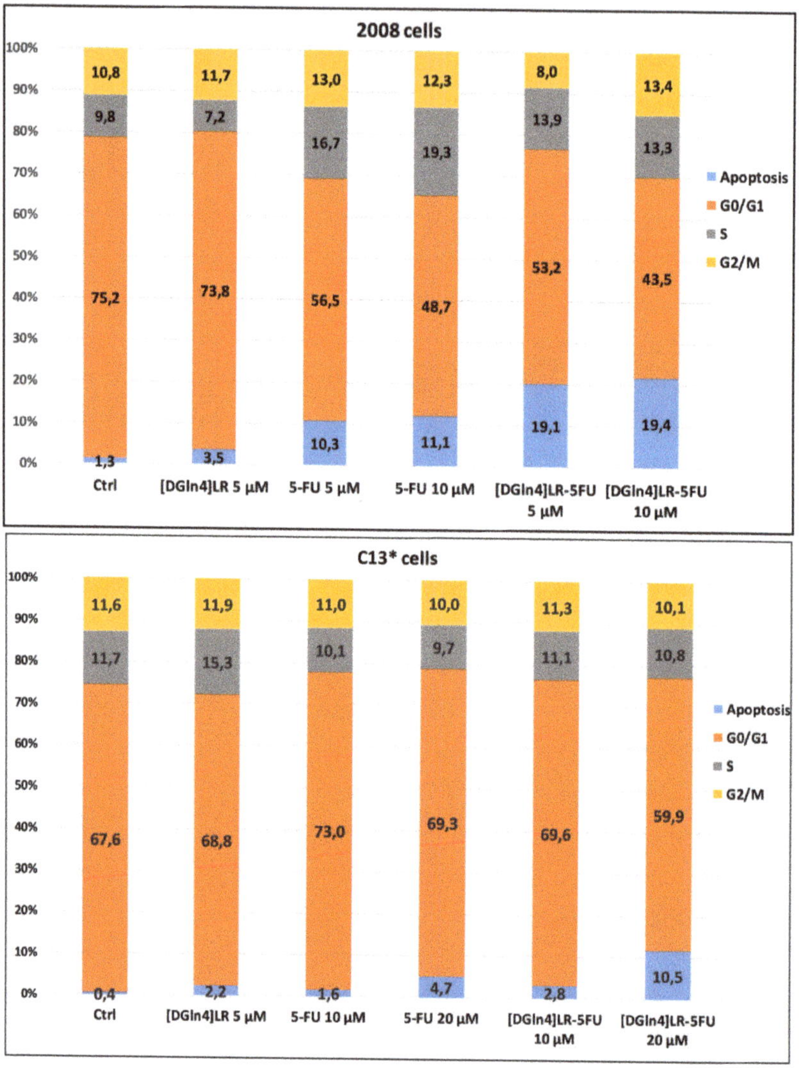

Figure 5. Effect of the [DGln4]LR peptide and 5-FU alone and in combination on the cell cycle phase distribution of 2008 and C13* cells by cytofluorimetric analysis of the DNA content by PI staining. After a 72 exposure to 5 µM [DGln4]LR and to the indicated concentrations of 5-FU alone or in concurrent combinations. Cells were processed according to methods described in Section 4. The inserted numbers indicate the percentages of cells in the different phases of the cell cycle and are the mean of two/three experiments. The error bars are omitted for a clearer visualization.

2.3. *Liposome Characterization*

Next, we move on, and pass from the SAINT-PhD delivery system to the engineered PEGylated liposomes, which represent an advancement towards a more drug-like delivery system. The PEGylated liposomes strategy is deemed necessary to implement further in vivo studies of the [D-Gln^4]LR peptide in combination with standard drugs.

Liposomes showed a size of about 200 nm with a good dimensional homogeneity, as shown by the low polydispersity index (PDI) value (Table 1). Moreover, the PDI did not show any significant alteration after drug addition, indicating that neither the homogeneity nor the stability of the liposomes were affected by the inclusion of the drug. The same occurred with the surface charge.

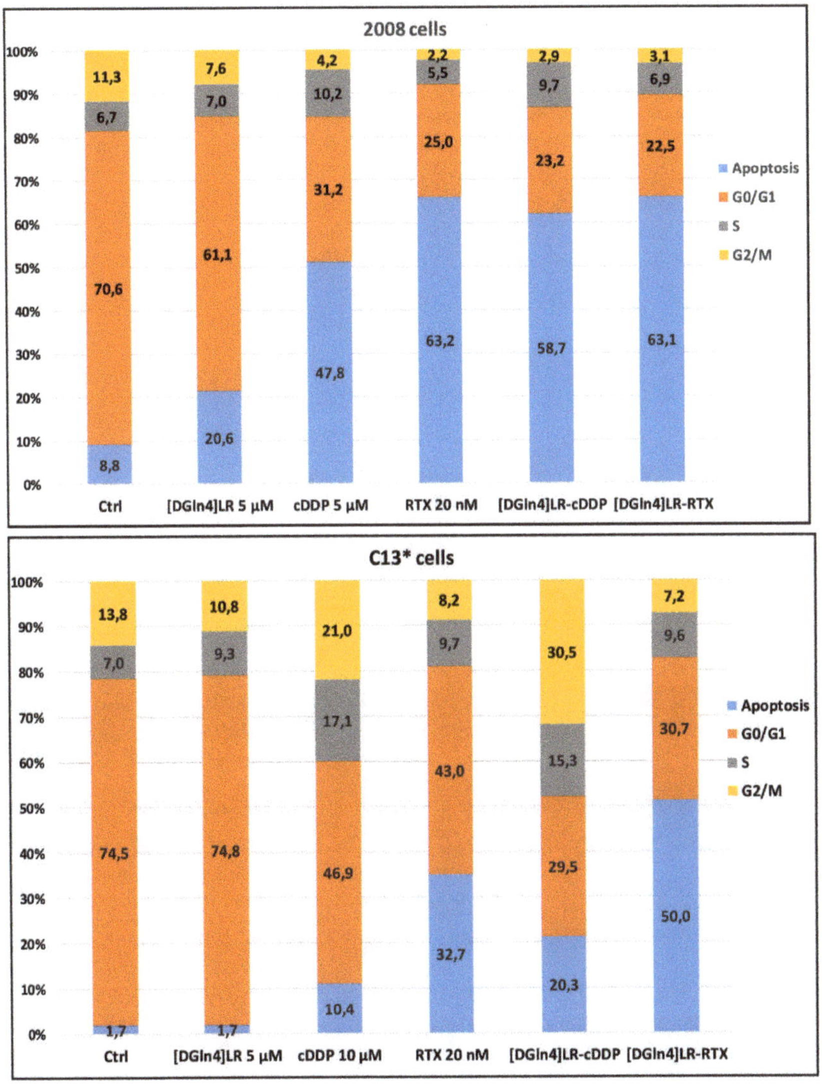

Figure 6. Effect of the [DGln4]LR peptide and cDDP alone and in combination on the cell cycle phase distribution of 2008 and C13* cells by cytofluorimetric analysis of the DNA content by PI staining. After a 72 exposure to 5 µM [DGln4]LR and 5 µM (2008 cells) or 10 µM (C13* cells) cDDP or 20 nM RTX alone and in concurrent combinations, cells were processed according to Section 4. The inserted numbers indicate the percentages of cells in the different phases of the cell cycle and are the mean of two/three experiments. The error bars were omitted for a clearer visualization.

Table 1. Physicochemical characterization of the optimized unloaded and [D-Gln4]LR-loaded liposomes. Each value represents the mean ± SD; PDI: polydispersity index; DL: drug loading; EE: encapsulation efficiency.

Sample	Z-Average (nm)	PDI	Z-Potential (mV)	DL (µg/mg)	EE (%)
DOPE:CHEMS:DSPE_PEG (PpHL)	198 ± 30	0.279 ± 0.077	−14.54 ± 2.41	-	-
[D-Gln⁴]LR_DOPE:CHEMS:DSPE_PEG ([D-Gln⁴]LR-PpHL)	209 ± 19	0.289 ± 0.072	−14.28 ± 2.19	21.03 ± 1.66	42 ± 3

The peptide was efficiently encapsulated in the liposomes. The drug loading (DL) was 21.03 ± 1.66 µg peptide/mg of lipid and the encapsulation efficiency approximately 40%. Consistently with the two-fold drug/lipid ratio used in the preparation, this formulation exhibited a DL twice that of the already reported homologue formulation [26]. The obtained loaded liposomes efficiently retained the peptide: after an 8 h incubation, only 30% of the initially encapsulated peptide was released and no further escape was observed in the next 16 h (Figure 7).

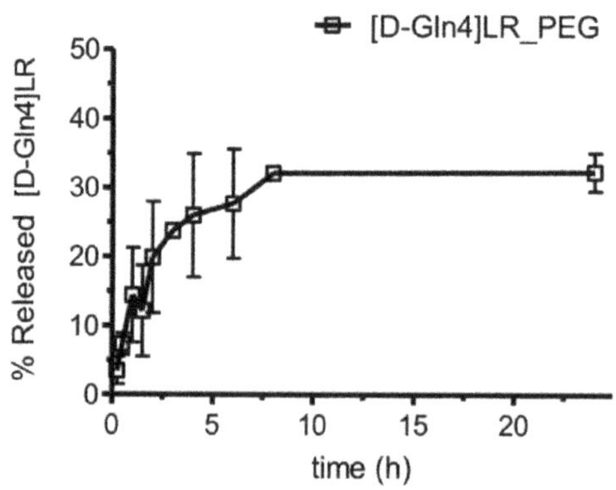

Figure 7. Time course of the [D-Gln4]LR release percentage from [D-Gln4]LR-PEGylated liposomes in phosphate buffer saline (PBS; pH 7.4). Error bars, SD; where not visible, error bars did not exceed symbol size.

2.4. *Cytotoxicity of Peptide-Loaded Liposomes*

The antiproliferative activity of the peptide encapsulated in these optimized liposomes (PpHL) was tested by the MTT assay on three OC cell lines, the cDDP-resistant C13* cells and their sensitive counterparts, the 2008, and the IGROV-1 cell lines, the latter sensitive to cDDP too.

In the first experiment, the cells were exposed to increasing concentrations of the peptide-loaded liposomes ([D-Gln^4]LR-PpHL) according to the optimised protocol reported, using the unloaded carriers as a control. Moreover, the free [D-Gln^4]LR peptide at the equivalent concentration was used for comparison. The elaborated results are reported in Figure 8. The unloaded liposomes (PpHL) were not toxic towards all cell lines since cell viabilities higher than 85% were observed for all the liposome concentrations and cells considered (Figure 8A).

Concerning the effect of [D-Gln^4]LR-PpHL, a marked difference in cytotoxicity between loaded and unloaded carriers was observed, particularly with the cDDP-sensitive 2008 cell line. Indeed, at the higher liposome concentration, 0.25 mg/mL, a 50% viability was obtained with these cells, while a slightly higher survival, 63%, was exhibited by the cDDP-resistant C13* cells. On the other hand, IGROV-1 cells, despite their cDDP-sensitivity, proved quite resistant to the peptide-loaded liposomes, exhibiting cell viabilities about 80% with all the amounts of liposomes employed. It should be noticed that IGROV1 cells are known to exhibit a different behaviour to drug treatment. Despite being sensitive to cisplatin in vitro, they are resistant to Asta Z, and present an intermediate drug response to adriamycin. Finally, concerning the effect of drug loading, its doubling caused only a moderate increase in cytotoxicity on the C13* cells, with a 63% survival vs a 70% survival measured with the original preparation [26], a finding likely due to the saturation of the intracellular target enzyme, hTS.

The importance of these liposomes as vehicles for the peptide internalization into cells was confirmed by the inability of the free [D-Gln[4]]LR peptide to interfere with the growth of all three cell lines [26].

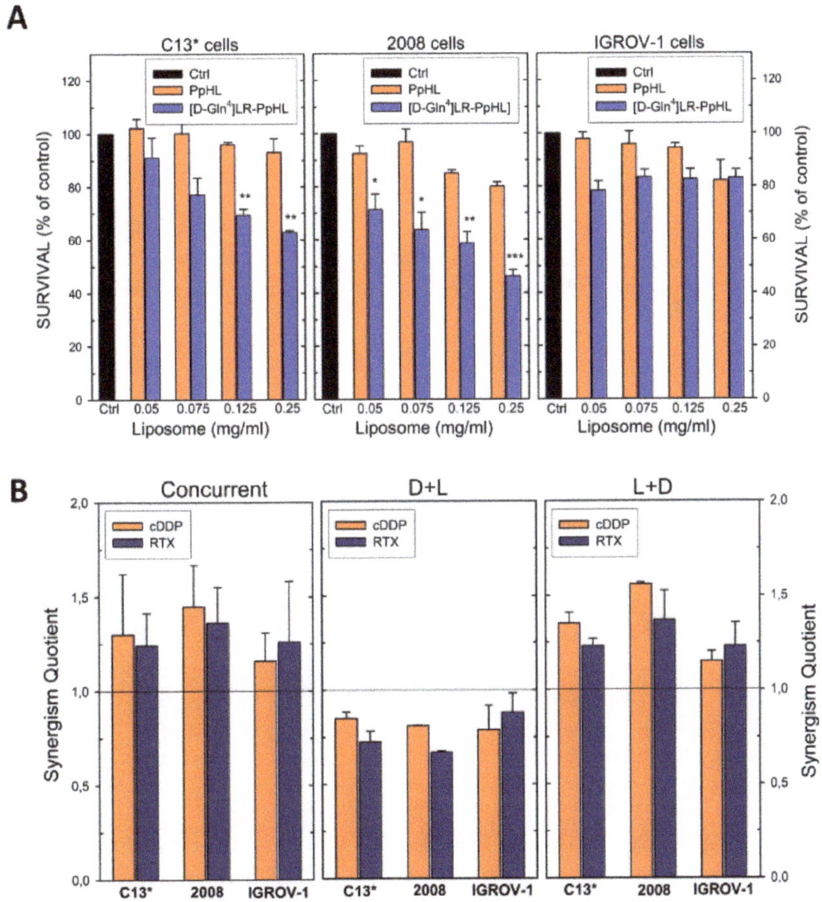

Figure 8. MTT test on C13*, 2008 and IGROV-1 cell lines. Cells were incubated with increasing amounts of [D-Gln4]LR-PpHL and PpHL (**A**) for 15 h, followed by a further 48-h incubation after the removal of the treatment. The results were expressed as percentage of cell growth with respect to the control (untreated cells), set at 100% of viability. Error bars, SD. Comparison between groups was performed by the ANOVA one-way test. Statistical significance levels were defined as * $p < 0.05$, ** $p < 0.01$ and *** $p < 0.005$. (**B**) The synergism of cell growth inhibition is reported as synergism quotient (SQ). The Concurrent chart corresponds to simultaneous liposomes + drug-exposure; D+L chart corresponds to sequential exposure in which the drugs (cDDP or RTX) was given 24 h before liposomes; L+D chart corresponds to the reversed sequential exposure. Error bars, SD.

2.5. Sequence-Dependent Synergistic Antiproliferative Effect of Peptide-Loaded Liposomes in Combination with RTX or cDDP

The peptide-loaded liposomes [D-Gln^4]LR-PpHL at a concentration of 0.125 mg/mL, corresponding to 2.12 μM overall extracellular peptide concentration, was combined with RTX and cDDP at different concentrations, according to the cell line sensitivities to these drugs, 10 nM RTX and 5 μM cDDP for C13*, 10 nM RTX and 2.5 μM cDDP for IGROV-1 and 2008 cells, respectively. Peptide-loaded liposomes combined with the two anticancer drugs showed greater efficacy against both cDDP-sensitive and -resistant cell lines when administered concurrently or sequentially (liposome, L+drug, D) (sequences I and II, respectively), while the reversed schedule (D+L, sequence III) produced an antagonistic effect; the combination sequences leading to the synergistic antiproliferative effect are the same observed with the SAINT-PhD delivery system (Figures 1–4). Noteworthy, sequences I and II synergistically killed even IGROV-1 cells, i.e., the least responsive to the peptide-loaded liposomes alone (Figure 8A). The SQ values obtained are shown in Figure 8B.

3. Discussion

The [D-Gln4]LR peptide and its lead, LR, have exhibited cancer cell-growth inhibitory activity by mainly reducing the abundance of the active form of hTS, and, unlike 5-FU and PMX, without inducing overexpression of the enzyme [19,21], but even by down-modulating the expression of other folate pathway genes, DHFR and AICAR transformylase (ATIC) [22]. Cells that acquire resistance to classical TS inhibitors because of an enhanced TS expression exhibit general cross-resistance with platinum-based drugs [3] and display cross-resistance to antifolates such as RTX [27,28].

Antifolates targeting hTS are not well known in OC therapy. All those inhibitors bind at the protein active site and this cause the loss of the translational control and TS levels regulations [29]. Our hypothesis was that if TS levels are reduced, drug resistance mechanisms will be limited or prevented. So, we propose a change of paradigm in TS inhibition based on new drugs that, unlike the well-known, traditional TS inhibitors (RTX, PMX, 5FU), bind at the protein interface. [22,30–32]. These compounds can be combined with platinum drugs, RTX, PMX or 5FU to keep TS levels low and contrast drug resistance development.

Owing to the intricate biochemical interconnections and the difference in biological pathway modulation between the peptides and the classical drugs that, in order to explore strategies to overcome tumour drug resistance and achieve greater therapeutic gains, the present in vitro study has investigated the effects of combining the novel TS inhibitor peptides with cDDP [1], as well as some classical antifolate agents. To provide a mechanistic interpretation of the results of our combination experiments, we first start from known individual drug effects as well as some drug-combination literature reports.

Combination of cDDP with classical folate cycle inhibitors, such as 5-FU, has been explored in experimental and clinical studies and has exhibited a pronounced activity against various types of human tumours, including chemo-resistant ovarian carcinoma [33,34]. However, the response and toxicity varied considerably according to the schedule and dose used [35,36]. The explanation of the synergistic response attained with 5-FU plus cDDP likely involves more than one mechanism. On one hand, it was correlated to increased TS inhibition related to higher levels of FdUMP, the active 5-FU metabolite. This, in turn, was associated with an increase in the intracellular reduced folate levels in the tumour cells by inhibition of the transport of L-methionine by cDDP, a well-known biochemical mechanism by which cDDP potentiates the cytotoxicity of 5-FU, [33,37]. However, a greater degree of fragmentation of both nascent and parent DNA was indicated as another possible mechanism of the 5-FU/cDDP synergism [38]. Finally, the better objective response to this drug combination exhibited by 5-FU-resistant patients has been explained by showing that TS inhibition by treatment with platinum drugs and 5-FU also occurs at the transcriptional level [39]. Overall, these explanations establish a causal relationship between hTS inhibition and a decreased efficiency of the mechanisms of repair and substitution of damaged DNA. This might apply to our case as well: the peptide delivered by the liposomes, before (sequence II) or concurrent (sequence I) with cDDP, causing the folate cycle enzyme inhibition, may hamper the repair and substitution of the DNA damaged by cDDP. Additional support to this mechanistic hypothesis comes from the previously reported evidence that treatment of OC cells with the TS inhibiting peptide [D-Gln^4]LR induces deregulation of such folate-related proteins as TS, DHFR and SHMT1 [21]. In particular, the idea that cDDP and the peptide may act synergistically on the folate pathway is reasonable since TS and DHFR operate in the same metabolic cycle [40–42].

Other nodal proteins of the folate pathway were found to be affected by the LR and [D-Gln^4]LR peptides. Among these, ATIC, is an enzyme involved in purine biosynthesis, and NME2, also known as nucleoside diphosphate kinase B, plays a major role in the synthesis of nucleoside triphosphates other than ATP [22]. Because both ATIC and NME2 are down-regulated following treatment with our peptidic inhibitors and are modulated at the gene level by c-Myc [43], a role of this proto-oncogene in the mechanism of action of the two peptides seems likely. Furthermore, the involvement of c-Myc in the mechanism of the observed synergistic effect of the peptides with cDDP and 5-FU is suggested by the known important role played by the downregulation of c-Myc in the cell response to cisplatin and to 5-FU [43–45].

The observed synergy of the peptide with RTX might be related to several different effects: RTX and/or its polyglutamated derivatives reduces the purine biosynthesis as suggested by the increase in the intracellular levels of phosphoribosylpyrophosphate (PRPP) following a 24 h exposure of colon carcinoma cells to RTX [46]. Because RTX is also a moderate direct inhibitor of DHFR, the observed peptide/RTX synergy could result from inhibition of this enzyme and the consequent increase of PRPP levels in the cell via inhibition of purinic biosynthesis [45], thereby potentiating the cellular inhibitory effects of the peptide.

In addition, the synergistic effect of combining the peptides with cDDP or the other drugs may derive from mis-incorporated genomic uracil, resulting from TS inhibition, which induces DNA damage and early cell-cycle arrest as a result of BER activity, and is a critical determinant of sensitivity to antifolate-based TS inhibition [46].

Our findings on the dependence of the effects of the combination of peptides and RTX on the administration schedule correlate with previous reports where synergistic effects, were obtained by a sequential exposure to the DHFR inhibitor methotrexate (MTX) followed by RTX [47]. Actually, the interaction between different antifolates or between antifolates and a Pt drug quite generally follows a preferential treatment schedule. In most cases, simultaneous and continuous administration of RTX and cisplatin, or the sequential administration of RTX followed by cisplatin, produce the highest cytotoxicity, while the reversed sequential administration produces antagonistic effects and thus resulting inappropriate [42].

In our case, the observed sequence-dependent synergy might be accounted for in connection with the finding that HSP90 and TRAP1, that are critical regulators of survival of tumour cells, are among the proteins down-regulated by the peptides. TRAP1 has been described as a mitochondrial chaperone of HSP90AA1. Effective cytoprotection may require TRAP-1 phosphorylation by the mitochondrial-localized kinase, PINK1, which associates with TRAP-1 in vivo. The high expression of TRAP-1 in cancer has been implicated in the inhibition of mitochondrial apoptosis, suppression of ROS production and acquisition of resistance to standard chemotherapeutics [48]. On the opposite hand, decreased TRAP1 expression leads to the accumulation of ubiquitinated/misfolded proteins and proteotoxic stress, a condition that makes cells more sensitive to apoptotic insults, including those caused by treatment with such drugs as cDDP [1]. HSP90 is a ubiquitously expressed molecular chaperone that is involved in the posttranslational folding and stability of multiple mutated, chimeric and over-expressed signalling proteins that promote the growth and/or survival of cancer cells [49]. Previous studies have shown that HSP90 is highly expressed in OC, and a sub-member, HSP90AA1, has been shown to be required for the survival and proliferation of human OC SKOV3 cells. High levels of HSP90AA1 can increase chemo-resistance to cisplatin of SKOV3 cells [50]. Moreover, the concomitant overexpression of ATIC and TRAP1 has been associated to chemoresistance to both cisplatin and

5-FU [51,52]. Therefore, inhibition of HSP90 could potentially act as a promising adjuvant to chemotherapies to overcome drug resistance. Inhibition of these two chaperones, HSP90AA1 and TRAP1 has been already reported following treatment with the antifolate drug MTX and yielded a synergistic effect by interfering with the synthesis of purines [53]. In this light, disabling of HSP90 cancer networks in their multiple subcellular compartments, mitochondrial and cytosolic, by our peptidic inhibitors, may cause irreversible collapse of mitochondria, degradation of HSP90 client proteins in the cytosol, and tumour cell killing by apoptosis and/or autophagy, thus favoring the action of both cDDP and RTX. Also, because RTX induces mitochondrial mediated apoptosis in SGC7901 human cancer cells by disrupting the PI3K/AKT/Hsp-90 cascade and mitochondrial integrity [54–56], a peptide-induced HSP90 disabling may explain the higher efficacy of a sequential treatment schedule.

The combination synergy of the peptide with paclitaxel, that like the other mentioned chemotherapeutic drugs, is frequently used in OC therapy, is justified by several experiments that report, already at the clinical level, on the positive effects of the combination of this microtubule stabilizer with folate-cycle inhibitors and/or platinum drugs [57–59].

4. Materials and Methods

4.1. Materials

Raltitrexed (RTX) and paclitaxel were purchased from Selleckchem (Houston, TX, USA) and were dissolved in DMSO immediately before addition to the cell cultures. Cisplatin (cDDP) was obtained from Santa Cruz Biotechnology. All other chemicals were purchased from Sigma–Aldrich S.r.L. (Milan, Italy), except otherwise indicated.

Reagents for liposome formulation: cholesteryl hemisuccinate (CHEMS) purchased from Sigma–Aldrich (St. Louis, MO, USA); 1,2-dioleoyl-sn-glycero-3-phosphoethanolamine (DOPE) and [N-(Cabonyl-methoxypolyethyleneglycol-2000)]-1,2-distearoyl-sn-glycero-3-phosphoethanolamine sodium salt (DSPE_PEG) purchased from Lipoid (Ludwigshafen, Germany). The active peptide [D-Gln^4]LR and the internal standard (IS) for quantitative analysis, LR-Ala7, were synthesized as previously described [21,22]. All the other reagents were of analytical grade.

4.2. Cell Lines

The five human cancer cell lines of ovarian origin, IGROV-1, A2780, A2780/CP, 2008 and C13*, were grown as monolayers in RPMI 1640 medium containing 10% heat-inactivated foetal bovine serum and 50 μg/mL gentamycin sulphate. The 2008 cell line was established from a patient with serous cystadenocarcinoma of the ovary, and their cisplatin (cDDP)-resistant variant C13* cell line, was developed by monthly exposure to cDDP, followed by chronic exposure to stepwise increases in cDDP concentration [60,61]. The human ovarian carcinoma A2780/CP cells are about 10 fold resistant to cDDP and derived from the parent A2780 cell line [62]. The IGROV-1 cell line, originating from an ovarian carcinoma of a 47 years old woman, was established in monolayer tissue culture. The IGROV-1 cell line exhibits an epithelial character, highly tumorigenic properties and a low doubling time. In addition, there are consistent cytogenetic markers in which oncogenic rearrangements can occur. IGROV-1 is therefore proposed as a model for experimental studies of human ovarian adenocarcinoma, including molecular and cell biology, preclinical pharmacology and experimental therapeutics [63]. The cells were incubated at 37 °C under 5% CO_2 for at least 24 h before treatment with the peptides. All cell media and serum were purchased from Lonza (Lonza Group Ltd., Basel, Switzerland). Cultures were equilibrated with humidified 5% CO_2 in air at 37 °C. All studies were performed in Mycoplasma negative cells, as routinely determined with the MycoAlert Mycoplasma detection kit (Lonza Group Ltd., Basel, Switzerland).

4.3. Cell Transfection with [D-Gln^4]LR Peptide by Means of the Delivery System SAINT-PhD

Treatment with peptide was performed according to the standard transfection protocol of the SAINT-PhD delivery system (Synvolux Therapeutics, Groningen, the Netherland). Complexes of peptide (µg) with SAINT-PhD (µL) were prepared at a ratio of 8 µg/20 µL (which corresponds to a concentration of 5 µM final). For each protein sample, complexes were prepared as follows: for the treatment of a 24-well, an appropriate amount of the peptide was diluted in HBS, then SAINT-PhD was pipetted into the solution without vortexing. The mixture was incubated for 5 min at room temperature, then filled with serum-free medium. The culture medium was aspirated from the cells, the SAINT-PhD/peptide complex was added to the wells and incubated 4 h (37 °C; 5% CO_2). After this time, complete RPMI was added to reach the appropriate volume.

4.4. Liposome Characterization

DOPE:CHEMS:DSPE_PEG (PpHL) liposomes were prepared using the Reverse Phase Evaporation technique followed by homogenization using Ultraturrax (Ika-euroturrax T 25 basic, IkaLabortechnik, Staufen, Germany). Briefly, phospholipid solutions in chloroform at fixed concentration, DOPE:CHEMS:DSPE (22.8: 15.2: 2 mM) for nPpHL liposomes and DOPE:CHEMS:DSPE_PEG (22.8: 15.2: 2 mM) for PpHL liposomes, were employed and an amount of 4 mg of [D-Gln^4]LR was loaded in the liposomes obtaining a final suspension of 15 mg/mL in HBS (Hepes Buffer Saline with 20 mM Hepes and 135 mM NaCl) [29]. Size, size homogeneity and surface charge of the liposomes diluted in water (1:5 v:v) were measured by Zetasizer Nano ZS analyser system (Zetasizer version 6.12; Malvern Instruments, Worcs, UK), equipped with a 4 mW He–Ne laser (633 nm) and a DTS software (Version 5.0). Measurements were performed in triplicate and each measurement was averaged over at least 12 runs.

Drug loading (DL), encapsulation efficiency (EE) and drug release were analysed by liquid chromatography (Agilent 1200 Series LC system; Agilent, Technologies, Milan, Italy) coupled with triple-quadrupole mass spectrometry (LC-MS/MS) (Agilent 6410 triple quadrupole-mass spectrometer, Agilent technologies, Milan, Italy), as described previously [26].

4.5. Cytotoxicity of Liposome-Drug Combinations by MTT Assay

The cytotoxicity assay was performed by the (3-(4,5-dimethylthiazol-2-yl)-2,5-diphenyl tetrazolium bromide; Sigma, Milan, Italy) assay (MTT assay) [64]. Briefly, C13*, 2008 and IGROV-1 cell lines were seeded at a density of 30,000 cells/well in 24-well plate in complete RPMI-1640 medium for at least 24 h. Immediately prior to the cell treatment, the culture medium was aspirated from each well and replaced with 500 µL of complete medium, containing either the unloaded or the [D-Gln^4]LR-loaded liposomes at concentrations of 0.05, 0.075, 0.125 and 0.250 mg/mL; moreover, cells were treated, for comparison, with equivalent amounts of naked peptide. Cells were then incubated in complete medium suspension in 5% CO_2 incubator at 37 °C for 15 h. Then, they were washed with PBS, added with 500 µL of complete RPMI-1640 medium and incubated for additional 48 h. After incubation, a final concentration of 0.5 mg/mL MTT was added to the medium and then incubated at 37 °C for 1 h. The medium was removed and the dark blue formazan crystals dissolved in DMSO were determined spectrophotometrically at 535 nm by the multi-plate reader Genios Pro (Tecan, Austria) with Magellan 6 software, and the results are expressed as percentage of cell growth with respect to the control (untreated cells).

4.6. Synergy Analysis

The nature of the combination between [D-Gln^4]LR-PEG liposomes and drugs (namely cDDP or RTX), combined at a fixed ratio either simultaneously or sequentially, was quantified by synergism quotient (SQ) [65,66]. SQ was defined as the net growth inhibitory effect of the analogue combination divided by the sum of the net individual analogue effects on growth inhibition. A quotient of > 1

indicates a synergistic effect, between 0.9 and 1.1 indicates an additive effect, while a quotient of < 1 indicates an antagonistic effect.

C13*, 2008 and IGROV-1 cell lines were exposed to the combinations using 3 different modalities. Simultaneous drug-exposure (sequence I): cDDP or RTX were given simultaneously with liposomes for 3 days; liposomes before (sequence II): [D-Gln^4]LR loaded liposomes was given 24 h before cDDP or RTX; drugs before: the drug (cDDP or RTX) was given 24 h before liposomes (sequence III).

All the incubation experiments were carried out according to the optimized protocol in which cells were exposed to liposome treatments for 15 h after the medium had been replaced with fresh one. Growth inhibition was assayed by MTT assay and the cytotoxicity of each combination was compared with the cytotoxicities of the drugs alone. Each experiment was performed at least 3 times.

The heatmap and clustering have been realized with the open source software R and Bioconductor repository, using ggplot2 and Heatplus packages (https://cran.r-project.org/ [67]; https://www.bioconductor.org/ [68]). For the clustering (Euclidean distance, complete linkage clustering), in order to highlight the distance between antagonism, addition and synergy values, the synergism quotient values were elaborated as follow: for synergism quotient values < 0.9, a value of 10 was subtracted; for synergism quotient values ≥ 1.1, a value of 10 was added.

4.7. Flow Cytometric Analysis of Cell Cycle

Quantitative measurements of the cell cycle phase distribution were performed by flow cytometry [69]. Cells were suspended in 0.5 mL of hypotonic fluorochrome solution (25 μg/mL PI, 0.1% sodium citrate, 0.1% Triton X-100). The samples were kept at 4 °C in the dark for at least 30 min, dispersed by repeated pipetting before flow cytometry analysis in a FACSCoulter Attune® NxT Acoustic Focusing Cytometer (Invitrogen, Milan, Italy) equipped with a single 488 nm argon laser. The percentage of nuclei in the different phases of the cell cycle (G0/G1, S and G2/M) was calculated with a DNA cell cycle analysis software (Attune® NxT Software version 1.0). A minimum of 1×10^4 cells/sample were analysed for each sample.

4.8. Statistical Analysis

A statistical analysis was performed using the one-way analysis of variance (ANOVA). The data are represented as means ± SD. A difference was considered statistically significant at * $p < 0.05$, ** $p < 0.01$ or *** $p < 0.005$.

5. Conclusions

This work has demonstrated that the TS allosteric inhibitor, the [D-Gln^4]LR peptide, showed a schedule-dependent synergistic antiproliferative effect against OC cells in combination with cDDP, RTX, 5-FU and paclitaxel. More precisely, simultaneous exposure as well as 24 h pre-treatment with the peptide, produced a synergistic cell killing, while a reversed schedule that consists in pre-treating with the non-peptidic drugs was always antagonistic. Furthermore, for the first time, we have provided evidence that this peptide can overcome resistance to cDDP, as its combination with this alkylating agent was synergistic also against the cDDP-resistant cell lines. Also, we have demonstrated a synergistic antiproliferative effect of the peptide in combination with 5-FU, or with RTX even against the C13* and A2780/CP cell lines that had been selected for their resistance to cDDP and cross-resistance to 5-FU and RTX, and feature amplified expression of the folate cycle enzymes [43]. Most of these effects paralleled with great alteration of cell cycle and additive or synergistic accumulation of apoptotic cells. Additionally, the present study indicates how liposome PEGylation, without changing the behaviour of the encapsulated drug, preserves its effectiveness and its therapeutic effect, making it comparable with that obtained with a specific peptide delivery system. Also, delivery by PEGylated liposomes allowed the carried peptide to be of effective formulations with other chemotherapy drugs. Notably, our results, in line with previous published evidence, show that, for a successful, synergistic functional combinatorial targeted therapy, careful attention must be paid to exposure, dose, schedule and target engagement. Thus, the results of this drug combination investigation concerning both in cytotoxicity and cell cycle studies imply the occurrence of complex biochemical interconnections between the peptide and the classical drugs used in this study. Many of these interactions involve apoptotic mechanisms accounting for the effects observed and discussed.

Overall, the results of this preclinical study suggest that the association of standard drugs with this novel peptidic hTS inhibitor encapsulated into PEGylated pH-sensitive liposomes is a promising strategy to impact the drug resistance problem and, combined with intracellular mechanistic analysis, may represent a rational basis for future clinical applications in ovarian tumor patients.

Supplementary Materials: Supplementary materials can be found at http://www.mdpi.com/1422-0067/21/12/4452/s1. Table S1. IC50 values (μM or nM) for selected compounds obtained after 72 h treatment of 2008, C13*, A2780, A2780/CP, and IGROV1 human ovarian cancer cell lines. Table S2. The effect of 72h-exposure to [DGln4]LR and 5-FU alone and in combination on the cell cycle phase distribution of 2008 and C13* cell lines by cytofluorimetric analysis. Table S3. The effect of 48h- and 72h-exposure to [DGln4]LR and cDDP alone and in combination on the cell cycle phase distribution of 2008 and C13* cell lines by cytofluorimetric analysis. Figure S1. Cell cycle-related analysis of the 2008 cells after treatment with the indicated compounds. Figure S2. Cell cycle-related analysis of the C13* cells after treatment with the indicated compounds. Figure S3. The effect of [DGln4]LR and cDDP alone and in combination on the cell cycle phase distribution of 2008 and C13* cells by cytofluorimetric analysis of the DNA content by PI staining.

Author Contributions: Data acquisition, data analysis and manuscript writing G.M.; participation in data analysis and revision of the manuscript G.G., E.M., A.L., L.S., F.S., S.F., and E.L.; contribution to the revision of the manuscript L.L., and S.P.; design of the entire study, data analysis and manuscript writing G.P., M.P.C., and D.D. All authors have read and agreed to the published version of the manuscript.

Funding: This work was financially supported by the Italian Association for Cancer Research (AIRC-2010, IG 10474 and AIRC-2015, IG 16977 to M.P.C.).

Acknowledgments: The authors thank the Centro Interdipartimentale Grandi Strumenti (CIGS), University of Modena and Reggio Emilia, Italy.

Conflicts of Interest: The authors declare that there are no conflict of interest.

References

1. Bowtell, D.D.; Böhm, S.; Ahmed, A.A.; Aspuria, P.J.; Bast, R.C.; Beral, V.; Berek, J.S.; Birrer, M.J.; Blagden, S.; Bookman, M.A.; et al. Rethinking ovarian cancer II: Reducing mortality from high-grade serous ovarian cancer. *Nat. Rev. Cancer* **2015**, *15*, 668–679. [CrossRef]
2. Lai, G.M.; Ozols, R.F.; Smyth, J.F.; Young, R.C.; Hamilton, T.C. Enhanced DNA repair and resistance to cisplatin in human ovarian cancer. *Biochem. Pharmacol.* **1988**, *37*, 4597–4600. [CrossRef]
3. Scanlon, K.J.; Kashani-Sabet, M. Elevated expression of thymidylate synthase cycle genes in cisplatin-resistant human ovarian carcinoma A2780 cells. *Proc. Natl. Acad. Sci. USA* **1988**, *85*, 650–653. [CrossRef]
4. Costi, M.P.; Ferrari, S. Update on antifolate drugs targets. *Curr. Drug Targets* **2001**, *2*, 135–166. [CrossRef]
5. Boyer, Q.; Li, C.; Lee, J.Y.; Shepard, H.M. A novel approach to thymidylate synthase as a target for cancer chemotherapy. *Mol. Pharmacol.* **2001**, *59*, 446–452.
6. Ackland, S.P.; Clarke, S.J.; Beale, P.; Peters, G.J. Thymidylate synthase inhibitors. *Cancer Chemother. Biol. Response Modif.* **2002**, *20*, 1–36.
7. Longley, D.B.; Harkin, D.P.; Johnston, P.G. 5-fluorouracil: Mechanisms of action and clinical strategies. *Nat. Rev. Cancer* **2003**, *3*, 330–338. [CrossRef]
8. Jackman, A.L.; Taylor, G.A.; Gibson, W.; Kimbell, R.; Brown, M.; Calvert, A.H.; Judson, I.R.; Hughes, L.R. ICI D1694, a quinazoline antifolate thymidylate synthase inhibitor that is a potent inhibitor of L1210 tumor cell growth in vitro and in vivo: A new agent for clinical study. *Cancer Res.* **1991**, *51*, 5579–5586.
9. Cocconi, G.; Cunningham, D.; Van Cutsem, E.; Francois, E.; Gustavsson, B.; Van Hazel, G.; Kerr, D.; Possinger, K.; Hietschold, S.M. Open, randomized, multicenter trial of raltitrexed versus fluorouracil plus high-dose leucovorin in patients with advanced colorectal cancer. Tomudex Colorectal Cancer Study Group. *J. Clin. Oncol. Off. J. Am. Soc. Clin. Oncol.* **1998**, *16*, 2943–2952. [CrossRef] [PubMed]

10. Van Triest, B.; Pinedo, H.M.; Van Hensbergen, Y.; Smid, K.; Telleman, F.; Schoenmakers, P.S.; Van der Wilt, C.L.; Van Laar, J.A.; Noordhuis, P.; Jansen, G.; et al. Thymidylate synthase level as the main predictive parameter for sensitivity to 5-fluorouracil, but not for folate-based thymidylate synthase inhibitors, in 13 nonselected colon cancer cell lines. *Clin. Cancer Res. Off. J. Am. Assoc. Cancer Res.* **1999**, *5*, 643–654.
11. Farrugia, D.C.; Ford, H.E.R.; Cunningham, D.; Danenberg, K.D.; Danenberg, P.V.; Brabender, J.; McVicar, A.D.; Aherne, G.W.; Hardcastle, A.; McCarthy, K.; et al. Thymidylate synthase expression in advanced colorectal cancer predicts for response to raltitrexed. *Clin. Cancer Res. Off. J. Am. Assoc. Cancer Res.* **2003**, *9*, 792–801.
12. Welsh, S.J.; Titley, J.; Brunton, L.; Valenti, M.; Monaghan, P.; Jackman, A.L.; Aherne, G.W. Comparison of thymidylate synthase (TS) protein up-regulation after exposure to TS inhibitors in normal and tumor cell lines and tissues. *Clin. Cancer Res. Off. J. Am. Assoc. Cancer Res.* **2000**, *6*, 2538–2546.
13. Peters, G.J.; Backus, H.H.J.; Freemantle, S.; Van Triest, B.; Codacci-Pisanelli, G.; Van der Wilt, C.L.; Smid, K.; Lunec, J.; Calvert, A.H.; Marsh, S.; et al. Induction of thymidylate synthase as a 5-fluorouracil resistance mechanism. *Biochim. Biophys. Acta* **2002**, *1587*, 194–205. [CrossRef]
14. Rahman, L.; Voeller, D.; Rahman, M.; Lipkowitz, S.; Allegra, C.; Barrett, J.C.; Kaye, F.J.; Zajac-Kaye, M. Thymidylate synthase as an oncogene: A novel role for an essential DNA synthesis enzyme. *Cancer Cell* **2004**, *5*, 341–351. [CrossRef]
15. Chu, E.; Takechi, T.; Jones, K.L.; Voeller, D.M.; Copur, S.M.; Maley, G.F.; Maley, F.; Segal, S.; Allegra, C.J. Thymidylate synthase binds to c-myc RNA in human colon cancer cells and in vitro. *Mol. Cell. Biol.* **1995**, *15*, 179–185. [CrossRef] [PubMed]
16. Chu, E.; Copur, S.M.; Ju, J.; Chen, T.M.; Khleif, S.; Voeller, D.M.; Mizunuma, N.; Patel, M.; Maley, G.F.; Maley, F.; et al. Thymidylate synthase protein and p53 mRNA form an in vivo ribonucleoprotein complex. *Mol. Cell. Biol.* **1999**, *19*, 1582–1594. [CrossRef]
17. Benhattar, J.; Cerottini, J.P.; Saraga, E.; Metthez, G.; Givel, J.C. p53 mutations as a possible predictor of response to chemotherapy in metastatic colorectal carcinomas. *Int. J. Cancer* **1996**, *69*, 190–192. [CrossRef]
18. Giovannetti, E.; Backus, H.H.J.; Wouters, D.; Ferreira, C.G.; Van Houten, V.M.M.; Brakenhoff, R.H.; Poupon, M.F.; Azzarello, A.; Pinedo, H.M.; Peters, G.J. Changes in the status of p53 affect drug sensitivity to thymidylate synthase (TS) inhibitors by altering TS levels. *Br. J. Cancer* **2007**, *96*, 769–775. [CrossRef]
19. Cardinale, D.; Guaitoli, G.; Tondi, D.; Luciani, R.; Henrich, S.; Salo-Ahen, O.M.H.; Ferrari, S.; Marverti, G.; Guerrieri, D.; Ligabue, A.; et al. Protein-protein interface-binding peptides inhibit the cancer therapy target human thymidylate synthase. *Proc. Natl. Acad. Sci. USA* **2011**, *108*, E542–E549. [CrossRef]
20. Ponterini, G.; Martello, A.; Pavesi, G.; Lauriola, A.; Luciani, R.; Santucci, M.; Pelà, M.; Gozzi, G.; Pacifico, S.; Guerrini, R.; et al. Intracellular quantitative detection of human thymidylate synthase engagement with an unconventional inhibitor using tetracysteine-diarsenical-probe technology. *Sci. Rep.* **2016**, *6*, 27198. [CrossRef]
21. Pelà, M.; Saxena, P.; Luciani, R.; Santucci, M.; Ferrari, S.; Marverti, G.; Marraccini, C.; Martello, A.; Pirondi, S.; Genovese, F.; et al. Optimization of peptides that target human thymidylate synthase to inhibit ovarian cancer cell growth. *J. Med. Chem.* **2014**, *57*, 1355–1367. [CrossRef]
22. Genovese, F.; Gualandi, A.; Taddia, L.; Marverti, G.; Pirondi, S.; Marraccini, C.; Perco, P.; Pelà, M.; Guerrini, R.; Amoroso, M.R.; et al. Mass spectrometric/bioinformatic identification of a protein subset that characterizes the cellular activity of anticancer peptides. *J. Proteome Res.* **2014**, *13*, 5250–5261. [CrossRef]
23. Miller, D.S.; Blessing, J.A.; Krasner, C.N.; Mannel, R.S.; Hanjani, P.; Pearl, M.L.; Waggoner, S.E.; Boardman, C.H. Phase II evaluation of pemetrexed in the treatment of recurrent or persistent platinum-resistant ovarian or primary peritoneal carcinoma: A study of the Gynecologic Oncology Group. *J. Clin. Oncol. Off. J. Am. Soc. Clin. Oncol.* **2009**, *27*, 2686–2691. [CrossRef]
24. Sacchetti, F.; Marraccini, C.; D'Arca, D.; Pelà, M.; Pinetti, D.; Maretti, E.; Hanuskova, M.; Iannuccelli, V.; Costi, M.P.; Leo, E. Enhanced anti-hyperproliferative activity of human thymidylate synthase inhibitor peptide by solid lipid nanoparticle delivery. *Colloids Surf. B Biointerfaces* **2015**, *136*, 346–354. [CrossRef]
25. Hama, S.; Itakura, S.; Nakai, M.; Nakayama, K.; Morimoto, S.; Suzuki, S.; Kogure, K. Overcoming the polyethylene glycol dilemma via pathological environment-sensitive change of the surface property of nanoparticles for cellular entry. *J. Control. Release Off. J. Control. Release Soc.* **2015**, *206*, 67–74. [CrossRef] [PubMed]

26. Sacchetti, F.; Marverti, G.; D'Arca, D.; Severi, L.; Maretti, E.; Iannuccelli, V.; Pacifico, S.; Ponterini, G.; Costi, M.P.; Leo, E. pH-promoted release of a novel anti-tumour peptide by "stealth" liposomes: Effect of nanocarriers on the drug activity in cis-platinum resistant cancer cells. *Pharm. Res.* **2018**, *35*, 206. [CrossRef]
27. Righetti, S.C.; Perego, P.; Corna, E.; Pierotti, M.A.; Zunino, F. Emergence of p53 mutant cisplatin-resistant ovarian carcinoma cells following drug exposure: Spontaneously mutant selection. *Cell Growth Differ. Mol. Biol. J. Am. Assoc. Cancer Res.* **1999**, *10*, 473–478.
28. Marverti, G.; Ligabue, A.; Paglietti, G.; Corona, P.; Piras, S.; Vitale, G.; Guerrieri, D.; Luciani, R.; Costi, M.P.; Frassineti, C.; et al. Collateral sensitivity to novel thymidylate synthase inhibitors correlates with folate cycle enzymes impairment in cisplatin-resistant human ovarian cancer cells. *Eur. J. Pharmacol.* **2009**, *615*, 17–26. [CrossRef] [PubMed]
29. Chu, E.; Koeller, D.M.; Johnston, P.G.; Zinn, S.; Allegra, C.J. Regulation of thymidylate synthase in human colon cancer cells treated with 5-fluorouracil and interferon-gamma. *Mol. Pharmacol.* **1993**, *43*, 527–533.
30. Severi, L.; Losi, L.; Fonda, S.; Taddia, L.; Gozzi, G.; Marverti, G.; Magni, F.; Chinello, C.; Stella, M.; Sheouli, J.; et al. Proteomic and Bioinformatic Studies for the Characterization of Response to Pemetrexed in Platinum Drug Resistant Ovarian Cancer. *Front. Pharmacol.* **2018**, *9*, 454. [CrossRef]
31. Carosati, E.; Tochowicz, A.; Marverti, G.; Guaitoli, G.; Benedetti, P.; Ferrari, S.; Stroud, R.M.; Finer-Moore, J.; Luciani, R.; Farina, D.; et al. Inhibitor of Ovarian Cancer Cells by Virtual Screening: A New Hydroxy-Thiazole Derivative Targeting Human Thymidylate Synthase. *J. Med. Chem.* **2012**, *55*, 10272–10276, (Brief Article). [CrossRef] [PubMed]
32. Taddia, L.; D'Arca, D.; Ferrari, S.; Marraccini, C.; Severi, L.; Ponterini, G.; Assaraf, Y.G.; Marverti, G.; Costi, M.P. Inside the biochemical pathways of thymidylate synthase perturbed by anticancer drugs: Novel strategies to overcome cancer chemoresistance. *Drug Resist. Updates* **2015**, *23*, 20–54. [CrossRef] [PubMed]
33. Wilson, P.M.; LaBonte, M.J.; Lenz, H.J.; Mack, P.C.; Ladner, R.D. Inhibition of dUTPase induces synthetic lethality with thymidylate synthase-targeted therapies in non-small cell lung cancer. *Mol. Cancer Ther.* **2012**, *11*, 616–628. [CrossRef] [PubMed]
34. Jackman, A.L.; Kelland, L.R.; Kimbell, R.; Brown, M.; Gibson, W.; Aherne, G.W.; Hardcastle, A.; Boyle, F.T. Mechanisms of acquired resistance to the quinazoline thymidylate synthase inhibitor ZD1694 (Tomudex) in one mouse and three human cell lines. *Br. J. Cancer* **1995**, *71*, 914–924. [CrossRef]
35. Scanlon, K.J.; Newman, E.M.; Lu, Y.; Priest, D.G. Biochemical basis for cisplatin and 5-fluorouracil synergism in human ovarian carcinoma cells. *Proc. Natl. Acad. Sci. USA* **1986**, *83*, 8923–8925. [CrossRef] [PubMed]
36. Lacave, A.J.; Barón, F.J.; Antón, L.M.; Estrada, E.; De Sande, L.M.; Palacio, I.; Esteban, E.; Gracia, J.M.; Buesa, J.M.; Fernández, O.A. Combination chemotherapy with cisplatin and 5-fluorouracil 5-day infusion in the therapy of advanced gastric cancer: A phase II trial. *Ann. Oncol. Off. J. Eur. Soc. Med. Oncol.* **1991**, *2*, 751–754. [CrossRef]
37. Ahlgren, J.D.; Trocki, O.; Gullo, J.J.; Goldberg, R.; Muir, W.A.; Sisk, R.; Schacter, L. Protracted infusion of 5-FU with weekly low-dose cisplatin as second-line therapy in patients with metastatic colorectal cancer who have failed 5-FU monotherapy. *Cancer Investig.* **1991**, *9*, 27–33. [CrossRef]
38. Chung, Y.S.; Yamashita, Y.; Inoue, T.; Matsuoka, T.; Nakata, B.; Onoda, N.; Maeda, K.; Sawada, T.; Kato, Y.; Shirasaka, T.; et al. Continuous infusion of 5-fluorouracil and low dose cisplatin infusion for the treatment of advanced and recurrent gastric adenocarcinoma. *Cancer* **1997**, *80*, 1–7. [CrossRef]
39. Shirasaka, T.; Shimamoto, Y.; Ohshimo, H.; Saito, H.; Fukushima, M. Metabolic basis of the synergistic antitumor activities of 5-fluorouracil and cisplatin in rodent tumor models in vivo. *Cancer Chemother. Pharmacol.* **1993**, *32*, 167–172. [CrossRef]
40. Nishiyama, M.; Yamamoto, W.; Park, J.S.; Okamoto, R.; Hanaoka, H.; Takano, H.; Saito, N.; Matsukawa, M.; Shirasaka, T.; Kurihara, M. Low-dose cisplatin and 5-fluorouracil in combination can repress increased gene expression of cellular resistance determinants to themselves. *Clin. Cancer Res. Off. J. Am. Assoc. Cancer Res.* **1999**, *5*, 2620–2628.
41. Yeh, K.H.; Cheng, A.L.; Wan, J.P.; Lin, C.S.; Liu, C.C. Down-regulation of thymidylate synthase expression and its steady-state mRNA by oxaliplatin in colon cancer cells. *Anticancer Drugs* **2004**, *15*, 371–376. [CrossRef] [PubMed]
42. Radivoyevitch, T. Folate system correlations in DNA microarray data. *BMC Cancer* **2005**, *5*, 95. [CrossRef] [PubMed]

43. Liu, Y.C.; Li, F.; Handler, J.; Huang, C.R.L.; Xiang, Y.; Neretti, N.; Sedivy, J.M.; Zeller, K.I.; Dang, C.V. Global regulation of nucleotide biosynthetic genes by c-Myc. *PLoS ONE* **2008**, *3*, e2722. [CrossRef] [PubMed]
44. Biroccio, A.; Benassi, B.; Amodei, S.; Gabellini, C.; Del Bufalo, D.; Zupi, G. c-Myc down-regulation increases susceptibility to cisplatin through reactive oxygen species-mediated apoptosis in M14 human melanoma cells. *Mol. Pharmacol.* **2001**, *60*, 174–182. [CrossRef]
45. Kugimiya, N.; Nishimoto, A.; Hosoyama, T.; Ueno, K.; Enoki, T.; Li, T.S.; Hamano, K. The c-MYC-ABCB5 axis plays a pivotal role in 5-fluorouracil resistance in human colon cancer cells. *J. Cell. Mol. Med.* **2015**, *19*, 1569–1581. [CrossRef] [PubMed]
46. Longo, G.S.; Izzo, J.; Chang, Y.M.; Tong, W.P.; Zielinski, Z.; Gorlick, R.; Chou, T.C.; Bertino, J.R. Pretreatment of colon carcinoma cells with Tomudex enhances 5-fluorouracil cytotoxicity. *Clin. Cancer Res. Off. J. Am. Assoc. Cancer Res.* **1998**, *4*, 469–473.
47. Kano, Y.; Akutsu, M.; Tsunoda, S.; Suzuki, K.; Yazawa, Y.; Furukawa, Y. Schedule-dependent synergism and antagonism between raltitrexed ("Tomudex") and methotrexate in human colon cancer cell lines in vitro. *Jpn. J. Cancer Res. GANN* **2001**, *92*, 74–82. [CrossRef]
48. Kano, Y.; Akutsu, M.; Suzuki, K.; Yazawa, Y.; Tsunoda, S. Schedule-dependent interactions between raltitrexed and cisplatin in human carcinoma cell lines in vitro. *Jpn. J. Cancer Res. GANN* **2000**, *91*, 424–432. [CrossRef]
49. Wilson, P.M.; Danenberg, P.V.; Johnston, P.G.; Lenz, H.J.; Ladner, R.D. Standing the test of time: Targeting thymidylate biosynthesis in cancer therapy. *Nat. Rev. Clin. Oncol.* **2014**, *1*, 282–298. [CrossRef]
50. Altieri, D.C.; Stein, G.S.; Lian, J.B.; Languino, L.R. TRAP-1, the mitochondrial Hsp90. *Biochim. Biophys. Acta* **2012**, *1823*, 767–773. [CrossRef]
51. Banerji, U. Heat shock protein 90 as a drug target: Some like it hot. *Clin. Cancer Res. Off. J. Am. Assoc. Cancer Res.* **2009**, *15*, 9–14. [CrossRef] [PubMed]
52. Chu, S.H.; Liu, Y.W.; Zhang, L.; Liu, B.; Li, L.; Shi, J.Z.; Li, L. Regulation of survival and chemoresistance by HSP90AA1 in ovarian cancer SKOV3 cells. *Mol. Biol. Rep.* **2013**, *40*, 1–6. [CrossRef] [PubMed]
53. Kim, H.K.; Choi, I.J.; Kim, C.G.; Kim, H.S.; Oshima, A.; Michalowski, A.; Green, J.E. A gene expression signature of acquired chemoresistance to cisplatin and fluorouracil combination chemotherapy in gastric cancer patients. *PLoS ONE* **2011**, *6*, e16694. [CrossRef] [PubMed]
54. Macleod, K.; Mullen, P.; Sewell, J.; Rabiasz, G.; Lawrie, S.; Miller, E.; Smyth, J.F.; Langdon, S.P. Altered ErbB receptor signaling and gene expression in cisplatin-resistant ovarian cancer. *Cancer Res.* **2005**, *65*, 6789–6800. [CrossRef] [PubMed]
55. French, J.B.; Zhao, H.; An, S.; Niessen, S.; Deng, Y.; Cravatt, B.F.; Benkovic, S.J. Hsp70/Hsp90 chaperone machinery is involved in the assembly of the purinosome. *Proc. Natl. Acad. Sci. USA* **2013**, *110*, 2528–2533. [CrossRef] [PubMed]
56. Xue, S.; Chen, Y.X.; Qin, S.K.; Yang, A.Z.; Wang, L.; Xu, H.J.; Geng, H.Y. Raltitrexed induces mitochondrial-mediated apoptosis in SGC7901 human gastric cancer cells. *Mol. Med. Rep.* **2014**, *10*, 1927–1934. [CrossRef]
57. Kang, G.H.; Kim, G.S.; Lee, H.R.; Yuh, Y.J.; Kim, S.R. A phase II trial of paclitaxel, 5-fluorouracil (5-FU) and cisplatin in patients with metastatic or recurrent gastric cancer. *Cancer Res. Treat. Off. J. Korean Cancer Assoc.* **2008**, *40*, 106–110. [CrossRef]
58. Wang, F.; Wang, Z.; Zhou, N.; An, X.; Xu, R.; He, Y.; Li, Y. Phase II study of biweekly paclitaxel plus infusional 5-fluorouracil and leucovorin as first-line chemotherapy in patients with advanced gastric cancer. *Am. J. Clin. Oncol.* **2011**, *34*, 401–405. [CrossRef]
59. Cho, B.C.; Kim, J.H.; Kim, C.B.; Sohn, J.H.; Choi, H.J.; Lee, Y.C.; Ahn, J.B. Paclitaxel and leucovorin-modulated infusional 5-fluorouracil combination chemotherapy for metastatic gastric cancer. *Oncol. Rep.* **2006**, *15*, 621–627. [CrossRef]
60. Andrews, P.A.; Murphy, M.P.; Howell, S.B. Differential potentiation of alkylating and platinating agent cytotoxicity in human ovarian carcinoma cells by glutathione depletion. *Cancer Res.* **1985**, *45*, 6250–6253.
61. Korch, C.; Spillman, M.A.; Jackson, T.A.; Jacobsen, B.M.; Murphy, S.K.; Lessey, B.A.; Craig, J.V.; Bradford, A.P. DNA profiling analysis of endometrial and ovarian cell lines reveals misidentification, redundancy and contamination. *Gynecol. Oncol.* **2012**, *127*, 241–248. [CrossRef] [PubMed]
62. Andrews, P.A.; Jones, J.A. Characterization of binding proteins from ovarian carcinoma and kidney tubule cells that are specific for cisplatin modified DNA. *Cancer Commun.* **1991**, *3*, 93–102. [CrossRef] [PubMed]

63. Bénard, J.; Da Silva, J.; De Blois, M.C.; Boyer, P.; Duvillard, P.; Chiric, E.; Riou, G. Characterization of a human ovarian adenocarcinoma line, IGROV1, in tissue culture and in nude mice. *Cancer Res.* **1985**, *45*, 4970–4979. [PubMed]
64. Mosmann, T. Rapid colorimetric assay for cellular growth and survival: Application to proliferation and cytotoxicity assays. *J. Immunol. Methods* **1983**, *65*, 55–63. [CrossRef]
65. Marverti, G.; Ligabue, A.; Guerrieri, D.; Paglietti, G.; Piras, S.; Costi, M.P.; Farina, D.; Frassineti, C.; Monti, M.G.; Moruzzi, M.S. Spermidine/spermine N1-acetyltranferase modulation by novel folate cycle inhibitors in cisplatin-sensitive and -resistant human ovarian cancer cell lines. *Gynecol. Oncol.* **2010**, *117*, 202–210. [CrossRef]
66. Cho, Y.S.; Cho-Chung, Y.S. Antisense protein kinase A RIalpha acts synergistically with hydroxycamptothecin to inhibit growth and induce apoptosis in human cancer cells: Molecular basis for combinatorial therapy. *Clin. Cancer Res. Off. J. Am. Assoc. Cancer Res.* **2003**, *9*, 1171–1178.
67. Software R. Available online: https://cran.r-project.org/ (accessed on 6 March 2017).
68. Bioconductor. Available online: https://www.bioconductor.org/ (accessed on 31 October 2017).
69. Dolbeare, F.; Gratzner, H.; Pallavicini, M.G.; Gray, J.W. Flow cytometric measurement of total DNA content and incorporated bromodeoxyuridine. *Proc. Natl. Acad. Sci. USA* **1983**, *80*, 5573–5577. [CrossRef]

International Journal of
Molecular Sciences

MDPI

Article

Phosphodiesterase *SMPDL3B* Gene Expression as Independent Outcome Prediction Marker in Localized Prostate Cancer

Frank Waldbillig [1,*], Katja Nitschke [1], Abdallah Abdelhadi [1], Jost von Hardenberg [1], Philipp Nuhn [1], Malin Nientiedt [1], Cleo-Aron Weis [2], Maurice Stephan Michel [1], Philipp Erben [1] and Thomas Stefan Worst [1]

1 Department of Urology and Urosurgery, University Medical Centre Mannheim, University of Heidelberg, 68167 Mannheim, Germany; katja.nitschke@medma.uni-heidelberg.de (K.N.); aabdelhadi92@gmail.com (A.A.); jost.vonhardenberg@medma.uni-heidelberg.de (J.v.H.); Philipp.Nuhn@medma.uni-heidelberg.de (P.N.); malin.nientiedt@umm.de (M.N.); maurice-stephan.michel@umm.de (M.S.M.); philipp.erben@medma.uni-heidelberg.de (P.E.); Thomas.Worst@umm.de (T.S.W.)

2 Institute of Pathology, University Medical Centre Mannheim, University of Heidelberg, 68167 Mannheim, Germany; Cleo-Aron.Weis@umm.de

* Correspondence: frank.waldbillig@umm.de; Tel.: +49-621-383-2201

Received: 25 May 2020; Accepted: 17 June 2020; Published: 19 June 2020

Abstract: Current outcome prediction markers for localized prostate cancer (PCa) are insufficient. The impact of the lipid-modifying Sphingomyelin Phosphodiesterase Acid Like 3B (SMPDL3B) in PCa is unknown. Two cohorts of patients with PCa who underwent radical prostatectomy ($n = 40$, $n = 56$) and benign prostate hyperplasia (BPH) controls ($n = 8$, $n = 11$) were profiled for *SMPDL3B* expression with qRT-PCR. Publicly available PCa cohorts (Memorial Sloane Kettering Cancer Centre (MSKCC; $n = 131$, $n = 29$ controls) and The Cancer Genome Atlas (TCGA; $n = 497$, $n = 53$ controls)) served for validation. SMPDL3B's impact on proliferation and migration was analyzed in PC3 cells by siRNA knockdown. In both cohorts, a Gleason score and T stage independent significant overexpression of *SMPDL3B* was seen in PCa compared to BPH ($p < 0.001$ each). A lower expression of *SMPDL3B* was associated with a shorter overall survival (OS) ($p = 0.005$) in long term follow-up. A *SMPDL3B* overexpression in PCa tissue was confirmed in the validation cohorts ($p < 0.001$ each). In the TCGA patients with low SMPDL3B expression, biochemical recurrence-free survival ($p = 0.011$) and progression-free interval ($p < 0.001$) were shorter. Knockdown of *SMPDL3B* impaired PC3 cell migration but not proliferation ($p = 0.0081$). In summary, *SMPLD3B* is highly overexpressed in PCa tissue, is inversely associated with localized PCa prognosis, and impairs PCa cell migration.

Keywords: cancer cell migration; prognosis; biomarker; extracellular vesicles; lipid metabolism

1. Introduction

Prostate cancer (PCa) is the most common solid tumor entity in men in developed countries [1,2]. Most tumors are detected in early stages and then, tend to show low aggressiveness with a slow tumor growth. Some of these patients probably do not need radical therapy. Aggressive PCa tumors on the other hand quickly metastasize to regional lymph nodes and the skeleton. This results in a fatal disease state that requires a long and demanding therapy. Therefore, it is of paramount priority to identify those patients in need of definite local treatment and those which can be spared from overtreatment.

Reliable PCa markers are needed both for therapy decision-making and risk prediction of tumor recurrence after curatively intended therapy, as the currently available tools still have significant deficiencies. In a recent study, we identified a number of potential protein biomarkers for high-risk PCa by proteomic profiling of PCa cells and their extracellular vesicles (EVs) combined with a screening of publicly available databases [3]. EVs have shown to be of great relevance for multiple tumor-associated processes such as local tumor invasion, induction of neoangiogenesis, and premetastatic niche formation in different tumor entities including PCa [4,5]. Screening of potential markers revealed that the SMPDL3B (acid sphingomyelinase-like phosphodiesterase 3B) protein is associated with PCa-derived EVs. Gonzales et al. have already detected the SMPDL3B protein among hundreds of others in human urinary EVs [6]. Principe et al. could detect the protein in analyses of EV's isolated from expressed prostate secretions in human urine [7]. Furthermore, our database analyses indicated an overexpression of the coding SMPDL3B gene in PCa tissue [3].

The SMPDL3B protein is an important enzyme for the lipid-modulation of the cell membrane and thus influences, e.g., the membrane fluidity [8]. So far, nothing has been published about the role of SMPDL3B in PCa. Recently, it was reported that SMPDL3B has impact on podocyte function in renal glomeruli. Thus, the SMPDL3B expression rate in vitro and in vivo correlates with podocyte damage in diabetic kidney disease (DKD) [9]. Apparently, insulin receptor signaling in podocytes in DKD is disturbed via the SMPDL3B-mediated reduction of ceramide-1-phosphate [10]. Watanabe et al. identified SMPDL3B as potential marker for therapeutic response in rituximab-based immuno-suppressive therapy in pediatric patients with intractable kidney disease with proteinuria [11]. This approach is mainly based on the findings of Heinz et al., who first described SMPDL3B as a negative regulator of innate immunity via reduction of Toll-like receptor function on macrophages [8].

Based on our previous findings we aimed to elucidate the role of SMPDL3B on the transcription level. Therefore, SMPDL3B expression and its influence on clinical outcome was analyzed by real-time quantitative polymerase chain reaction (qRT-PCR) in two cohorts of patients with benign prostate hyperplasia (BPH), localized and locally advanced PCa. Furthermore, these results were validated in current, publicly available PCa expression data sets, and in vitro analyses were performed in a PCa cell line.

2. Results

2.1. qRT-PCR Analyses in Patients

In both cohorts analyzed with qRT-PCR a significant overexpression of SMPDL3B could be seen in tumor samples compared to BPH samples (Tissue scan: 6.46× and Mannheim: 37.81×, Mann–Whitney for both $p < 0.001$). This was also seen after stratification of tumor samples for T stage (both $p < 0.001$, Figure 1a,b) and Gleason score (both $p < 0.001$, Figure 1c,d). In the tissue scan cohort, multiple comparison showed significant overexpression of SMPDL3B in both locally confined (T1/2: 5.42×, $p = 0.001$) and locally advanced (T3/4: 8.94×, $p < 0.001$) PCa and in different Gleason groups (≤ 6: 7.01×, $p = 0.002$; 7: 6.00×, $p < 0.001$; ≥ 8: 7.14×, $p = 0.004$) compared to BPH. No significant differences were seen between tumor groups. Similar results were seen in the Mannheim cohort: T1/2 vs. BPH: 46.6× ($p < 0.001$), T3/4 vs. BPH: 32.37 ($p < 0.001$), Gleason ≤ 6 vs. BPH: 50.19 ($p < 0.001$). Controversially, no significant overexpression in Gleason 7 or Gleason ≥ 8 compared to BPH was seen, while the expression was also significantly higher in Gleason ≤ 6 tumors compared to Gleason 7 ($p = 0.045$) and Gleason ≥ 8 ($p = 0.030$) tumors. In the Mannheim cohort, SMPDL3B expression did not correlate with the serum PSA level (Spearman $r = -0.103$, $p = 0.453$). For the tissue scan cohort, no serum PSA data were available.

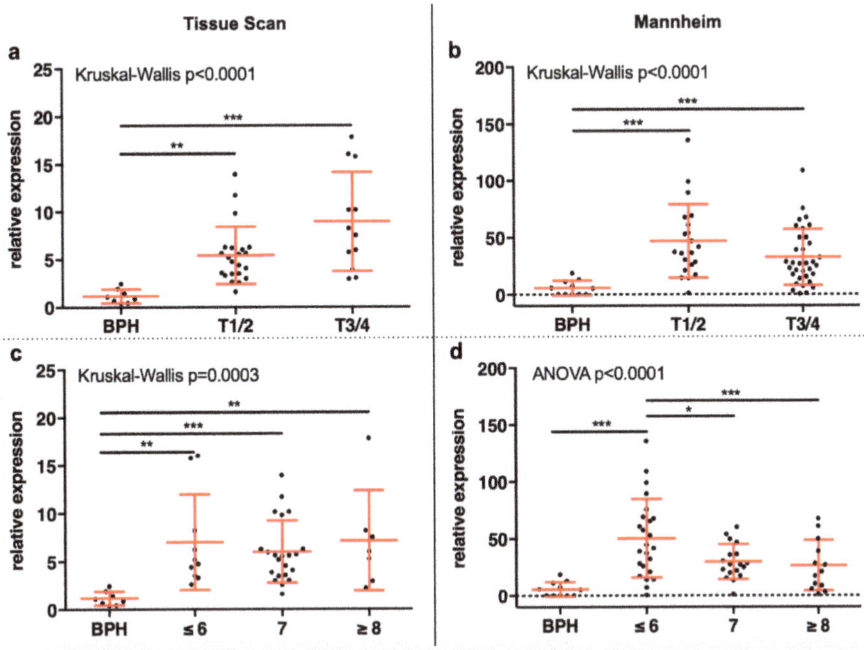

Figure 1. Expression of Sphingomyelin Phosphodiesterase Acid Like 3B (SMPDL3B) was analyzed by qRT-PCR. (**a**) Both locally confined and locally advanced tumors showed an overexpression of SMPDL3B compared to benign prostate hyperplasia (BPH). (**b**) In the Mannheim cohort, T1/2 and T3/4 tumors had an SMPDL3B overexpression. (**c**) All three Gleason groups had an SMPDL3B overexpression in the tissue scan cohort. (**d**) In the Mannheim cohort, Gleason 6 tumors had an overexpression of SMPDL3B both compared to BPH and to Gleason 7 and ≥ 8 tumors. (* $p > 0.05$; ** $p > 0.01$; *** $p > 0.001$)

Interestingly, after a 50:50 division of the Mannheim cohort into two groups by SMPDL3B expression, a lower expression of SMPDL3B in tumor samples correlated with a shorter OS (Figure 2a, $p = 0.005$) in long-term follow-up (average follow-up time: 168 months). Using the same cutoff, no significant difference was seen for BCR (Figure 2b).

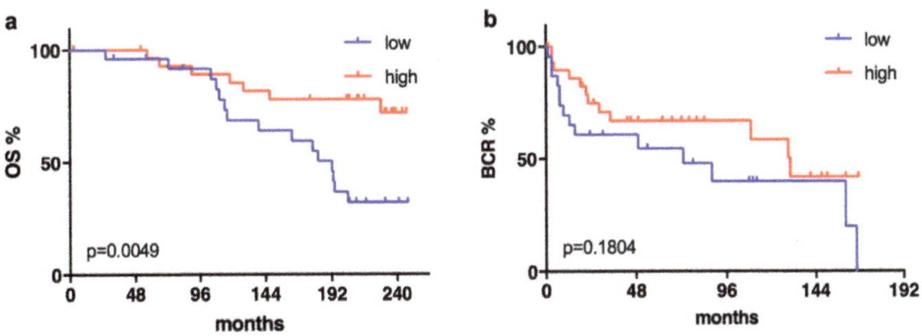

Figure 2. (**a**) In the Mannheim cohort, a low expression of SMPDL3B correlated with a significantly shorter overall survival (OS) but (**b**) not with a shorter biochemical recurrence (BCR)-free survival of patients with localized prostate cancer (PCa) who underwent RP.

2.2. In Silico Validation

In silico analyses confirmed the overexpression of SMPDL3B in PCa tissue in the MSKCC dataset ($p < 0.001$) and in the TCGA dataset ($p < 0.001$). In the MSKCC dataset, this was both seen for T2 and T3/4 tumors (both $p < 0.001$), with no differences in expression between T2 and T3/4 tumors (Figure 3a). In the TCGA cohort, besides a higher expression in T2 and T3/4 compared to BPH (both $p < 0.001$), interestingly, also a significantly higher expression in T2 compared to T3/4 tumors was seen ($p < 0.001$, Figure 3b).

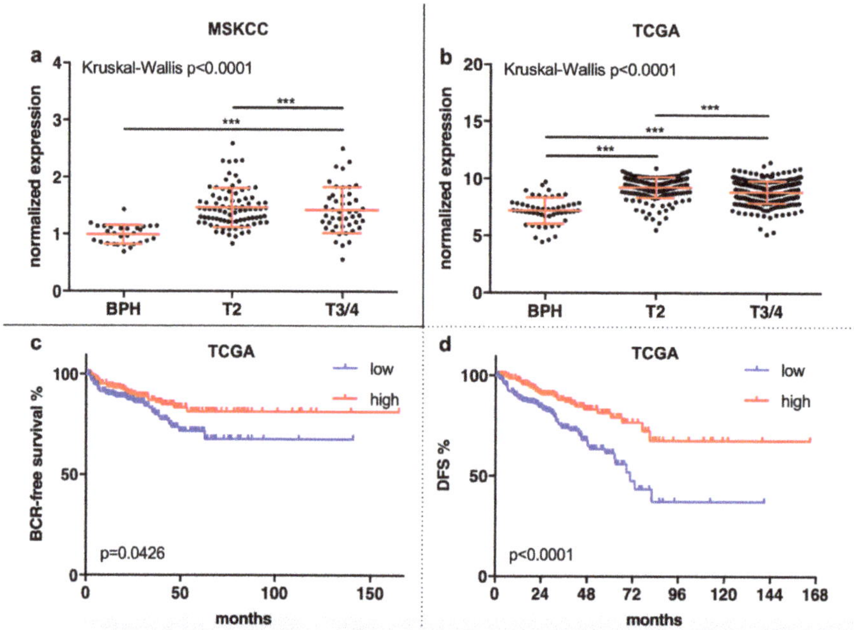

Figure 3. (**a**) and (**b**) Both in the Memorial Sloane Kettering Cancer Centre (MSKCC) and in The Cancer Genome Atlas (TCGA) cohort, the expression of SMPDL3B was higher in T2 and in T3/4 tumors compared to BPH. In the TCGA cohort, the expression was also significantly higher in T2 compared to T3/4 tumors. (**c**) and (**d**) A low expression of SMPDL3B was associated with a significantly shorter BCR-free survival and progression-free interval (PFI) in the PCa patients from the TCGA cohort. (*** $p > 0.001$)

In the TCGA cohort, the SMPDL3B expression correlated positively with the expression of KLK3, the gene which codes for PSA, ($r = 0.287$, $p < 0.001$) and correlated negatively with the expression of *AR* ($r = -0.148$, $p < 0.001$) and the Gleason sum score ($r = -0.336$, $p < 0.001$). Again, with a 50:50 SMPDL3B expression cut-off in the MSKCC cohort, no difference was seen for BCR-free survival and OS. Yet, the number of events in this cohort is quite small (only 26 cases with BCR and only 7 deceased patients). In the TCGA cohort, also no significant difference was seen for OS (only $n = 10$ patients deceased in total), but a low expression in tumor samples correlated with a significantly shorter BCR-free survival ($p = 0.043$) and a shorter progression-free interval ($p < 0.001$) (Figure 3c,d).

2.3. Functional Experiments

The siRNA knockdown of SMPDL3B significantly impaired migration of PC3 cells compared to nontargeting siCtrl ($n = 16$). In brief, 24 h after application of the defect, the wound area of the cells treated with siSMPDL3B was 0.666 ± 0.107 compared to siCtrl at 0.556 ± 0.236 ($p = 0.98$). After 48 h of growth, the wound area of the cells treated with siSMPDL3B was 0.264 ± 0.116 compared to siCtrl at 0.079 ± 0.054, (p= 0.0081, Figure 4a,b).

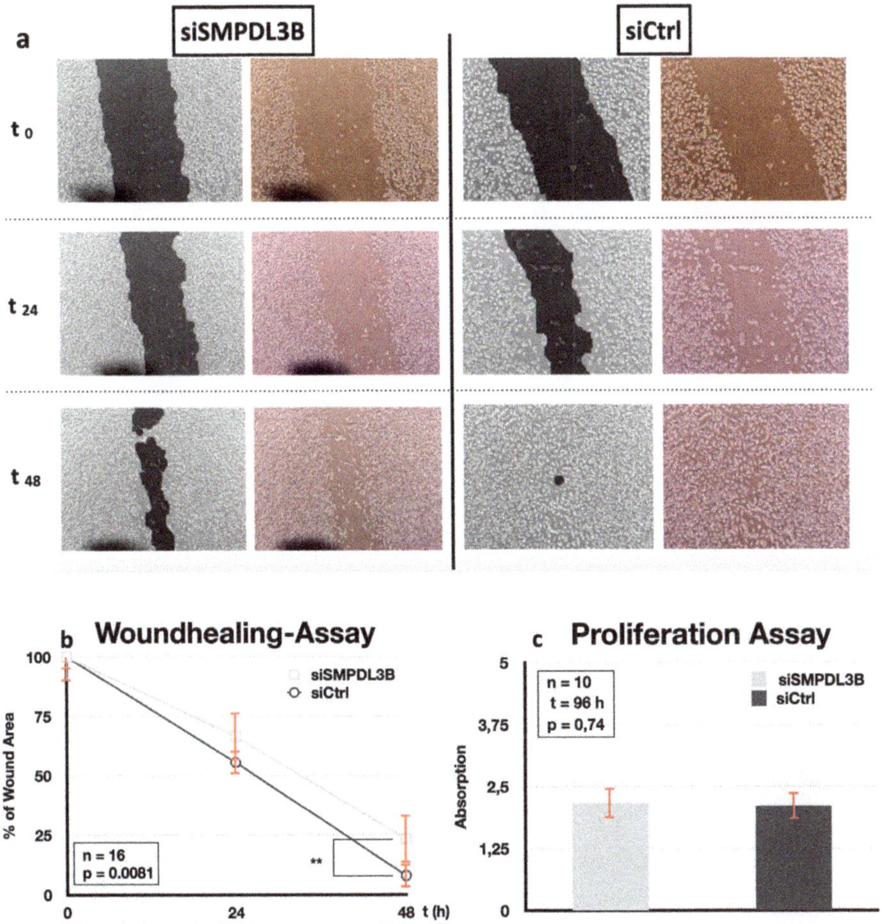

Figure 4. (**a**) Wound-healing migration assay in PC3 cells after siRNA knockdown of SMPDL3B (n = 16). Images were taken at 0, 24, and 48 h after scratch. Calculation of the wound area was performed using TScratch software® (black and white images), with a percentage of the wound area being given for each image. (**b**) Quantitative results of the migration assay which showed a significant reduction of the migration of siSMPDL3B-transfected cells compared to siCtrl after 48 h (p = 0.0081). (** $p > 0.01$) (**c**) PC3 cell proliferation after siRNA knockdown was not significantly altered after knockdown of SMPDL3B (n = 10, p = 0.74).

However, the siRNA knockdown of SMPDL3B had no significant effect on PC3 cell proliferation after 96 h in n = 10 replicates compared to siCtrl (2.165 ± 0.312 vs. 2.103 ± 0.273, respectively) (Figure 4c).

3. Discussion

PCa development is highly heterogenetic and a multifactorial process with intracellular dysregulations on several molecular levels [12]. Although these mechanisms are gaining a better understanding, it is probably the PCa diversity that has prevented potential tumor progression markers from being reliable. A little-noticed factor that plays an important role in cancer signaling is the lipid-based cell membrane [13]. In addition to its central role as boundary and stabilizer of cells, cell membranes can present different cancer-specific surface structures such as proteins and carbohydrates. In this study, we focused on the GPI membrane-anchored phosphodiesterase SMPDL3B. This, to date, little described protein is an enzymatic component of the lipid rafts of cell membranes and appears to be involved in a wide variety of molecular processes including the sphingomyelin metabolism [8]. Besides its relevance for membrane fluidity and lipid composition, SMPDL3B has been identified as a negative regulator of the innate immune signaling [14]. In addition, SMPDL3B overexpression reduces cell apoptosis and strengthens the actin skeleton.

Yet, little is known about the role of SMPDL3B in malignant diseases. In context of PCa, it has only been described in one study so far. By coincidence, Zhu et al. found SMPDL3B overexpression in docetaxel-resistant PC3 cells in a microarray analysis [15].

To our knowledge, this is the first study in which a significant overexpression of SMPDL3B in PCa tissue compared to nonmalignant prostate tissue could be demonstrated in two different patient cohorts. Since this overexpression was independent of the serum PSA level, SMDPL3B could potentially serve as an additional PCa screening marker. Curiously, we were able to show that a lower SMPDL3B expression in PCa was associated with a shorter OS but not with a shorter BCR-free survival. In silico validation in two freely available PCa cohorts (MSKCC, TCGA) confirmed a general SMPDL3B overexpression in patients with PCa. While the MSKCC cohort also showed no correlation of SMPDL3B expression with BCR-free survival, the TCGA cohort indicated that low SMPDL3B expression was associated with shorter BCR-free survival. This is in line with our findings of a stage-dependent, higher SMPDL3B expression in T2 tumors compared to T3/4 tumors. Interestingly, recent studies show that an excess of SMPDL3B leads to a decrease in ceramide-1-phosphate (C1P), a bioactive sphingolipid metabolite regulating key physiologic cell functions. In turn, C1P has been shown to interfere in cancer cell growth, migration, and survival [16]. One gene expression study described a downregulation of SMPDL3B during the mesenchymal to epithelial transition [17]. This might partly explain the inverse association of SMPDL3B expression with PCa patient prognosis.

Both the mechanisms regulating SMPDL3B expression and function and its main substrates remain unclear. However, SMPDL3B can cleave choline-containing substrates including CDP-choline and release phosphate from ATP and ADP. Furthermore, amino acid residues close to the active center of SMPDL3B have been identified. These can recognize endogenous molecules that modulate macrophage function [14]. Studies could show that decreased SMPDL3B levels lead to membrane disorder and a TLR-moderated hyperinflammatory cell phenotype [14]. One could speculate that high concentrations of SMPDL3B lead to a higher membrane stability and thus less disruption of cytoskeleton-lipid raft assemblies promoting carcinogenesis [18]. At the same time, its higher occurrence would reduce the inflammatory environmental reaction and thus reduce the cancer-specific immune defense.

The siRNA knockdown of SMPDL3B lead to a decrease in tumor cell migration. No effects on the proliferation rate were found. This may be explained by the important role of SMPDL3B for membrane stability. However, additional functional studies are required to confirm these results.

To the best of our knowledge, the present study for the first time showed that SMPDL3B is overexpressed in PCa independent of PSA, with higher expression levels in early stages. Its correlation with other clinical parameters promotes its potential applicability in screening and outcome prediction of PCa.

Excitingly, rituximab (RTX), a well-established monoclonal antibody against cell membrane-based CD20, which is used, e.g., in the treatment of malignant hematological diseases, also binds to SMPDL3B [19]. Different case reports have already shown a potential clinical use of RTX in the treatment of metastatic PCA [20–22]]. RTX binds to SMPDL3B in a CD20-independent mechanism [23], which leads to a stabilization of the SMPDL3B concentration in the raft domains and thus, to the stabilization of the cell membrane. Based on this, it could be shown that RTX-treated cells are less radiation sensitive [24]. Considering a modification of the antibody-structure presupposed, one could possibly revert this effect for a combination with PCa radiotherapy. Presumably, an antibody-mediated decrease in membrane-anchored SMPDL3B would lower the membrane stability and could render PCa cells more susceptible to external radiation.

SMPDL3B is a promising biomarker candidate, with potential value for screening and outcome prediction in PCa and at the same time could be an easily accessible target structure in PCa therapy.

4. Materials and Methods

4.1. Patient Characteristics

The expression of SMPDL3B was analyzed by qRT-PCR in a complementary deoxyribonucleic acid (cDNA) array (OriGene, Rockville, MD, USA; $n = 40$ PCa patients and $n = 8$ benign control samples, patient characteristics Table S1) and a cohort of 55 patients who underwent radical prostatectomy (RP) in the Department of Urology and Urosurgery of the Mannheim Medical Center between 1998 and 2001 (patient characteristics in Table S2). Tumor-free prostate tissue from patients who underwent cystoprostatectomy or transurethral resection of the prostate served as controls ($n = 11$). All experiments conducted in this retrospective analysis were in accordance with the institutional ethics review board (ethics approval date 05.12.2013, ethics number: 2013-845R-MA). The study includes retrospective analysis of patient data and qRT-PCR analysis of patient tissue samples obtained during surgery for routine pathologic examination. All analyses and experiments were approved by the local ethic committee (Medical Ethics Committee II of the Medical Faculty Mannheim; ethic approval: 2013-845R-MA).

The Memorial Sloane Kettering Cancer Centre (MSKCC, $n = 131$, $n = 29$ controls, Table S3) [25] and The Cancer Genome Atlas (TCGA, $n = 497$, $n = 53$ controls, Table S4) [26] were obtained from cBioPortal [27], and the University of California, Santa Cruz Xena Browser [28] served for in silico validation. SMPDL3B RNA expression was stratified by tumor characteristics, i.e., serum prostate specific antigen (PSA) levels and was correlated with biochemical recurrence (BCR)-free survival, progression-free survival (PFS), and overall survival (OS).

4.2. RNA Extraction and qRT-PCR Analyses

Hematoxylin- and eosin-stained sections of tumor-bearing or tumor-free formalin-fixed paraffin-embedded (FFPE) prostate tissue specimen of the Mannheim cohort were reviewed by a trained uropathologist. Areas with at least 70% of tumor or tumor-free areas from control patients were marked and dissected from subsequent unstained 10 μm tissue sections. Ribonucleic acid (RNA) was extracted using the bead-based XTRAKT FFPE Kit (Stratifyer, Cologne, Germany), as recommended by the manufacturer and recently published [29].

Multiplexed specific cDNA synthesis with equimolar pooling of transcript-specific reverse PCR primers was used. Superscript III (Life Technologies, Darmstadt, Germany) was used as reverse transcriptase at 55 °C for 120 min, followed by an incubation at 70 °C for 15 min. cDNA was immediately used for qPCR or stored at −20 °C.

In the cDNA array and the Mannheim cohort, the expression of SMPDL3B was determined in relation to the housekeeping gene calmodulin 2 (CALM2) on a StepOnePlus qRT-PCR Cycler (Applied Biosystems, Waltham, MA, USA) (all sequences in Table S5). Relative expression was calculated with the $2^{(-\Delta\Delta CT)}$ method [30]. The methodology was recently published elsewhere [29]. The sequences of primers and probes are given in Supplementary Table S1.

4.3. Small Interfering RNA (siRNA) Knockdown

The impact of SMPDL3B siRNA knockdown on proliferation and migration was analyzed in humane metastatic PC3 prostate cancer cells. These were obtained from ATCC (Wesel, Germany) and grown under standard conditions in DMEM medium (Life Technologies, Carlsbad, CA, USA) supplemented with 10% FCS (Sigma Aldrich, St. Louis, LA, USA). siGENOME pooled and individual siRNAs against SMPDL3B were transfected using DharmaFECT I transfection reagent (Dharmacon,

Lafayette, CO, USA) (siSMPDL3B). Dharmacon nontargeting siRNAs were used as negative control (siCtrl). Briefly, cells were detached, harvested, spun down, and diluted to the desired concentration. Meanwhile siRNAs were diluted to a target concentration of 30 nMol in pure RPMI (Life Technologies) and incubated for 10 min at room temperature. DharmaFECT I was diluted 1:1000 in RPMI medium. After 10 min, diluted siRNA and transfection reagent were mixed (1:1) and again incubated at room temperature for 30 min. Hereafter, cell suspension was added to the transfection mix (3:1) and incubated at 37 °C.

qRT-PCR was conducted to validate knockdown SMPDL3B. RNA extraction was performed using the RNeasy Mini Kit (Qiagen, Hilden, Germany) as recommended by the manufacturer. cDNA synthesis was performed as described previously [31]: in brief, 40 µL of diluted RNAs were mixed with 4 µL of 5 mg/mL pdN6 random primers, 4 µL of 10 mM dNTP Mix, 16 µL of 5× M-MLV buffer, 8 µL of 0.1 M RNase inhibitor, 4 µL of 0.1 M DTT, and 4 µL of M-MLV reverse transcriptase (all from Roche Diagnostics, Basel, Switzerland). After an incubation for 2 h at 37 °C and a deactivation step of 5 min at 65 °C, cDNA was directly used for qPCR or stored at −20 °C. qRT-PCR analyses were performed using the same primers, reagents, and PCR protocol as described for tissue sample analyses.

4.4. Proliferation Assay

PC3 cells were seeded and transfected in 96-well plates (4500 cells in 100 µL/well). After 24 h, the supernatant was replaced by 100 µL of fresh growth medium. After further 96 h of incubation, 10 µL of MTT-reagent (stock concentration 1.585 mg/mL, product number G3582, Promega, Mannheim, Germany) was added to each well and incubated for 3 h at 37 °C. Absorption measurement at 570 nm was done with an Infinite M1000 PRO plate reader (Tecan, Männedorf, Switzerland).

4.5. Migration Assay

Using the same transfection protocol, a wound-healing migration assay of PC3 cells was performed. The cells were seeded in 24-well plates (250,000 cells in 1 mL of DMEM with 10% FCS per well). The medium was changed 24 h after transfection. Again 24 h later, a defined scratch was introduced in the center of the well with a sterile 200 µL pipette tip, and the medium was changed again. The scratch was photographed at 10× magnification. Subsequent images were acquired after 24 and 48 h. The cell-free space in the scratch area was calculated with the open source software TScratch (ETH Zürich, Switzerland) [32]. The free area at 24 and 48 h after scratch was normalized to the initial scratch size.

4.6. Statistics

Statistical calculations were performed using Prism 7 (GraphPad, La Jolla, CA, USA). Different subgroups were tested for normal distribution using the D'Agostino-Pearson test for normal distribution. In case of a normal distribution, subsequent group comparisons were performed by two-sided t-test or parametric ANOVA with subsequent Turkey's post hoc multiple comparison tests. In case of non-normal distribution Mann–Whitney test or Kruskal–Wallis test with Dunn's post hoc multiple comparison tests were performed. Outcome correlations in Kaplan–Meier analyses were done using the log-rank test. Correlations with the PSA serum level were performed using Spearman correlation. p-values ≤ 0.05 were deemed significant.

Supplementary Materials: Supplementary materials can be found at http://www.mdpi.com/1422-0067/21/12/4373/s1. Table S1: Characteristics of the purchased cohort; Table S2: Characteristics of the Mannheim cohort; Table S3: Characteristics of the MSKCC cohort; Table S4: Characteristics of the TCGA cohort; Table S5: Primers & Probes.

Author Contributions: Conceptualization, T.S.W., J.v.H., and P.E.; methodology, T.S.W., J.v.H., P.E., and K.N.; formal analysis, T.S.W. and F.W.; investigation, T.S.W. and F.W.; resources, M.S.M. and P.E.; data curation, F.W., T.S.W., A.A., and K.N.; writing—original draft preparation, T.S.W. and F.W.; writing–review and editing, all authors; visualization, T.S.W., M.N., and F.W.; supervision, M.S.M., P.N., and P.E.; validation, C.-A.W.; and funding acquisition, T.S.W. All authors have read and agreed to the published version of the manuscript.

Funding: This work was supported by the foundation on cancer and scarlet research of the University of Heidelberg. T.S.W. was supported by a Ferdinand Eisenberger scholarship of the German Society of Urology. Funding number WoT1/FE-14; funding date: 01.09.2014-31.08.2015.

Acknowledgments: The technical help of Annette Steidler in experiment planning and conduction is greatly acknowledged.

Conflicts of Interest: The authors declare no conflict of interest.

References

1. Jemal, A.; Bray, F.; Center, M.M.; Ferlay, J.; Ward, E.; Forman, D. Global cancer statistics. *CA Cancer J. Clin.* **2011**, *61*, 69–90. [CrossRef] [PubMed]
2. Siegel, R.L.; Miller, K.D.; Jemal, A. Cancer statistics, 2016. Ca. *Cancer J. Clin.* **2016**, *66*, 7–30. [CrossRef] [PubMed]
3. Worst, T.S.; von Hardenberg, J.; Gross, J.C.; Erben, P.; Schnoelzer, M.; Hausser, I.; Bugert, P.; Michel, M.S.; Boutros, M. A database-augmented, exosome-based mass spectrometry approach exemplarily identifies circulating claudin 3 as biomarker in prostate cancer. *Mol. Cell. Proteom. Mcp* **2017**, *16*, 998–1008. [CrossRef] [PubMed]
4. Théry, C.; Witwer, K.W.; Aikawa, E.; Alcaraz, M.J.; Anderson, J.D.; Andriantsitohaina, R.; Antoniou, A.; Arab, T.; Archer, F.; Atkin-Smith, G.K.; et al. Minimal information for studies of extracellular vesicles 2018 (MISEV2018): A position statement of the International Society for Extracellular Vesicles and update of the MISEV2014 guidelines. *J. Extracell. Vesicles* **2018**, *7*, 1535750. [CrossRef]
5. Vlaeminck-Guillem, V. Extracellular Vesicles in Prostate Cancer Carcinogenesis, Diagnosis, and Management. *Front. Oncol.* **2018**, *8*, 222. [CrossRef]
6. Gonzales, P.A.; Pisitkun, T.; Hoffert, J.D.; Tchapyjnikov, D.; Star, R.A.; Kleta, R.; Wang, N.S.; Knepper, M.A. Large-scale proteomics and phosphoproteomics of urinary exosomes. *J. Am. Soc. Nephrol. Jasn* **2009**, *20*, 363–379. [CrossRef]
7. Principe, S.; Jones, E.E.; Kim, Y.; Sinha, A.; Nyalwidhe, J.O.; Brooks, J.; Semmes, O.J.; Troyer, D.A.; Lance, R.S.; Kislinger, T.; et al. In-depth proteomic analyses of exosomes isolated from expressed prostatic secretions in urine. *Proteomics* **2013**, *13*, 1667–1671. [CrossRef]
8. Heinz, L.X.; Baumann, C.L.; Köberlin, M.S.; Snijder, B.; Gawish, R.; Shui, G.; Sharif, O.; Aspalter, I.M.; Müller, A.C.; Kandasamy, R.K.; et al. The Lipid-Modifying Enzyme SMPDL3B Negatively Regulates Innate Immunity. *Cell Rep.* **2015**, *11*, 1919–1928. [CrossRef]
9. Yoo, T.-H.; Pedigo, C.E.; Guzman, J.; Correa-Medina, M.; Wei, C.; Villarreal, R.; Mitrofanova, A.; Leclercq, F.; Faul, C.; Li, J.; et al. Sphingomyelinase-Like Phosphodiesterase 3b Expression Levels Determine Podocyte Injury Phenotypes in Glomerular Disease. *J. Am. Soc. Nephrol. Jasn* **2015**, *26*, 133–147. [CrossRef]
10. Mitrofanova, A.; Mallela, S.K.; Ducasa, G.M.; Yoo, T.H.; Rosenfeld-Gur, E.; Zelnik, I.D.; Molina, J.; Varona Santos, J.; Ge, M.; Sloan, A.; et al. SMPDL3b modulates insulin receptor signaling in diabetic kidney disease. *Nat. Commun.* **2019**, *10*, 2692. [CrossRef]
11. Watanabe, S.; Tsugawa, K.; Tsuruga, K.; Imaizumi, T.; Tanaka, H. Urinary excretion of sphingomyelinase phosphodiesterase acid-like 3b in children with intractable nephrotic syndrome. *Pediatr. Int. Off. J. Jpn. Pediatr. Soc.* **2017**, *59*, 1112–1115. [CrossRef] [PubMed]
12. Tolkach, Y.; Kristiansen, G. The Heterogeneity of Prostate Cancer: A Practical Approach. *Pathobiol. J. Immunopathol. Mol. Cell. Biol.* **2018**, *85*, 108–116. [CrossRef] [PubMed]
13. Ogretmen, B. Sphingolipid metabolism in cancer signalling and therapy. *Nat. Rev. Cancer* **2018**, *18*, 33–50. [CrossRef]
14. Gorelik, A.; Heinz, L.X.; Illes, K.; Superti-Furga, G.; Nagar, B. Crystal Structure of the Acid Sphingomyelinase-like Phosphodiesterase SMPDL3B Provides Insights into Determinants of Substrate Specificity. *J. Biol. Chem.* **2016**, *291*, 24054–24064. [CrossRef] [PubMed]
15. Zhu, S.; Min, Z.; Qiao, X.; Chen, S.; Yang, J.; Zhang, X.; Liu, X.; Ran, W.; Lv, R.; Lin, Y.; et al. Expression profile-based screening for critical genes reveals S100A4, ACKR3 and CDH1 in docetaxel-resistant prostate cancer cells. *Aging* **2019**, *11*, 12754–12772. [CrossRef] [PubMed]
16. Hait, N.C.; Maiti, A. The Role of Sphingosine-1-Phosphate and Ceramide-1-Phosphate in Inflammation and Cancer. *Mediat. Inflamm.* **2017**, *2017*, 4806541. [CrossRef] [PubMed]

17. Shah, A.D.; Inder, K.L.; Shah, A.K.; Cristino, A.S.; McKie, A.B.; Gabra, H.; Davis, M.J.; Hill, M.M. Integrative Analysis of Subcellular Quantitative Proteomics Studies Reveals Functional Cytoskeleton Membrane-Lipid Raft Interactions in Cancer. *J. Proteome Res.* **2016**, *15*, 3451–3462. [CrossRef]
18. Gröger, C.J.; Grubinger, M.; Waldhör, T.; Vierlinger, K.; Mikulits, W. Meta-Analysis of Gene Expression Signatures Defining the Epithelial to Mesenchymal Transition during Cancer Progression. *PLoS ONE* **2012**, *7*, e51136. [CrossRef]
19. Fornoni, A.; Sageshima, J.; Wei, C.; Merscher-Gomez, S.; Aguillon-Prada, R.; Jauregui, A.N.; Li, J.; Mattiazzi, A.; Ciancio, G.; Chen, L.; et al. Rituximab targets podocytes in recurrent focal segmental glomerulosclerosis. *Sci. Transl. Med.* **2011**, *3*, 85ra46. [CrossRef]
20. Dalgleish, A.; Featherstone, P.; Vlassov, V.; Rogosnitzky, M. Rituximab for treating CD20+ prostate cancer with generalized lymphadenopathy: A case report and review of the literature. *Invest. New Drugs* **2014**, *32*, 1048–1052. [CrossRef]
21. Bindal, P.; Jalil, S.A.; Holle, L.M.; Clement, J.M. Potential role of rituximab in metastatic castrate-resistant prostate cancer. *J. Oncol. Pharm. Pract. Off. Publ. Int. Soc. Oncol. Pharm. Pract.* **2019**, *25*, 1509–1511. [CrossRef] [PubMed]
22. Pirozzi, A.; Cartenì, G.; Scagliarini, S.; Fusco, M.; Riccardi, F. Incidental finding of non-Hodgkin's lymphoma in a patient affected by castration-sensitive prostate cancer: A case report. *Medicine (Baltimore)* **2019**, *98*, e14805. [CrossRef]
23. Perosa, F.; Favoino, E.; Caragnano, M.A.; Dammacco, F. Generation of biologically active linear and cyclic peptides has revealed a unique fine specificity of rituximab and its possible cross-reactivity with acid sphingomyelinase-like phosphodiesterase 3b precursor. *Blood* **2006**, *107*, 1070–1077. [CrossRef] [PubMed]
24. Ahmad, A.; Mitrofanova, A.; Bielawski, J.; Yang, Y.; Marples, B.; Fornoni, A.; Zeidan, Y.H. Sphingomyelinase-like phosphodiesterase 3b mediates radiation-induced damage of renal podocytes. *FASEB J. Off. Publ. Fed. Am. Soc. Exp. Biol.* **2017**, *31*, 771–780. [CrossRef] [PubMed]
25. Taylor, B.S.; Schultz, N.; Hieronymus, H.; Gopalan, A.; Xiao, Y.; Carver, B.S.; Arora, V.K.; Kaushik, P.; Cerami, E.; Reva, B.; et al. Integrative genomic profiling of human prostate cancer. *Cancer Cell* **2010**, *18*, 11–22. [CrossRef] [PubMed]
26. Cancer Genome Atlas Research Network The Molecular Taxonomy of Primary Prostate Cancer. *Cell* **2015**, *163*, 1011–1025. [CrossRef]
27. Cerami, E.; Gao, J.; Dogrusoz, U.; Gross, B.E.; Sumer, S.O.; Aksoy, B.A.; Jacobsen, A.; Byrne, C.J.; Heuer, M.L.; Larsson, E.; et al. The cBio cancer genomics portal: An open platform for exploring multidimensional cancer genomics data. *Cancer Discov.* **2012**, *2*, 401–404. [CrossRef]
28. Goldman, M.; Craft, B.; Hastie, M.; Repečka, K.; McDade, F.; Kamath, A.; Banerjee, A.; Luo, Y.; Rogers, D.; Brooks, A.N.; et al. The UCSC Xena platform for public and private cancer genomics data visualization and interpretation. *bioRxiv* **2019**, 326470. [CrossRef]
29. Worst, T.S.; Waldbillig, F.; Abdelhadi, A.; Weis, C.-A.; Gottschalt, M.; Steidler, A.; von Hardenberg, J.; Michel, M.S.; Erben, P. The EEF1A2 gene expression as risk predictor in localized prostate cancer. *BMC Urol.* **2017**, *17*, 86. [CrossRef]
30. Livak, K.J.; Schmittgen, T.D. Analysis of Relative Gene Expression Data Using Real-Time Quantitative PCR and the $2-\Delta\Delta CT$ Method. *Methods* **2001**, *25*, 402–408. [CrossRef]
31. Worst, T.S.; Meyer, Y.; Gottschalt, M.; Weis, C.-A.; von Hardenberg, J.; Frank, C.; Steidler, A.; Michel, M.S.; Erben, P. RAB27A, RAB27B and VPS36 are downregulated in advanced prostate cancer and show functional relevance in prostate cancer cells. *Int. J. Oncol.* **2017**, *50*, 920–932. [CrossRef] [PubMed]
32. Ashby, W.J.; Zijlstra, A. Established and novel methods of interrogating two-dimensional cell migration. *Integr. Biol. (Camb.)* **2012**, *4*, 1338–1350. [CrossRef] [PubMed]

International Journal of
Molecular Sciences

MDPI

Article

Hedgehog Signal Inhibitor GANT61 Inhibits the Malignant Behavior of Undifferentiated Hepatocellular Carcinoma Cells by Targeting Non-Canonical GLI Signaling

Kensuke Harada, Ryuya Ohashi, Kyoko Naito and Keita Kanki *

Department of Biomedical Engineering, Faculty of Engineering, Okayama University of Science, 1-1 Ridai-cho, Kita-ku, Okayama 700-0005, Japan; t15s047hk@ous.jp (K.H.); t16s011or@ous.jp (R.O.); t20mm02nk@ous.jp (K.N.)
* Correspondence: kkanki@bme.ous.ac.jp

Received: 11 March 2020; Accepted: 27 April 2020; Published: 28 April 2020

Abstract: The Hedgehog (HH)–GLI pathway plays an important role in cell dedifferentiation and is therefore pivotally involved in the malignant transformation of cancer cells. GANT61, a selective inhibitor of GLI1 and GLI2, was reported as a promising treatment for cancer in various tissues; however, the biological impact of GANT61 in hepatocellular carcinoma (HCC), especially in undifferentiated HCC cells, remains unclear. In this study, we investigated the antitumor effect of GANT61 using two undifferentiated hepatoma cell lines: HLE and HLF. Quantitative PCR and RT-PCR analyses revealed that these cells express *GLI* transcripts, showing mesenchymal phenotypes characterized by the loss of epithelial and hepatic markers and specific expression of epithelial–mesenchymal transition (EMT)-related genes. GANT61 significantly reduced the proliferation and cell viability after drug treatment using 5-FU and Mitomycin C. We showed that *GLI* transcript levels were down-regulated by the MEK inhibitor U0126 and the Raf inhibitor sorafenib, suggesting that non-canonical signaling including the Ras–Raf–MEK–ERK pathway is involved. Sphere formation and migration were significantly decreased by GANT61 treatment, and it is suggested that the underlying molecular mechanisms are the down-regulation of stemness-related genes (*Oct4, Bmi1, CD44,* and *ALDH*) and the EMT-related gene *Snail1*. The data presented here showed that direct inhibition of GLI might be beneficial for the treatment of dedifferentiated HCC.

Keywords: GANT61; non-canonical Hedgehog-GLI signaling; hepatocellular carcinoma

1. Introduction

Hepatocellular carcinoma (HCC), the most frequently diagnosed primary liver cancer, is an aggressive disease caused by chronic hepatitis B or C virus infection, excessive alcohol intake, non-alcoholic steatohepatitis, or exposure to aflatoxin B [1]. The prognosis for patients with advanced stage HCC is poor because of its high rates of recurrence and intrahepatic metastasis. Consequently, there is an essential requirement for the development of an improved therapeutic strategy that aims to inhibit tumor progression. The malignant transformation of tumor cells is related to dedifferentiation of the cells [2]. A loss of cellular specific functions, increased proliferation, and enhanced glycolytic activity are the common features of tumor cells. Additional genetic mutations often cause aberrant activation of intracellular signaling pathways, leading to a loss of cell–cell adhesion, increased cell motility, and anchorage-independent cell proliferation, which are the crucial malignant phenotypes that trigger invasiveness and metastasis [3].

The Hedgehog (HH) pathway, originally identified as an important pathway during embryonic development, is known to play a role in carcinogenesis in various tissues including the liver [4]. In recent years, inappropriate HH signaling was demonstrated in more than 30% of human cancers [5]. Canonical activation of HH signaling is triggered by the binding of HH ligands (SHH, IHH, and DHH) to their receptor Patched-1 (PTCH) on the surface of target cells. Ligand-bound PTCH lost its inhibitory regulation on Smoothened (SMO); thus, active SMO causes the downstream glioma-associated homologue (GLI) transcription factors GLI1 and GLI2 to be activated via nuclear translocation [6,7]. GLI target genes include factors involved in cell proliferation, survival, self-renewal, and invasiveness; therefore, aberrant activation of this signal pathway has a positive correlation with poor prognosis [8,9]. Small-molecule inhibitors of the HH-GLI signal, such as SMO inhibitors, were developed to increase the clinical benefit of patients. Some have entered clinical trials as promising therapies for advanced cancers, including basal cell carcinoma and medulloblastoma [10–12]. However, several patients were reported to be resistant to the SMO inhibitor because of an acquired mutation in SMO that disrupts drug binding and mutations in genes downstream of SMO, such as SUFU mutations [13–15]. Moreover, an alternative activation pathway of GLI, namely the non-canonical HH pathway, has been identified [16]. These problems of drug resistance led to the development of new small molecules that function as direct inhibitors of GLI transcription.

GANT61 is a selective inhibitor of GLI1- and GLI2-mediated gene transactivation. Preclinical studies were performed for numerous cancer types, including rhabdomyosarcoma, neuroblastoma, and leukemia, as well as colon, pancreas, cervix, skin, lung, head-neck, and gastric cancers [5]. Although substantial promising results were reported, the biological impact of GANT61 in HCC, in particular undifferentiated HCC cells, has not been clarified. In this study, we investigated the antitumor effect of GANT61 in undifferentiated HCC cells, which show greater expression of GLI transcription factors than well-differentiated HCC cells. The presented data show that GANT61 significantly reduced cell proliferation and viability after drug treatment. Malignant phenotypes, such as sphere formation and migration, were significantly decreased with reduced expression of stemness-related and epithelial–mesenchymal transition (EMT)-related genes. We also found that the gene expression of *GLI* is regulated, in part, by non-canonical signaling, including the Ras–Raf–MEK–ERK pathway, in these cells. Our data suggest that the application of a direct inhibitor of GLI transcription might be beneficial for the treatment of dedifferentiated HCC.

2. Results

2.1. Preferential Expression of GLI Genes in Undifferentiated HCC Cell Lines

To determine the intracellular status of GLI-mediated signaling in hepatoma cell lines, a gene expression analysis of *GLI1*, *GLI2*, and *GLI3* was performed on a panel of hepatoma cells including three differentiated (HepG2, HuH1, and HuH7) and two undifferentiated (HLE and HLF) types of HCC cell. Quantitative PCR revealed that *GLI1* mRNA is expressed in both differentiated and undifferentiated HCC cells, being highly expressed in undifferentiated cells (Figure 1a). Among HCC cells, HepG2, a typical well-differentiated type of HCC cell showed the lowest expression of the *GLI1* gene. The expression of *GLI2* and *GLI3* was positively detected in undifferentiated cells but not in differentiated HCC cells. Morphologically, HepG2 showed an epithelial-like shape characterized by tight cell adhesion, while HLE and HLF showed mesenchymal morphology characterized by loose cell contact and an irregular cell shape (Figure 1b). RT-PCR analysis revealed that HLE and HLF lack the expression of *E-cadherin* and hepatic markers (*ALB* and *HNF4α*), but these cells express mesenchymal markers such as *vimentin*, *CD44*, *snail1*, and *twist1* (Figure 1c). These data suggest that GLI-mediated signaling is activated in undifferentiated HCC cell lines showing the mesenchymal phenotype. Therefore, we chose HLE and HLF cells to investigate the effect of the HH signaling inhibitor GANT61.

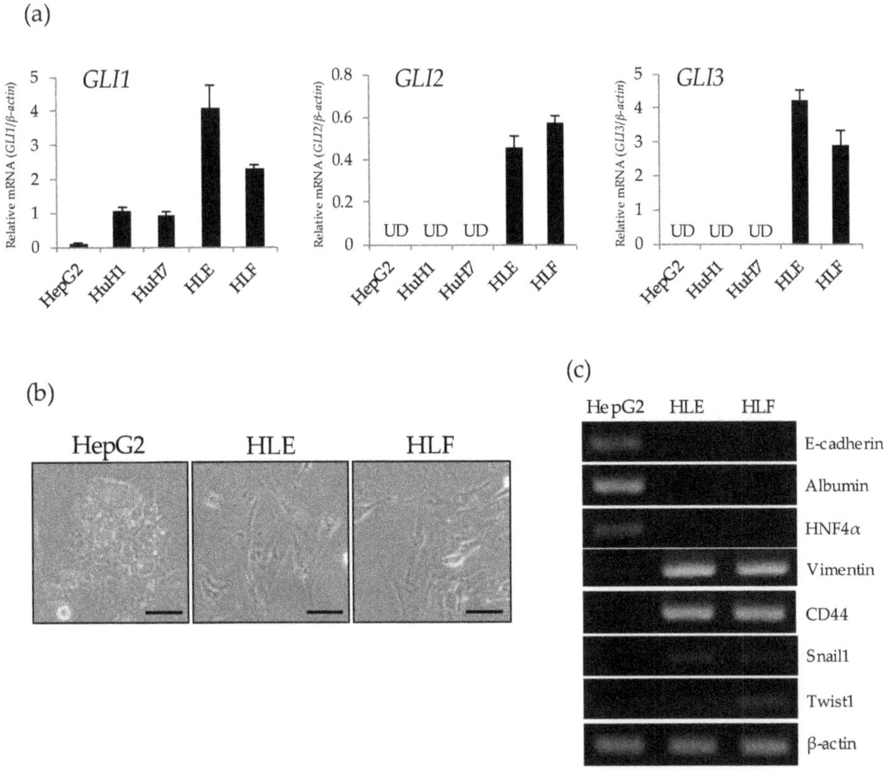

Figure 1. Preferential expression of *GLI* genes in undifferentiated hepatocellular carcinoma (HCC) cell lines. (**a**) Relative gene expression levels of *GLI1*, *GLI2*, and *GLI3* compared to *β-actin* were determined in hepatoma cell lines by qRT-PCR analysis (UD; undetectable). (**b**) Microscopic observation of HepG2 (left), HLE (center), and HLF (right). Scale bar = 20 μm. (**c**) RT-PCR analysis of cell-type specific markers including *E-cadherin* (epithelial), *Albumin*, *HNF4α* (hepatic), *Vimentin*, *CD44*, *Snail1*, *Twist1* (mesenchymal), and *β-actin* as a house keeping gene.

2.2. Antitumor Effect of GANT61 on the Proliferation of Undifferentiated HCC Cells

To evaluate the antitumor potential of GANT61, cell proliferation was investigated in HLE and HLF cells treated with 0–10 μM of GANT61. No statistically significant effect on cell proliferation was observed during the first two days of treatment (Figure 2a). In both types of cell, significant inhibition of cell proliferation was observed on day 3 of 10 μM treatment (Figure 2b). Treatment with a concentration greater than 10 μM, for example, with 20 or 30 μM, markedly decreased cell viability at days 2 and 3 of the experiment (Figure 2c). Therefore, we selected a treatment with 10 μM of GANT61, a quantity which did not largely affect the viability of HCC cells, for further experiments.

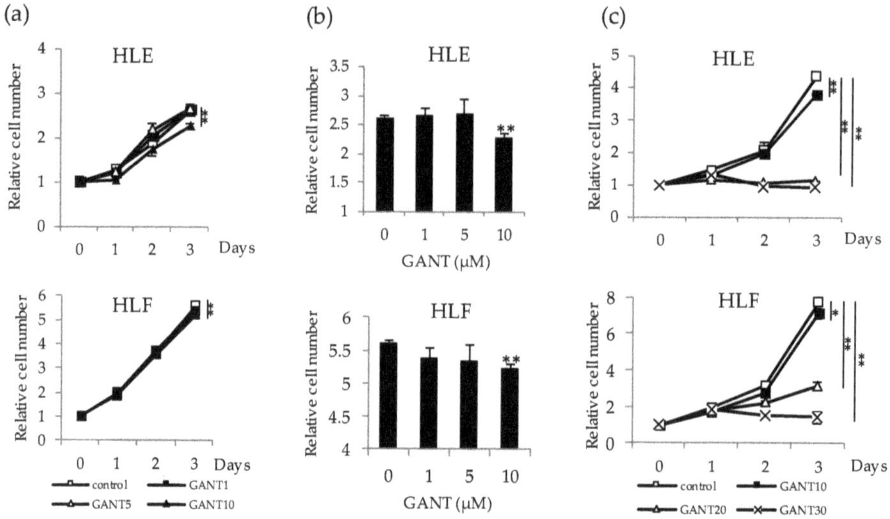

Figure 2. Antitumor effect of GANT61 on the cell proliferation of undifferentiated HCC cells. (**a**) HLE and HLF cells were treated with 0, 1, 5, and 10 μM of GANT61 for 3 days for cell proliferation assays. Cell viability was determined by WST assay and expressed as the relative number of viable cells compared to day 0. (**b**) Column graphs showing the values at day 3 of the experiment. (**c**) HLE and HLF cells were treated with 0, 10, 20, and 30 μM of GANT61 for 3 days and analyzed as described in (a). * $p < 0.05$, ** $p < 0.01$ vs. the vehicle control (Dimethylsulfoxide; DMSO).

2.3. Drug-Specific Enhancing Effect of Cytotoxicity by GANT61 and Involvement of the Ras–Raf–MEK–ERK Pathway in the Regulation of GLI Expression in Undifferentiated HCC Cells

Next, to examine the effect on drug sensitivity, HLE and HLF cells were treated with a series of concentrations of Mitomycin C (0–50 μg/mL), 5-FU (0–500 μg/mL), and sorafenib (0–50 μg/mL) in combination with GANT61 for 48 h. GANT61 was used at 10 μM which showed no inhibitory effect on the proliferation of HLE and HLF cells up to 48 h after the treatment. The three anticancer drugs significantly reduced the cell viability of HLE and HLF cells in a dose-dependent manner; however, the effect of combination treatment with GANT61 differed between the drugs. GANT61 significantly reduced the viability of HCC cells treated with relatively high concentrations of mitomycin C and 5-FU. In contrast, this enhancement of drug sensitivity was barely observed in the combination with sorafenib (Figure 3a,b). To understand the mechanism of the drug-specific effect of GANT61, we next examined the expression level of GLI mRNA in sorafenib-treated cells. Sorafenib is known as a multi-kinase inhibitor that targets the kinase activity of B-Raf, vascular endothelial growth factor receptors (VEGFRs), and platelet-derived growth factor receptor β (PDGFR-β). We first examined the amount of the active forms of MEK1 and ERK1/2, downstream molecules of the Ras–Raf pathway, in HLE cells treated with the inhibitors. Phosphorylated forms of MEK1 (p-MEK1) and ERK1/2 (p-ERK1/2) were shown to be decreased by treatment with the MEK inhibitor U0126 (1 μM) and sorafenib (10 μM) (Figure 3c,d). Quantitative PCR revealed that these treatments significantly decreased the expression levels of *GLI1*, *GLI2*, and *GLI3* mRNA (Figure 3e). Altogether, these data suggest that GANT61 has an antitumor effect by decreasing cell proliferation and cell viability when it is combined with anticancer drugs in undifferentiated HCC cells. The effect of combination treatment on drug sensitivity differed between the drugs due to the non-canonical regulation of GLI-mediated signaling by the Ras–Raf–MEK–ERK pathway, a critical target of sorafenib.

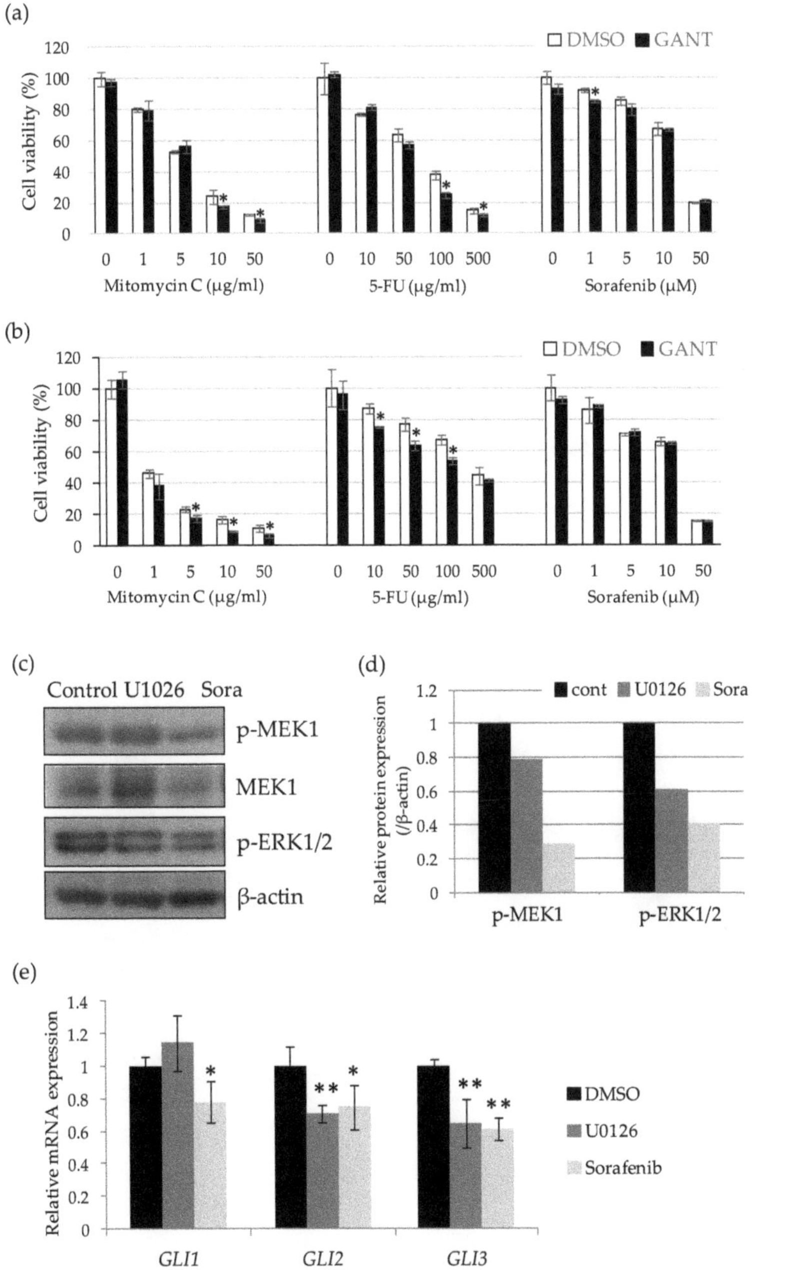

Figure 3. Drug-specific enhancing effect of cytotoxicity by GANT61 and involvement of the Ras–Raf–MEK–ERK pathway in the regulation of *GLI* expression in undifferentiated HCC cells. Cell viability assays of HLE (**a**) and HLF (**b**) cells treated with 10 μM of GANT61 in combination with various concentrations of anticancer drugs. After 48 h, cell viability was measured by WST assay and is expressed as the percentage of drug-free control. (**c**) Western blot analysis of phospho-MEK

(Ser2017/221), total MEK, and phospho-ERK (Thr202/Tyr204) in the lysate extracted from the cells treated with U1026 (1 μM) and sorafenib (Sora) (10 μM) for 24 h. β-Actin served as a control of protein loading. (**d**) Relative protein expression levels were calculated from the band intensity with ImageJ software (NIH). (**e**) Relative gene expression levels of *GLI1*, *GLI2*, and *GLI3* in U1026- and sorafenib-treated cells compared to the control (DMSO). * $p < 0.05$, ** $p < 0.01$ vs. the vehicle control (DMSO).

2.4. GANT61 Inhibits the Sphere Forming Capacity of Undifferentiated HCC Cells

To examine the effect of GANT61 on the cancer stem cell (CSC)-like phenotype, sphere formation in suspension culture was performed in the presence or absence of GANT61. HLE and HLF cells were able to form large spheres in a low adherence plate with media containing the vehicle. In contrast, large spheres were barely observed in the presence of 10 μM of GANT61 (Figure 4a,b). To understand the possible molecular mechanism underlying this, we next examined the expression of stemness-related genes in control and GANT61-treated cells. Quantitative PCR revealed that *Oct4*, a transcription factor related to pluripotency and self-renewal, was significantly down-regulated in HLE cells treated with GANT61. In HLF cells, GANT61 treatment showed a tendency to reduce the expression of three genes—*Oct4*, *Bmi1* and *CD44*—although the difference was not statistically significant (Figure 4c). Two subtypes of aldehyde dehydrogenase, *ALDH1A1* and *ALDH1A3*, were significantly down-regulated by GANT61 in HLF and HLE cells, respectively. These data suggest that GANT61 inhibits sphere formation by down-regulating stemness-related genes in undifferentiated HCC cells.

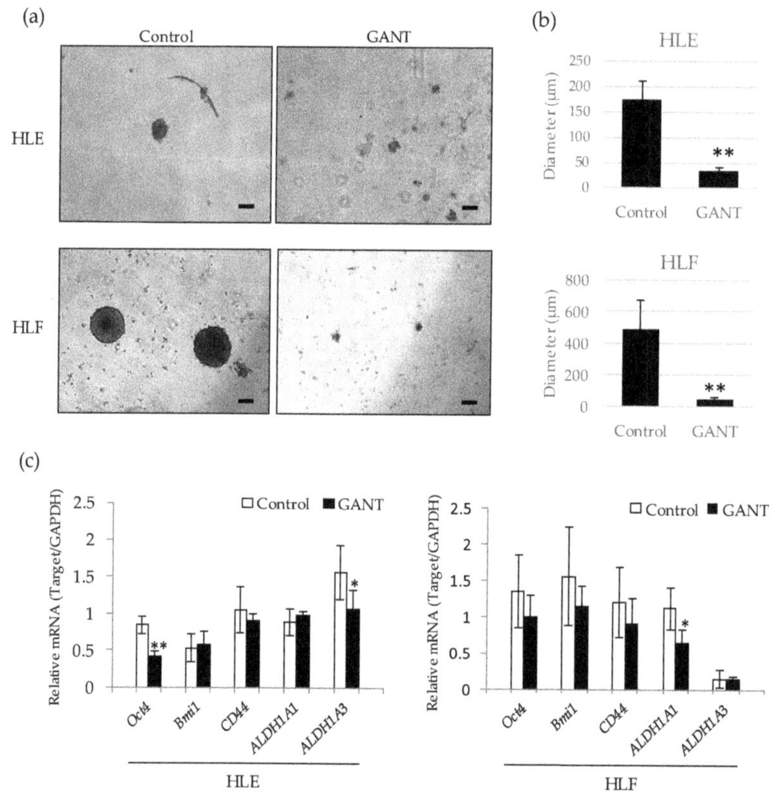

Figure 4. GANT61 inhibits sphere formation by down-regulating stemness-related genes.

(**a**) Morphological appearance of spheres formed in the presence or absence of 10 μM GANT61. HLE (upper) and HLF (lower) cells were cultured in an ultra-low-attachment culture plate for 10 days and photographed. Scale bar = 100 μm. (**b**) The diameter of the sphere was measured in five representative spheres of each group. (**c**) Quantitative PCR of stemness-related genes *Oct4*, *Bmi1*, *CD44*, *ALDH1A1*, and *ALDH1A3* in HLE (left) and HLF (right) cells treated with 10 μM of GANT61. *GAPDH* mRNA was used as an internal control. * $p < 0.05$, ** $p < 0.01$ vs. the vehicle control (DMSO).

2.5. GANT61 Inhibits Cell Migration by Down-Regulating Snail1 Gene Expression

To investigate the effect of GANT61 on cell migration ability, a wound healing assay was performed by scratching monolayer-cultured cells. To exclude the effect on cell proliferation, assays were performed under the condition of serum deprivation (1% FBS or serum-free) using 5 and 10 μM of GANT61. Twenty-four hours after scratching, GANT61-treated cells showed slower migration than control cells (Figure 5a). Quantification of cell migrating activity was carried out by measuring a closed area after 24 h, and this revealed a significant and dose-dependent decrease in cell migrating activity in HLE and HLF cells (Figure 5b).

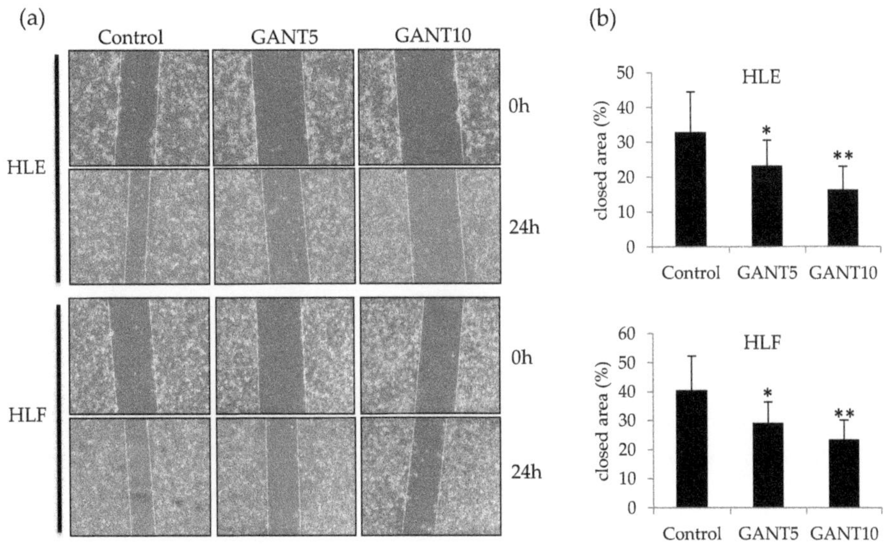

Figure 5. Wound healing assay of undifferentiated HCC cells treated with GANT61. (**a**) Scratched area of monolayer-cultured HLE (upper) and HLF (lower) cells treated with 5 μM (GANT5) and 10 μM (GANT10) of GANT61 under the serum-free condition. (**b**) A closed area was used to evaluate cell migration activity. * $p < 0.05$, ** $p < 0.01$ vs. the vehicle control (DMSO).

To confirm the inhibitory effect of GANT61 on cell migration, a transwell assay was performed. As shown in Figure 6, GANT61 lowered the migration of HLE cells in a dose-dependent manner (Figure 6a,b). Furthermore, an increase in the epithelial marker E-cadherin and a decrease in the mesenchymal marker vimentin were observed by Western blotting of the lysates extracted from GANT61-treated cells (Figure 6c). Quantitative PCR revealed that the mesenchymal transcription factor *Snail1* was down-regulated by GANT61 treatment compared to the control (Figure 6d). Snail knockdown by small-interference RNA (siRNA) significantly reduced the migration ability of HLE cells (Figure 6e–g). Together, these data suggest that GANT61 inhibits cell migration by attenuating the mesenchymal phenotype of undifferentiated HCC cells.

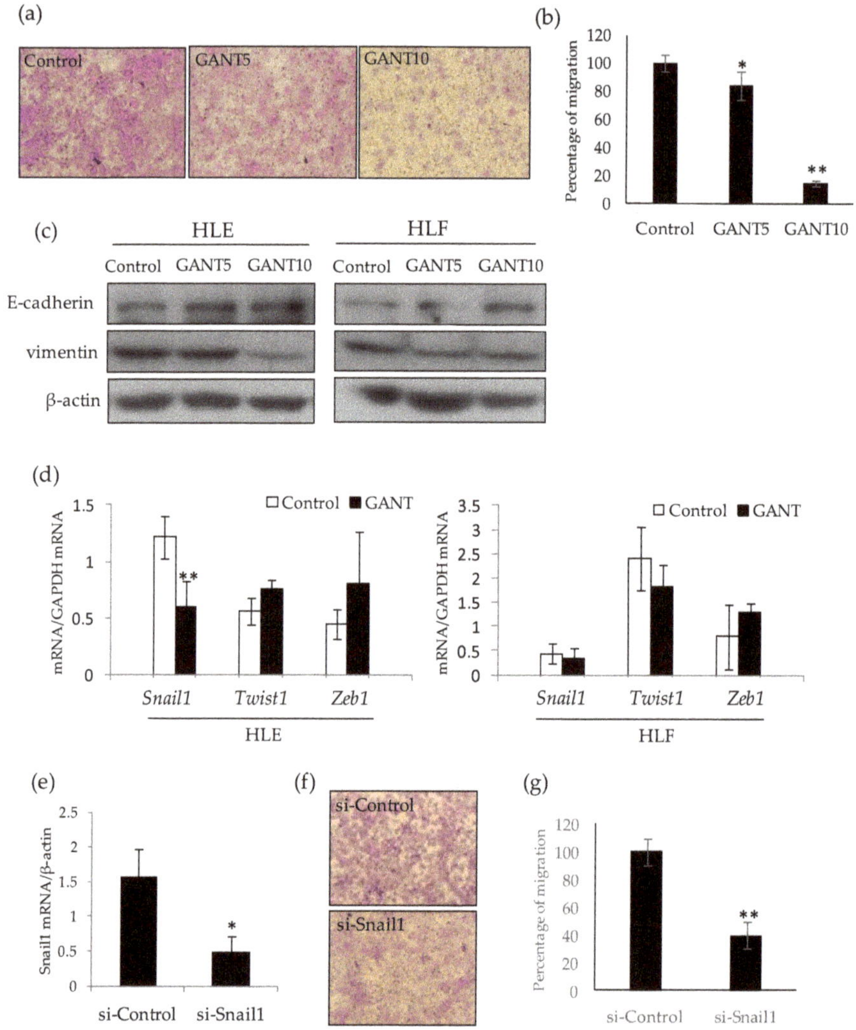

Figure 6. GANT61 inhibits cell migration by attenuating the mesenchymal phenotype of HLE cells. (**a**) Transwell migration assay of HLE cells treated with 5 μM (GANT5) and 10 μM (GANT10) of GANT61. Migrated cells were stained with crystal violet. (**b**) The migration activity of the cells is expressed as a percentage of the cell number compared with the control. (**c**) Western blotting of E-cadherin, vimentin, and β-actin as an loading control. (**d**) Quantitarive PCR analysis of EMT-related transcription factors *Snail1*, *Twist1*, and *Zeb1* in HLE (left) and HLF (right) cells treated with 10 μM GANT61. (**e**) Quantitative PCR analysis of Snail mRNA in HLE cells treated with control (si-Control) and Snail (si-Snail) siRNA. (**f**,**g**) Transwell migration assay of siRNA-treated HLE cells and their migration activity. * $p < 0.05$, ** $p < 0.01$ vs. vehicle control (DMSO).

3. Discussion

The HH-GLI signaling pathway was shown to play a critical function during development by regulating cellular proliferation and differentiation. In wound healing and regeneration processes, this pathway plays important roles in cellular differentiation, migration, and maintenance of tissue progenitor cells [17]. Therefore, aberrant activation of this signal in cancer cells leads to malignant transformations such as dedifferentiation, acquisition of stemness features, and migration potency. In this study, we investigated the biological significance of GLI-mediated signaling in undifferentiated HCC cell lines using the GLI specific inhibitor GANT61. We found preferential expression of *GLI* mRNAs in undifferentiated HCC cell lines that showed the mesenchymal phenotype. Consistent with previous reports, HLE and HLF were characterized by increased mesenchymal markers and the loss of epithelial and hepatic markers [18,19]. Among *GLI* genes, *GLI2* and repressor *GLI3* were specifically expressed in undifferentiated HCC cells. Generally, GLI1 acts mainly transcriptional activation in cancer cells, whereas GLI2 requires N-terminal processing for full activation [20,21]. GLI2 is reported to induce genomic instability by inhibiting apoptosis, which is responsible for eliminating transformed aberrant cells [22]. These reports and our data suggest that *GLI2* expression, in addition to that of *GLI1*, contributes to the promotion of cell dedifferentiation and confers HCC cells with mesenchymal phenotypes.

To determine the biological impact of GANT61 in undifferentiated HCC cells, we investigated the cell proliferation and sensitivity to cell death induced by anticancer drugs in GANT61-treated HLE and HLF cells. Although we observed a positive antitumor effect of GANT61 on cell viability, the impact seemed to be relatively weak compared with that observed in other type of cancers. For instance, previous studies reported that only 10%–40% of cells remained viable after treatment with 10–20 μM of GANT61 in lung, melanoma, prostate, glioma, and breast cancer cells [23–29]. These reports and our observation allow us to hypothesize that the GANT61 target may be a small subpopulation in HLE and HLF cells. In combination treatment, there was little enhancing effect on drug sensitivity when GANT61was used with low concentrations of mitomycin C and 5-FU; however, a significant decrease in cell viability was observed when GANT61 was used in combination with relatively high concentrations of these drugs. This result suggests that a small population of drug-resistant cells were targeted by GANT61. Previous reports showed that HH-signal inhibitors, such as the SMO inhibitor and direct GLI inhibitors, have antitumor effects by targeting CSC-like cell populations [30,31]. Therefore, it is reasonable to hypothesize that a small subpopulation of CSC-like cells which depends on GLI signaling would be a target of GANT61.

Interestingly, we observed no enhancement of cell death in combination with sorafenib, a multi-kinase inhibitor targeting the serine–threonine kinases Raf-1 and B-Raf and the receptor tyrosine kinase activity of VEGFRs and PDGFR-β. A possible mechanism underlying the drug-specific response might be explained by the observation that sorafenib down-regulated the expression of *GLI* genes. GLI-dependent transcription is known to be activated in an HH ligand-independent manner, namely non-canonical HH signaling, which includes oncogenic intracellular signaling pathways such as PI3–AKT, TGFβ–smad, and Ras–Raf–MEK–ERK [16]. Therefore, in undifferentiated HCC, the GLI-mediated HH pathway is suggested to be regulated, at least in part, by the Ras–Raf–MEK–ERK pathway.

Increasing evidence is suggesting that the acquisition of stemness features in tumor cells is correlated with poor clinical outcomes in various type of cancers. These malignantly transformed cells often show enhanced resistance to chemotherapy, anchorage-independent survival, and migration potency that causes postoperative recurrence and metastasis. As well as well-known oncogenic intracellular signaling components such as Wnt/β-catenin, Notch, TGF-β, and AKT, HH-GLI signaling was described as an important pathway for maintaining the stemness of undifferentiated and/or immature cells [3]. In normal liver development, HH-GLI signaling contributes to the proliferation of immature progenitor cells, and its activation progressively decreases during liver epithelial cell maturation [32]. HH-GLI signal activation re-increases when the tissue regenerates after injury to promote the proliferation of liver progenitor cells. Therefore, HH-GLI signaling is considered to promote dedifferentiation and confer cancer cells to have stemness features, making it a promising target for cancer therapy. In this study, we observed significant decreases in sphere forming activity and cell migration by GANT61. Gene expression analysis revealed that GLI inhibition significantly down-regulated *Oct4* in HLE cells and in HLF cells, showing a tendency to reduce the expression of stemness-related genes (*Oct4*, *Bmi1*, and *CD44*) which play important roles in anchorage-independent cell growth. We also observed significant down-regulation of *ALDH* mRNA expression by GANT61. ALDH was shown to play an important role in self-renewal in various cancers including HCC and was reported to be regulated by GLI signaling in melanomas [33,34]. Therefore, these stemness-related genes may be crucial targets of GLI signaling in undifferentiated HCC. In the sphere assay, the rate of sphere formation was found to be about 1% for the cells seeded in a low-attachment dish (data not shown). In future research, identification and isolation of a subpopulation that has sphere forming activity is required to assess the critical effect of GANT61.

A significant decrease in cell motility by GANT61 was confirmed in a wound healing assay and transwell assay. These assays and the Western blot analysis suggest that the mesenchymal features of undifferentiated HCC cells are attenuated by GANT61 treatment. In terms of the molecular mechanism, *Snail1*, a pivotal transcriptional regulator for migration potency, was found to be down-regulated in HLE cells. Although *Snail1* is known as a key regulator of EMT, the impact of GANT61 on mRNA expression differed among two cell lines (Figure 6d). Moreover, GANT61 treatment was associated with greater inhibition of migration activity than *Snail* knockdown, despite the weaker down-regulation effect of *Snail* mRNA compared with that of si-Snail (Figure 6b,g). Together, these results suggest that other genes, in addition to *Snail*, responsible for cell migration may be affected by GANT61. A thorough study of target genes including coding and non-coding genes should be required to understand the cell-type-specific action of GAN61. An anti-migrative effect was related to HH signaling inhibition by the SMO inhibitors NVP-LDE-225 and GDC-0449, both of which were approved for clinical application [35–37]. However, these drugs may be ineffective for the non-canonical GLI-activation pathway observed in this study. In addition to the SMO inhibitor, the use of a direct inhibitor of GLI and/or inhibitors of signaling pathways involved in the non-canonical HH-GLI pathway may be required for effective treatment.

As depicted in Figure 7, the results of our study demonstrate that the HH-GLI signal inhibitor GANT61 has an antitumor effect on the malignant character of undifferentiated HCC cells with the mesenchymal phenotype. Direct inhibition of GLI was suggested to be beneficial in the case of dedifferentiated HCC which is involved in non-canonical HH-GLI signaling.

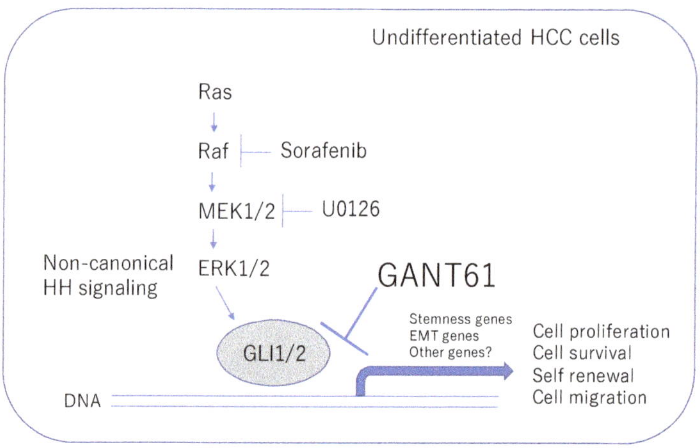

Figure 7. Schematic view of the antitumor effect of GANT61 in undifferentiated HCC cells. GANT61 inhibits the malignant behavior of undifferentiated HCC cells by targeting non-canonical GLI signaling. Arrows denote activation and T-shaped lines indicate inhibition.

4. Materials and Methods

4.1. Cell Culture and Reagents

Hepatoma cell lines (HepG2, HuH1, HuH7, HLE and HLF) were purchased from RIKEN BRC. Cells were maintained in DMEM containing 10% FBS, 4.5 mg/mL glucose, and 2 mM L-glutamine in a humidified incubator at 37 °C and 5% CO_2. GANT61 (Santa Cruz Biotechnology; sc-202630) and sorafenib (ChemScene LLC; Monmouth Junction, NJ, USA) were dissolved in DMSO at a concentration of 10 mM and stored in a freezer at –20 °C.

4.2. Cell Viability Assays

Cell viability was determined by the WST assay using the Cell Counting Kit-8 (Dojin Chemical Co., Ltd.; Kumamoto, Japan). Briefly, cells were cultured in a 96-well plate and subjected to the experiment. To determine cell viability, a 10% volume of CCK-8 reagent was added to each well, and the plate was incubated for 1 h at 37 °C. Cell viability was calculated by measuring the absorbance at a wavelength of 450 nm. For cell proliferating assays, cells were seeded at 3×10^3 cells/well and measured at 24, 48, and 72 h after the treatment. The relative number of viable cells was calculated from OD450 to create proliferation curves.

4.3. Gene Expression Analysis

Total RNA was isolated using the TRIZOL reagent in accordance with the manufacturer's instructions (Invitrogen; Carlsbad, CA, USA). One microgram of RNA was used to synthesize cDNA using GeneAce Reverse Transcriptase (Nippon Gene; Tokyo, Japan). A gene expression analysis was performed using PowerUp SYBR Green Master Mix (Thermo Fisher Scientific; Waltham, MA, USA) and the ABI qPCR system (Thermo Fisher Scientific). The primers used in this study are listed in Supplementary Table S1.

4.4. Sphere Forming Assay

A sphere assay was performed by seeding HLE and HLF cells in a 12-well low-attachment plate. Cells were treated with DMSO or 10 µM of GANT61 and incubated for 10 days. Half of the medium was changed every three days. Formed spheres were photographed and their diameters were measured.

4.5. Cell Migration Assays

Cells were grown at 90%–100% confluency and subjected to the experiment. Monolayer cells were scratched using microtips and photographed at day 0. Cells were treated with serum-free DMEM containing DMSO or 10 µM of GANT61 for 24 h and photographed at day 1. A closed area was used to evaluate cell migration activity.

4.6. Transwell Migration Assays

HLE cells were pre-treated with DMSO or GANT61 (5 and 10 µM) in a serum-starved condition for 24 h. A total of 5×10^4 cells were seeded into the upper chambers of transwell inserts with 8 µM pores (Corning Inc.; Corning, NY, USA) with DMEM but without serum. The bottom well contained 10% FBS DMEM to induce cell migration. After 24 h, the cells that migrated to the underside of the membrane were fixed and stained with crystal violet containing 10% methanol and counted to quantify the migration ability. A knockdown experiment was performed by transfection of Stealth RNAi Pre-Designed siRNAs (Thermo Fisher Scientific) for *Snail* mRNA and a negative control for 24 h. Cells treated with siRNA were subjected to the transwell assay as described above.

4.7. Western Blot Analysis

Cell lysates were prepared from the cells subjected to the experiment by using Cell Lysis Buffer (Cell Signaling Technology; Danvers, MA, USA) and stored in a deep freezer until use. The protein concentration was determined by the Pierce™ BCA Protein Assay Kit (Thermo Fisher Scientific). Equal amounts of the protein sample were separated on a SDS-PAGE and immunoblotted with antibodies against human E-cadherin (Santa Cruz Biotechnology), vimentin (Santa Cruz Biotechnology), MEK1 (Novus Biologicals, Centennial CO, USA), phospho-MEK1/2 (Ser2017/221) (Cell Signaling Technology), phospho-ERK1/2 (Thr202/Tyr204) (Cell Signaling Technology), and beta-actin (Cell Signaling Technology). HRP-conjugated secondary antibodies were used to detect primary antibodies and were visualized with ECL detection reagent (GE Healthcare UK; Buckinghamshire, UK) and a light-sensitive X-ray film (Fuji film; Tokyo, Japan).

4.8. Statistical Analysis

Statistical comparisons for in vitro experiments were performed using Student's *t*-tests. $p < 0.05$ was considered statistically significant.

Supplementary Materials: Supplementary materials can be found at http://www.mdpi.com/1422-0067/21/9/3126/s1.

Author Contributions: K.H. and K.K. designed the experiments. K.H. performed the cell viability assay and sphere forming assay. R.O. performed cell migration assays. K.N. performed the siRNA experiment. K.K. performed the gene expression analysis and Western blotting. K.K. wrote the manuscript and acquired funding. All authors have read and agreed to the published version of the manuscript.

Funding: This research was supported by Grants-in-Aid for Scientific Research (KAKENHI, 18K07214), Japan Society for the Promotion of Science (JSPS) and The Okayama Medical Foundation.

Conflicts of Interest: The authors declare no conflict of interest.

Abbreviations

HCC	Hepatocellular carcinoma
HH	Hedgehog
GLI	Glioma-associated homologue
SMO	Smoothened
EMT	Epithelial–mesenchymal transition
RT-PCR	Reverse Transcription-Polymerase Chain Reaction
HNF4α	Hepatocyte nuclear factor 4α
DMSO	Dimethyl sulfoxide
siRNA	Small-interference RNA
ALDH	Aldehyde dehydrogenase
CSC	Cancer Stem Cells

References

1. Yang, J.D.; Hainaut, P.; Gores, G.J.; Amadou, A.; Plymoth, A.; Roberts, L.R. A global view of hepatocellular carcinoma: Trends, risk, prevention and management. *Nat. Rev. Gastroenterol. Hepatol.* **2019**, *16*, 589–604. [CrossRef] [PubMed]
2. Gupta, P.B.; Pastushenko, I.; Skibinski, A.; Blanpain, C.; Kuperwasser, C. Phenotypic Plasticity: Driver of Cancer Initiation, Progression, and Therapy Resistance. *Cell Stem Cell* **2019**, *24*, 65–78. [CrossRef] [PubMed]
3. Flores-Téllez, T.N.; Villa-Treviño, S.; Piña-Vázquez, C. Road to stemness in hepatocellular carcinoma. *World J. Gastroenterol.* **2017**, *23*, 6750–6776. [CrossRef] [PubMed]
4. Della Corte, C.M.; Viscardi, G.; Papaccio, F.; Esposito, G.; Martini, G.; Ciardiello, D.; Martinelli, E.; Ciardiello, F.; Morgillo, F. Implication of the Hedgehog pathway in hepatocellular carcinoma. *World J. Gastroenterol.* **2017**, *23*, 4330–4340. [CrossRef]

5. Gonnissen, A.; Isebaert, S.; Haustermans, K. Targeting the Hedgehog signaling pathway in cancer: Beyond Smoothened. *Oncotarget* **2015**, *6*, 13899–13913. [CrossRef]
6. Rohatgi, R.; Milenkovic, L.; Scott, M.P. Patched1 regulates hedgehog signaling at the primary cilium. *Science* **2007**, *317*, 372–376. [CrossRef]
7. Kovacs, J.J.; Whalen, E.J.; Liu, R.; Xiao, K.; Kim, J.; Chen, M.; Wang, J.; Chen, W.; Lefkowitz, R.J. Beta-arrestin-mediated localization of smoothened to the primary cilium. *Science* **2008**, *320*, 1777–1781. [CrossRef]
8. Katoh, Y.; Katoh, M. Hedgehog target genes: Mechanisms of carcinogenesis induced by aberrant hedgehog signaling activation. *Curr. Mol. Med.* **2009**, *9*, 873–886. [CrossRef]
9. Gupta, S.; Takebe, N.; Lorusso, P. Targeting the Hedgehog pathway in cancer. *Ther. Adv. Med. Oncol.* **2010**, *2*, 237–250. [CrossRef]
10. Sekulic, A.; Migden, M.R.; Oro, A.E.; Dirix, L.; Lewis, K.D.; Hainsworth, J.D.; Solomon, J.A.; Yoo, S.; Arron, S.T.; Friedlander, P.A.; et al. Efficacy and safety of vismodegib in advanced basal-cell carcinoma. *N. Engl. J. Med.* **2012**, *366*, 2171–2179. [CrossRef]
11. Rodon, J.; Tawbi, H.A.; Thomas, A.L.; Stoller, R.G.; Turtschi, C.P.; Baselga, J.; Sarantopoulos, J.; Mahalingam, D.; Shou, Y.; Moles, M.A.; et al. A phase I, multicenter, open-label, first-in-human, dose-escalation study of the oral smoothened inhibitor Sonidegib (LDE225) in patients with advanced solid tumors. *Clin. Cancer Res.* **2014**, *20*, 1900–1909. [CrossRef] [PubMed]
12. Jimeno, A.; Weiss, G.J.; Miller, W.H., Jr.; Gettinger, S.; Eigl, B.J.; Chang, A.L.; Dunbar, J.; Devens, S.; Faia, K.; Skliris, G.; et al. Phase I study of the Hedgehog pathway inhibitor IPI-926 in adult patients with solid tumors. *Clin. Cancer Res.* **2013**, *19*, 2766–2774. [CrossRef] [PubMed]
13. Chang, A.L.; Oro, A.E. Initial assessment of tumor regrowth after vismodegib in advanced Basal cell carcinoma. *Arch. Dermatol.* **2012**, *148*, 1324–1325. [CrossRef]
14. Yauch, R.L.; Dijkgraaf, G.J.; Alicke, B.; Januario, T.; Ahn, C.P.; Holcomb, T.; Pujara, K.; Stinson, J.; Callahan, C.A.; Tang, T.; et al. Smoothened mutation confers resistance to a Hedgehog pathway inhibitor in medulloblastoma. *Science* **2009**, *326*, 572–574. [CrossRef] [PubMed]
15. Kool, M.; Jones, D.T.; Jäger, N.; Northcott, P.A.; Pugh, T.J.; Hovestadt, V.; Piro, R.M.; Esparza, L.A.; Markant, S.L.; Remke, M.; et al. Genome sequencing of SHH medulloblastoma predicts genotype-related response to smoothened inhibition. *Cancer Cell* **2014**, *25*, 393–405. [CrossRef] [PubMed]
16. Pietrobono, S.; Gagliardi, S.; Stecca, B. Non-canonical Hedgehog Signaling Pathway in Cancer: Activation of GLI Transcription Factors Beyond Smoothened. *Front. Genet.* **2019**, *10*, 556. [CrossRef]
17. Singh, B.N.; Koyano-Nakagawa, N.; Donaldson, A.; Weaver, C.V.; Garry, M.G.; Garry, D.J. Hedgehog Signaling during Appendage Development and Regeneration. *Genes* **2015**, *6*, 417–435. [CrossRef]
18. Okabe, H.; Ishimoto, T.; Mima, K.; Nakagawa, S.; Hayashi, H.; Kuroki, H.; Imai, K.; Nitta, H.; Saito, S.; Hashimoto, D.; et al. CD44s signals the acquisition of the mesenchymal phenotype required for anchorage-independent cell survival in hepatocellular carcinoma. *Br. J. Cancer* **2014**, *110*, 958–966. [CrossRef]
19. Mima, K.; Okabe, H.; Ishimoto, T.; Hayashi, H.; Nakagawa, S.; Kuroki, H.; Watanabe, M.; Beppu, T.; Tamada, M.; Nagano, O.; et al. CD44s regulates the TGF-beta-mediated mesenchymal phenotype and is associated with poor prognosis in patients with hepatocellular carcinoma. *Cancer Res.* **2012**, *72*, 3414–3423. [CrossRef]
20. Carpenter, R.L.; Lo, H.W. Hedgehog pathway and GLI1 isoforms in human cancer. *Discov. Med.* **2012**, *13*, 105–113.
21. Speek, M.; Njunkova, O.; Pata, I.; Valdre, E.; Kogerman, P. A potential role of alternative splicing in the regulation of the transcriptional activity of human GLI2 in gonadal tissues. *BMC Mol. Biol.* **2006**, *7*, 13. [CrossRef] [PubMed]
22. Pantazi, E.; Gemenetzidis, E.; Trigiante, G.; Warnes, G.; Shan, L.; Mao, X.; Ikram, M.; The, M.T.; Lu, Y.J.; Philpott, M.P. GLI2 induces genomic instability in human keratinocytes by inhibiting apoptosis. *Cell Death Dis.* **2014**, *5*, e1028. [CrossRef] [PubMed]
23. Huang, L.; Walter, V.; Hayes, D.N.; Onaitis, M. Hedgehog-GLI signaling inhibition suppresses tumor growth in squamous lung cancer. *Clin. Cancer Res.* **2014**, *20*, 1566–1575. [CrossRef] [PubMed]
24. Vlčková, K.; Réda, J.; Ondrušová, L.; Krayem, M.; Ghanem, G.; Vachtenheim, J. GLI inhibitor GANT61 kills melanoma cells and acts in synergy with obatoclax. *Int. J. Oncol.* **2016**, *49*, 953–960. [CrossRef] [PubMed]

25. Tong, W.; Qiu, L.; Qi, M.; Liu, J.; Hu, K.; Lin, W.; Huang, Y.; Fu, J. GANT-61 and GDC-0449 induce apoptosis of prostate cancer stem cells through a GLI-dependent mechanism. *J. Cell Biochem.* **2018**, *119*, 3641–3652. [CrossRef]
26. Li, J.; Cai, J.; Zhao, S.; Yao, K.; Sun, Y.; Li, Y.; Chen, L.; Li, R.; Zhai, X.; Zhang, J.; et al. GANT61, a GLI inhibitor, sensitizes glioma cells to the temozolomide treatment. *J. Exp. Clin. Cancer Res.* **2016**, *35*, 184. [CrossRef]
27. Riaz, S.K.; Ke, Y.; Wang, F.; Kayani, M.A.; Malik, M.F.A. Influence of SHH/GLI1 axis on EMT mediated migration and invasion of breast cancer cells. *Sci. Rep.* **2019**, *9*, 6620. [CrossRef]
28. Riaz, S.K.; Khan, J.S.; Shah, S.T.A.; Wang, F.; Ye, L.; Jiang, W.G.; Malik, M.F.A. Involvement of hedgehog pathway in early onset, aggressive molecular subtypes and metastatic potential of breast cancer. *Cell Commun. Signal.* **2018**, *16*, 3. [CrossRef]
29. Réda, J.; Vachtenheim, J.; Vlčková, K.; Horák, P.; Vachtenheim, J., Jr.; Ondrušová, L. Widespread Expression of Hedgehog Pathway Components in a Large Panel of Human Tumor Cells and Inhibition of Tumor Growth by GANT61: Implications for Cancer Therapy. *Int. J. Mol. Sci.* **2018**, *19*, 2682. [CrossRef]
30. Nanta, R.; Shrivastava, A.; Sharma, J.; Shankar, S.; Srivastava, R.K. Inhibition of sonic hedgehog and PI3K/Akt/mTOR pathways cooperate in suppressing survival, self-renewal and tumorigenic potential of glioblastoma-initiating cells. *Mol. Cell Biochem.* **2019**, *454*, 11–23. [CrossRef]
31. Miyazaki, Y.; Matsubara, S.; Ding, Q.; Tsukasa, K.; Yoshimitsu, M.; Kosai, K.; Takao, S. Efficient elimination of pancreatic cancer stem cells by hedgehog/GLI inhibitor GANT61 in combination with mTOR inhibition. *Mol. Cancer* **2016**, *15*, 49. [CrossRef] [PubMed]
32. Sicklick, J.K.; Li, Y.X.; Melhem, A.; Schmelzer, E.; Zdanowicz, M.; Huang, J.; Caballero, M.; Fair, J.H.; Ludlow, J.W.; McClelland, R.E.; et al. Hedgehog signaling maintains resident hepatic progenitors throughout life. *Am. J. Physiol. Gastrointest. Liver Physiol.* **2006**, *290*, G859–G870. [CrossRef] [PubMed]
33. Ma, S.; Chan, K.W.; Lee, T.K.; Tang, K.H.; Wo, J.Y.; Zheng, B.J.; Guan, X.Y. Aldehyde dehydrogenase discriminates the CD133 liver cancer stem cell populations. *Mol Cancer Res.* **2008**, *6*, 1146–1153. [CrossRef] [PubMed]
34. Santini, R.; Vinci, M.C.; Pandolfi, S.; Penachioni, J.Y.; Montagnani, V.; Olivito, B.; Gattai, R.; Pimpinelli, N.; Gerlini, G.; Borgognoni, L.; et al. Hedgehog-GLI signaling drives self-renewal and tumorigenicity of human melanoma-initiating cells. *Stem Cells* **2012**, *30*, 1808–1818. [CrossRef]
35. Nanta, R.; Kumar, D.; Meeker, D.; Rodova, M.; Van Veldhuizen, P.J.; Shankar, S.; Srivastava, R.K. NVP-LDE-225 (Erismodegib) inhibits epithelial-mesenchymal transition and human prostate cancer stem cell growth in NOD/SCID IL2Rγ null mice by regulating Bmi-1 and microRNA-128. *Oncogenesis* **2013**, *2*, e42. [CrossRef]
36. Tu, Y.; Niu, M.; Xie, P.; Yue, C.; Liu, N.; Qi, Z.; Gao, S.; Liu, H.; Shi, Q.; Yu, R.; et al. Smoothened is a poor prognosis factor and a potential therapeutic target in glioma. *Sci. Rep.* **2017**, *7*, 42630. [CrossRef]
37. Magistri, P.; Battistelli, C.; Strippoli, R.; Petrucciani, N.; Pellinen, T.; Rossi, L.; Mangogna, L.; Aurello, P.; D'Angelo, F.; Tripodi, M.; et al. SMO Inhibition Modulates Cellular Plasticity and Invasiveness in Colorectal Cancer. *Front. Pharmacol.* **2018**, *8*, 956. [CrossRef]

International Journal of
Molecular Sciences

MDPI

Article

Efficacy of Immune Checkpoint Inhibitor Monotherapy for Advanced Non-Small-Cell Lung Cancer with *ALK* Rearrangement

Yuko Oya [1,2], Hiroaki Kuroda [2,*], Takeo Nakada [2], Yusuke Takahashi [2], Noriaki Sakakura [2] and Toyoaki Hida [1]

1 Department of Thoracic Oncology, Aichi Cancer Center Hospital, 1-1 Kanokoden, Chikusa-ku, Nagoya, Aichi 464-8681, Japan; yshima@aichi-cc.jp (Y.O.); 107974@aichi-cc.jp (T.H.)

2 Department of Thoracic Surgery, Aichi Cancer Center Hospital, 1-1 Kanokoden, Chikusa-ku, Nagoya, Aichi 464-8681, Japan; t.nakada@aichi-cc.jp (T.N.); y.takahashi@aichi-cc.jp (Y.T.); nsakakura@aichi-cc.jp (N.S.)

* Correspondence: h-kuroda@aichi-cc.jp; Tel.: +81-52-762-6111; Fax: +81-52-764-2963

Received: 21 March 2020; Accepted: 7 April 2020; Published: 9 April 2020

Abstract: Programmed death-ligand 1 (PD-L1) expression is a predictor of immune checkpoint inhibitor (ICI) treatment efficacy. The clinical efficacy of ICIs for non-small-cell lung cancer (NSCLC) patients harboring major mutations, such as *EGFR* or *ALK* mutations, is limited. We genotyped 190 patients with advanced lung adenocarcinomas who received nivolumab or pembrolizumab monotherapy, and examined the efficacy in NSCLC patients with or without major mutations. Among the patients enrolled in the genotyping study, 47 patients harbored *EGFR* mutations, 25 patients had *KRAS* mutations, 5 patients had a *HER2* mutation, 6 patients had a *BRAF* mutation, and 7 patients had *ALK* rearrangement. The status of PD-L1 expression was evaluated in 151 patients, and the rate of high PD-L1 expression (≥50%) was significantly higher in patients with *ALK* mutations. The progression-free survival was 0.6 (95% CI: 0.2–2.1) months for *ALK*-positive patients and 1.8 (95% CI: 1.2–2.1) months for *EGFR*-positive patients. All patients with *ALK* rearrangement showed disease progression within three months from the initiation of anti-PD-1 treatment. Our data suggested that ICI treatment was significantly less efficacious in patients with *ALK* rearrangement than in patients with *EGFR* mutations, and PD-L1 expression was not a critical biomarker for ICI treatment for patients with one of these mutations.

Keywords: *ALK*; *EGFR*; non-small-cell lung cancer (NSCLC); immune checkpoint inhibitor (ICI); adenocarcinoma

1. Introduction

Anaplastic lymphoma kinase (*ALK*) is a tyrosine kinase that is constitutively activated in certain types of cancers due to genetic abnormalities, such as chromosomal translocations, gene amplification, and point mutations [1,2]. In 2007, Mano et al. [3] reported a small translocation within chromosome 2 in lung adenocarcinoma cells, which results in the generation of an echinoderm microtubule-associated protein-like 4 (*EML4*)–*ALK* fusion gene. The *ALK* is a receptor tyrosine kinase that is responsible for cell growth, but when it is fused with EML4, the enzyme activity is increased to become an EML4–ALK kinase that has strong oncogenic potential. Additionally, Mano et al. [4] reported the oncogenicity of the *EML4–ALK* fusion gene in genetically modified mice, which produced *EML4–ALK* kinase specific to lungs and developed multiple lung cancers soon after birth.

ALK gene rearrangement occurs in a small portion of patients with non-small-cell lung cancer (NSCLC) [5], but it accounts for about 30%–40% of lung adenocarcinoma in young individuals, and is present in many non-smokers and patients with epidermal growth factor receptor (*EGFR*)-mutated lung cancer [6–8]. *ALK* tyrosine kinase inhibitors (TKIs) are effective in NSCLC patients with *ALK* gene rearrangement. Crizotinib, a first-generation *ALK* inhibitor that was originally developed as a *c-Met* inhibitor, also potently inhibits *ALK* kinase [9,10]. Crizotinib showed clinical benefit for NSCLC patients with *ALK* rearrangement beyond cytotoxic chemotherapy in the PROFILE 1007 trial and PROFILE 1014 trial, and has been approved and used in clinical practice [9,10]. However, after the initial positive response, all patients developed resistance to crizotinib after approximately one year. The most frequent resistance mechanism is the L1196M gatekeeper mutation, but resistance mechanisms of *ALK* inhibitors are more diverse than those of *EGFR* TKIs [11–13]. Second-generation *ALK* inhibitors have been developed as effective drugs for *ALK*-positive NSCLC patients who have acquired resistance, such as alectinib, which has shown remarkable efficacy after acquired resistance to crizotinib in clinical trials [14]. To answer the question of whether sequential treatment with alectinib after crizotinib is better than alectinib for *ALK*-positive NSCLC as an initial treatment, alectinib and crizotinib were previously compared as a first-line treatment for *ALK*-positive NSCLC. The results showed that alectinib was superior to crizotinib as a first-line treatment for *ALK*-positive NSCLC [15,16]. However, resistance to alectinib eventually developed. Moreover, the resistance mechanism of alectinib has been reported to be more diverse than that of crizotinib [17]. Approximately 30% of resistance to *ALK* inhibitors is due to gene mutations, but the new acquired resistance mechanisms are diverse [18]. Attempting to overcome resistance by focusing on resistant mutated genes is not a common treatment strategy [19]. Therefore, cytotoxic chemotherapy and immune therapy are needed to treat patients who develop resistance to *ALK* inhibitors. Treatment after acquired resistance to *ALK* inhibitors include other *ALK* inhibitors, but an immune checkpoint inhibitor (ICI) + chemotherapy, which is the standard first-line treatment for NSCLC without a driver oncogene, is a candidate for *ALK* inhibitor resistance. However, there is less data on the efficacy of ICIs for *ALK*-rearranged NSCLC. Among the clinical trials for ICI + chemotherapy, those including *ALK*- and *EGFR*-positive cases have reported the use of atezorizumab + bevacizumab + carboplatin + paclitaxel (ABCP) (IMPOWER 150 trial). For *EGFR/ALK*-positive cases, ABCP showed a 70.6% objective response rate (ORR), and ABCP was superior to BCP in progression-free survival (PFS) and overall survival (OS) in a subgroup analysis (PFS—hazard ratio (HR): 0.59, 95% confidence interval (CI): 0.37–0.94; OS—HR: 0.54, 95% CI: 0.29–1.0). However, only 11 *ALK*-positive cases were included in the IMPOWER 150 trial [20]. Similarly, there are a few reports of *ALK*-positive NSCLC with ICI monotherapy [21,22]. Additionally, in many clinical trials, *ALK* and *EGFR* are administered in combination. Although the backgrounds of patients who have *EGFR*- and *ALK*-positive tumors commonly include non-smokers and young patients [6–8], it is unclear whether a similar clinical course is followed in immunotherapy. It has been reported that programmed death-ligand 1 (PD-L1) expression is high in *ALK*-positive lung cancer [23,24]. PD-L1 expression is one of the predictors of efficacy for ICI treatment [25,26]. However, there are a few reports of studies that have investigated the efficacy of PD-1 inhibitors in *ALK*-rearranged NSCLC patients [21,22]. Therefore, we compared the efficacies of ICIs for the treatment of lung adenocarcinoma with *ALK* rearrangement and other oncogene drivers, including *EGFR* mutations.

2. Results

2.1. Patients and Treatment

The median follow-up was 10.2 months (range: 0–51 months). The patient characteristics are summarized in Table 1. The median age of the patients was 66 years (range: 32–87 years). All patients had adenocarcinoma, and 141 patients were smokers. The Eastern Cooperative Oncology Group Performance Status (PS) was 0 to 1 for 166 patients and 2 to 3 for 24 patients. Anti-PD-1 therapy was the first-line treatment for 27 patients, the second-line treatment for 70 patients, and the third-line treatment for 93 patients. Nivolumab was used to treat 138 patients, and pembrolizumab was used to treat 52 patients. Forty-seven patients (24%) harbored *EGFR* mutations, 25 patients (13%) had v-Ki-ras2 Kirsten rat sarcoma viral oncogene homolog (*KRAS*) mutations, 5 patients (3%) had a human epidermal growth factor receptor type 2 (*HER2*) mutation, 6 patients (3%) had a B-Raf proto-oncogene, serine/threonine kinase (*BRAF*) mutation, and 7 patients (4%) had *ALK* rearrangement. The status of PD-L1 expression could be evaluated in 117 (62%) patients. Fifty-eight patients had high PD-L1 expression (≥50%), 52 patients had low PD-L1 expression (1%–49%), and 41 patients were PD-L1 negative.

Table 1. Patient characteristics.

Characteristics		N = 190
Age	Median (Range)	66 (32–87)
	<75 years	173 (91%)
	≤75 years	17 (9%)
Sex	Male	128 (67%)
	Female	62 (33%)
Stage	Relapse post-operation	32 (17%)
	III/IV	158 (83%)
ICI	Nivolumab	138 (73%)
	Pembrolizumab	52 (27%)
ECOG PS	0–1	166 (87%)
	2–3	24 (13%)
Histology	Adenocarcinoma	190 (100%)
Smoking	Smoker	141 (74%)
	Non-smoker	49 (26%)
Mutation	*EGFR*	47 (25%)
	ALK	7 (4%)
	Others	40 (21%)
	None	96 (51%)
PD-L1 status	Unknown	39 (21%)
	0	41 (22%)
	1%–49%	52 (27%)
	≥50%	58 (31%)

ICI, immune checkpoint inhibitor; ECOG: Eastern Cooperative Oncology Group; PS: Performance status; and PD-L1, program death-ligand 1.

When the database was closed, 23 of the patients were continuing anti-PD-1 treatment. For efficacy measurements, the objective response rate (ORR) and the median progression-free survival (PFS) in all patients (N = 190) were 22% and 2.4 (95% CI: 2.0–3.1) months, respectively (Figure 1A), and the OS of patients treated with anti-PD-1 therapy was 14.3 (95% CI: 10.0–19.3) months (Figure 1B). The PFS and OS data presented in this cohort are comparable to those presented in ICI monotherapy clinical trials [20,25–27].

2.2. *PD-L1 Expression for Each Mutation*

The PD-L1 expression for each mutation is shown in Figure 1C. The rate of high PD-L1 expression (≥50%) was significantly higher in patients with *ALK* mutations than in patients with other mutations, including *EGFR* mutations.

2.3. *Efficacy of Immune Checkpoint Inhibitors According to Oncogenic Subtype*

The PFS was 0.6 (95% CI: 0.2–2.1) months for *ALK*-positive patients, 1.8 (95% CI: 1.2–2.1) months for *EGFR*-positive patients, 3.2 (95% CI: 1.8–7.6) months for patients with other mutations, and 3.2 (95% CI: 2.3–4.8) months for patients without any mutations (Figure 2A). Among patients with *EGFR* or *ALK* mutations, the PFS following ICI treatment was significantly longer in patients with *EGFR* mutations than in those with *ALK* mutations (*ALK*: 0.6 (95% CI: 0.2–2.1) months, *EGFR*: 1.8 (95% CI: 1.2–2.1) months; $p < 0.01$) (Figure 2B). The predictive factors in the univariate and multivariate analyses are presented in Table 2. The good PS (0–1), PD-L1 status (positive), mutation (without *EGFR* or

ALK), and baseline C-reactive protein (CRP) value (<1.0 mg/dL) were significant factors for PFS of ICI-treated patients.

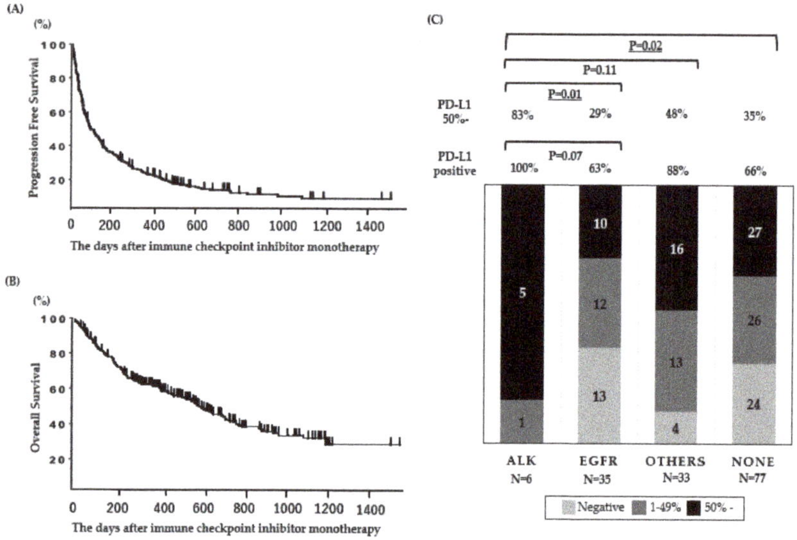

Figure 1. Clinical outcomes and PD-L1 expression in patients treated with ICI. (**A**) Progression-free survival and (**B**) overall survival (N = 190). (**C**) PD-L1 status according to oncogenic subtypes in patients assessed for PD-L1 expression (N = 151). Underlined values indicate a significant difference.

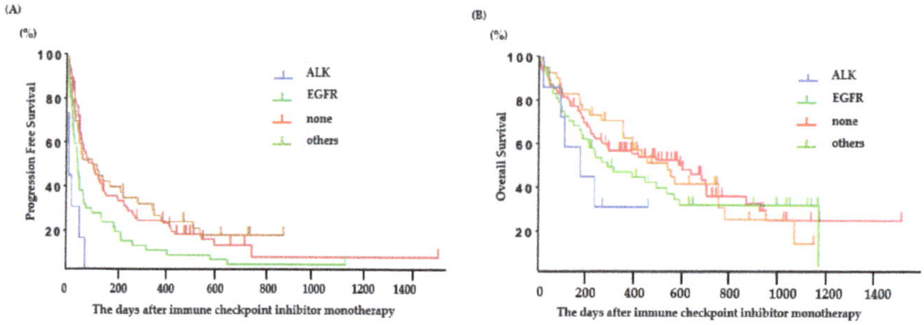

Figure 2. Clinical outcomes in patients treated with ICI. (**A**) Progression-free survival based on each oncogenic subtype and (**B**) overall survival based on each oncogenic subtype.

Table 2. Univariate and multivariate analyses for progression-free survival of ICI-treated patients.

	Univariate				Multivariate
Objectives		**Risk Ratio**	**95% CI**	***p*-Value**	
Sex	Male/Female	0.62	0.45–0.85	<0.01	0.37
Mutation	Without *EGFR*, *ALK*/*EGFR*, or *ALK*	0.54	0.39–0.75	<0.01	0.02
PS	0–1/2–3	0.24	0.15–0.38	<0.01	<0.01
PD-L1	Positive/negative	0.57	0.39–0.83	<0.01	0.01
Smoking	Smoker/non-smoker	0.59	0.42–0.83	<0.01	0.92
CRP	<1.0/≥1.0 mg/dL	0.65	0.48–0.88	<0.01	0.02
LDH	<245/≥245 U	0.59	0.43–0.81	<0.01	0.07
Albumin	<3.5/≥3.5 g/dL	0.68	0.48–0.98	0.04	0.8
NLR	<5.0/≥5.0	0.56	0.40–0.80	<0.01	0.33

PS, performance status; PD-L1, program death-ligand 1; CRP, C-reactive protein; LDH, lactate dehydrogenase; and NLR, neutrophil-to-lymphocyte ratio.

The PFS of ICI-treated patients with PD-L1 high expression (≥50%), PD-L1 low expression (1%–49%), and PD-L1-negative tumors was 5.2 (95% CI: 2.1–9.2) months, 2.9 (95% CI: 2.1–4.8) months, and 2.1 (95% CI: 1.4–3.2) months (Figure 3A), and the OS was 25.4 (95% CI: 15.5–32.0) months, 18.3 (95% CI: 8.3–23.0) months, and 10.9 (95% CI: 6.5–20.3) months, respectively (Figure 3B).

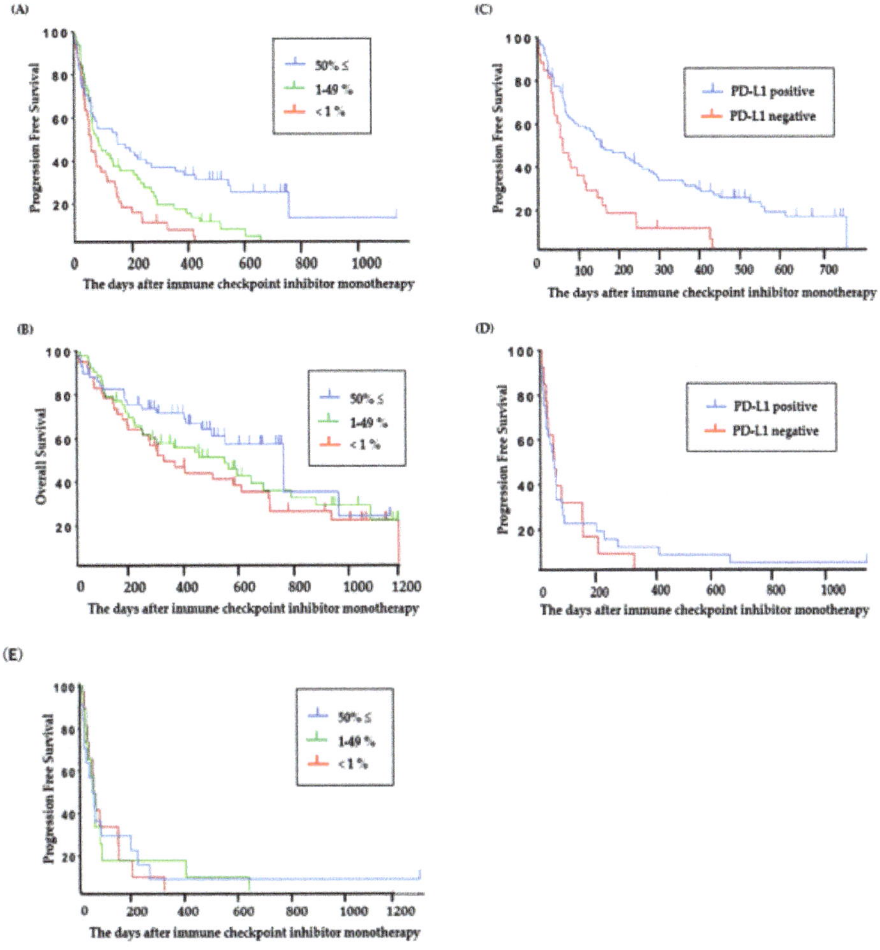

Figure 3. Clinical outcomes in patients treated with ICI. (**A**) Progression-free survival and (**B**) overall survival based on PD-L1 status in all patients viable for PD-L1 assessment (N = 151). (**C**) Progression-free survival based on PD-L1 status in patients without *EGFR* or *ALK* mutations (N = 141) and (**D**,**E**) with *EGFR* or *ALK* mutations (N = 54).

The PFS of ICI was significantly longer in patients who were PD-L1 positive than in those who were PD-L1 negative in the group without *EGFR* or *ALK* mutations (PD-L1 positive: 5.2 (95% CI: 3.1–8.6) months (Figure 3C), PD-L1 negative: 2.1 (95% CI: 1.4–3.8) months; $P < 0.01$), but for patients with *EGFR* or *ALK* mutations, there was no significant difference between the PFS of ICI-treated patients who were PD-L1 positive and those who were PD-L1 negative (PD-L1 positive: 1.8 (95% CI: 1.0–2.1) months, PD-L1 negative: 1.9 (95% CI: 0.9–5.1) months; $P = 0.83$) (Figure 3D). ICI-treated patients with PD-L1 high expression (≥50%) and patients with *ALK* or *EGFR* mutations showed short PFS. The PFS of patients with PD-L1 high expression (≥50%), PD-L1 low expression (1%–49%), and PD-L1-negative tumors in the group with *ALK* or *EGFR* mutations was 1.9 (95% CI: 0.9–5.1) months, 1.9 (95% CI: 0.9–2.7) months, and 1.7 (95% CI: 0.6–2.8) months, respectively (Figure 3E).

All patients with *ALK* rearrangement showed disease progression within three months from the initiation of anti-PD-1 treatment.

3. Discussion

We found that anti-PD-1 therapy was less efficacious in patients with *ALK* rearrangement than in patients with other mutations, including *EGFR* mutations. Recently, several studies that examined the correlation between oncogenic subtype and efficacy of ICIs have been reported [21,22,27–29]. Mazieres et al. reported the association between the efficacy of ICI and driver oncogenes and found that patients with actionable mutations (*EGFR*, *ALK*, and *ROS-1*) had poor ICI treatment outcomes [22,30]. The present study is the first to show the difference in ICI efficacy between patients with *EGFR* mutations and *ALK* rearrangement. TKIs are used as first-line treatment for patients with driver oncogenes, but after patients become resistant, ICI treatment is considered at any stage. The efficacy of nivolumab was found to be inferior to that of docetaxel in patients with *EGFR* mutations in the Checkmate 057 trial [27], but there is less data on the efficacies of ICIs for NSCLC with *ALK* rearrangement. Different *ALK* inhibitors are clinically used subsequent to the resistance to *ALK* inhibitors, but ICI + chemotherapy, which is the standard first-line treatment for NSCLC without a driver oncogene, is another option. However, little is known about the efficacy of these regimens in patients with an *EGFR* or *ALK* mutation. The IMPOWER 150 trial [20] reported ICI + chemotherapy efficacy in patients, including those with *ALK* and *EGFR* mutations, and ABCP was used for treatment in the trial. In the *EGFR*-positive cases, ABCP gave a 70.6% ORR, and the HR for BCP therapy was 0.61 (95% CI: 0.29–1.28). However, only 11 *ALK*-positive patients were included in the IMPOWER 150 trial, and the effects of treatment on those patients were not described [20].

In addition, it is not known whether the predictive factors of ICI treatment efficacy are equivalent between patients with and without driver oncogenes. In this study, a multivariate analysis detected that good PS (0–1), PD-L1 status (positive), mutation (without *EGFR* or *ALK*), and baseline C-reactive protein (CRP) value (<1.0 mg/dL) were significant factors for PFS following ICI treatment. There are some reports that ICI treatment has less efficacy in patients with a high CRP level. There are several reports, including our report, that ICI treatment has less efficacy in patients with poor PS or high CRP. It is thought that these are due to immunosuppression and the relationship with inflammatory cytokines [31–33], but the reasons are not clear. Therefore, CRP is not a clear biomarker. However, PD-L1 expression, one of the biomarkers of ICI treatment, is important when determining the course of treatment [24,25]. We do not know whether PD-L1 is a suitable biomarker for patients with *EGFR* or *ALK* mutations. In this study, we found that the correlation between PD-L1 expression and the effect of ICI treatment was not strongly associated in patients with *ALK* rearrangement, in spite of the small population included in the study.

Recently, several studies have shown that oncogenic signals derived from mutations or loss of tumor suppressor genes upregulate the expression of immune checkpoint molecules in cancer cells during immune escape [34]. In *EGFR*-mutated NSCLC, PD-L1 expression was reportedly enhanced, which resulted in the suppression of T-cell function via activation of the PD-1/PD-L1 pathway. Furthermore, it has been reported that the *EML4-ALK* rearrangement upregulated PD-L1 expression via HIF-1α and STAT3 in vitro [35].

The expression of PD-L1 occurs by two mechanisms. First, tumor-infiltrating lymphocytes (TIL) in the tumor microenvironment produce interferon-γ, thereby upregulating PD-L1. Thus, PD-L1 expression may be correlated with the existence of TILs, and PD-L1 may be used as a biomarker of the efficacy of ICI treatment [36–38]. Second, PD-L1 is upregulated by a pathway downstream of the driver oncogene [35,39]. In this case, the expression of PD-L1 was increased secondarily with no relationship with the presence of TILs or effects of ICI. In the present study, all patients with *ALK*-positive cases experienced exacerbation within three months after the initiation of ICI treatment, even patients with high expression of PD-L1. This result indicates that PD-L1 expression may not be an effective predictor in patients with lung cancer who have *ALK* or *EGFR* mutations (Figure 4).

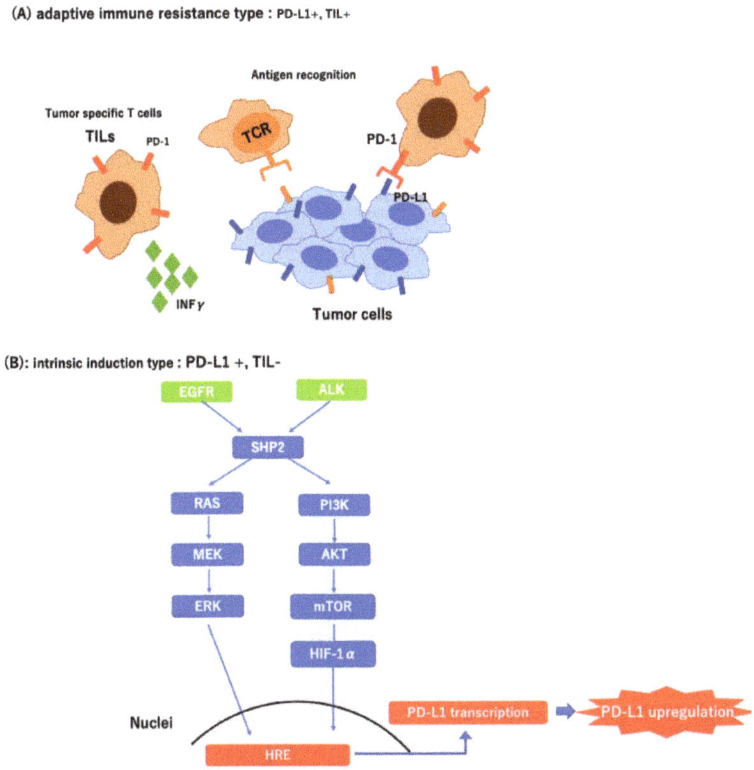

Figure 4. Different types of tumor microenvironment in PD-L1-positive tumors. (**A**) Adaptive immune resistance type: Tumor Infiltrating Lymphocyte (+); (**B**) Intrinsic induction type: Tumor Infiltrating Lymphocyte (−)

Another point to be considered in the treatment with ICI for *ALK*-positive patients is that early treatment with ICI could take away the next treatment opportunity. For example, some patients experience a unique pattern of progression disease called hyperprogression [40–42], which is characterized by rapid disease progression after the initiation of ICI. Patients who experience hyperprogression show poor prognosis, and they cannot take the next treatment. Some retrospective studies have also reported an increased risk of hepatotoxicity with the sequential use of ICIs and crizotinib [43]. These data are consistent with the unexpected severe AE rates reported for different combinations of *ALK* inhibitors and PD-1/PD-L1 inhibitors [44]. Patients who show sever AEs cannot take the next treatment until the AEs are reduced. Patients with *EGFR* or *ALK* mutations respond to TKIs dramatically. Even after they acquire resistance to TKIs, they are still subjected to certain cytotoxic chemotherapeutic effects. In particular, pemetrexed has been reported to have good antitumor activity in patients with *ALK* rearrangement [45]. Therefore, it is a great disadvantage that these patients cannot receive the next treatment because an ICI was used earlier.

In this study, the effect of ICI in patients with other mutations was almost equivalent to the effect in patients without mutations, and PD-L1 expression served as a predictive factor. These differences between *EGFR*, *ALK*, and other oncogenes were not biologically obvious. However, *EGFR* and *ALK* greatly contribute to carcinogenesis as driver oncogenes, and targeted therapy is effective. Other genes, including *KRAS*, may not be as carcinogenic as *EGFR* or *ALK*.

There were several limitations in this study. First, this was a retrospective study at a single institution. Bylicki et al. reported that there were no significantly difference in PFS between the *ALK* group (2.4 months) and the EGFR group (2.2 months). The PFS of *ALK*-positive patients were shorter in our study (0.6 months) compared with the Bylicki study [22]. There are some differences in the Bylicki study and our study, such as race, smoking status, and gender proportion, but these are not sufficient reasons to explain the difference. Therefore, these differences may be accidental due to the small sample size. Therefore, larger studies need to be performed to accurately validate the usefulness of ICI treatment in *ALK*-positive patients. Second, the treatment lines of ICI varied in this study. In lung cancer, it is unknown whether TKIs affect the efficacy of ICI treatment. Treatment with a TKI is recommended for the initial treatment of NSCLC patients with a driver oncogene, and almost all patients with a driver oncogene, except *EGFR* exon 20 insertion, were pre-treated with a TKI in this study. The effects of prior treatment with a TKI may be considered as a reason for poor efficacy of ICI treatment in patients with *EGFR* and *ALK* mutations. Given this limitation, prospective trials will be required to confirm the effect of driver oncogenes on the efficacy of ICI treatments. However, in patients with *EGFR* mutations or *ALK* rearrangement, TKI is firmly positioned as an initial treatment because of its high efficacy, and it is difficult to verify the effect of ICI with or without a treatment history of TKI. A single-arm phase 2 trial for *EGFR*-positive lung cancer in *EGFR* TKI-untreated patients had been initiated, but was stopped during the interim analysis because of poor efficacy [46]. *EGFR* itself may be a factor that lowers the effect of ICI, and the patients' sex may also be a factor. In addition, in lung cancer patients, pneumonitis frequently occurred in the combination therapy of pembrolizumab and osimertinib in the TATTON trial, and there were concerns about safety in the combination treatment of TKI and ICI [47]. In addition, there have been some reports of pneumonitis caused by treatment with osimertinib after ICI. Therefore, the sequential treatment strategy of TKI followed by ICI is difficult in lung cancer. In addition, it has been reported that the efficacy of ICI will be poor in late-treatment lines. Patients with *EGFR* or *ALK* mutations tend to receive ICI as a late-line treatment because other candidates are preferred for the treatment. However, in this study, there was no bias in the treatment line between patients with *EGFR* and *ALK* mutations and other cases. Third, this study did not include information on adverse effects. In melanoma patients, adverse effects, including vitiligo and rash, were reported to be good prognostic factors for patients treated with nivolumab [27,48].

4. Materials and Methods

4.1. Patients

We retrospectively analyzed 190 advanced lung adenocarcinoma patients who received nivolumab or pembrolizumab (ICIs) monotherapy from January 2015 to September 2018 at the Aichi Cancer Center Hospital. The database was closed on 1 May 2019; at this time, 117 of the 190 patients had died. This study was approved by the Institutional Review Board in our institution (ACC-2019-1-002). The informed consent was waived because of the retrospective nature of this study.

Data on patient characteristics, genetic characteristics (*EGFR*, *KRAS*, *ALK*, *HER2*, and *BRAF*), baseline laboratory data (C-reactive protein (CRP), lactate dehydrogenase (LDH), neutrophil-to-lymphocyte ratio (NLR), and serum albumin level), PD-L1 expression of the tumors, objective response, PFS of ICI, and OS were obtained. Serum LDH and CRP levels were measured just before initiation of treatment with ICI. The cut-off values for LDH, CRP, albumin, and NLR were determined from standard values and previous reports [18,41–43] and in this study, we used the following levels: serum LDH: <245 vs. ≥ 245 IU/L, serum CRP: <1.0 vs. ≥ 1.0 mg/dL, serum albumin: <3.5 vs. ≥ 3.5 mg/dL, and NLR: <5 vs. ≥ 5.

4.2. Analysis of Efficacy of ICI

The patients received at least one infusion of nivolumab (3 mg/kg every 2 weeks) or pembrolizumab (200 mg/body every 3 weeks) monotherapy. The patients were treated with an ICI until they showed disease progression or experienced unacceptable adverse events. In general, the patients underwent radiographic imaging every 2 months and were evaluated for tumor response according to the Response Evaluation Criteria in Solid Tumors, version 1.1. The ORR was calculated as the total percentage of patients with a complete response or a partial response.

4.3. Mutation Analyses of EGFR, ALK, KRAS, HER2, and BRAF

EGFR mutations (exons 18–21) were identified through the cycleave polymerase chain reaction (PCR) method. *HER2* (exon 20), *KRAS* (exons 2 and 3), and *BRAF* mutations (exons 11–15) were analyzed using fragment analysis, and the results were partially validated with direct sequencing, as previously reported [40]. *ALK* fusions were examined by reverse transcriptase (RT)-PCR, immunohistochemistry (IHC) analysis, or fluorescence in situ hybridization (FISH) assays (Vysis *ALK* Break Apart FISH Probe Kit; Vysis, Inc, Downers Grove, IL, USA). A tumor was considered to be *ALK* positive when two or more mutations were present or the RT-PCR, IHC, or FISH tests had positive results, as previously reported [49].

4.4. PD-L1 Expression Analysis

Tumor PD-L1 protein expression was evaluated during pre-treatment of (archival or recent) tumor biopsy or surgical resection specimens using an automated IHC assay. Depending on the drug, the antibody for companion diagnostic testing varied. The 28-8 pharmDx (Dako, North America, Carpinteria, CA, USA) and pembrolizumab use 22C3 pharmDx (Dako North America, Carpinteria, CA, USA). PD-L1 was evaluated, before initial treatment, with the 22C3 antibody for currently diagnosed cases. However, PD-L1 expression was evaluated by using the 28-8 antibody in cases treated with an ICI before pembrolizumab was approved. Therefore, two antibodies were used to examine some patients. In this study, PD-L1 expression was evaluated by using the values obtained by the antibodies used for companion diagnostic testing. The status of PD-L1 expression was measured as the proportion of PD-L1-expressing tumor cells in a section that included 100 or more tumor cells. In 190 patients, we identified 151 patients with tumor specimens that were evaluated for PD-L1 expression.

4.5. Statistical Analysis

All statistical analyses were performed using the JMP version 11 statistical software package (SAS Institute, Cary, NC, USA). Differences in the baseline characteristics between the groups were compared by Fisher's exact tests for categorical data. PFS was calculated from the date of therapy initiation to disease progression. OS was calculated from the date of nivolumab therapy initiation to death and censored at the date of the last visit for patients whose death could not be confirmed. The survival probabilities were estimated using the Kaplan–Meier method, where differences in the variables were calculated with the log-rank test. The multivariate regression analysis was performed according to the Cox proportional hazard model. Covariates with $P \leq 0.05$ in the univariate analysis were included in the multivariate model.

5. Conclusions

ICI treatment was significantly less efficacious in patients with *ALK* rearrangement than in patients with *EGFR* mutations, but there was no significant difference in efficacy between patients with other mutations and no mutations. ICIs have an important role in the treatment strategy for advanced NSCLC, but patients with *ALK* and *EGFR* mutations showed little benefit from ICIs. In addition, PD-L1 expression was not a critical biomarker for ICI treatment in patients with *EGFR* or *ALK* mutations.

This is important information when choosing the next treatment after TKIs. Further investigation of the association of each driver oncogene with the efficacy of ICI treatment is warranted.

Author Contributions: Conceptualization, Y.O. and T.H.; methodology, H.K.; software, Y.O. and Y.T.; validation, N.S., T.N. and T.H.; formal analysis, H.K.; investigation, Y.O.; resources, Y.O.; data curation, Y.O. and T.N.; writing—original draft preparation, Y.O.; writing—review and editing, H.K.; visualization, H.K.; supervision, T.H.; project administration, N.S. All authors have read and agreed to the published version of the manuscript.

Funding: This research received no external funding.

Conflicts of Interest: Toyoaki Hida has obtained research grants from Ono Pharmaceutical, Novartis Pharma, Chugai Pharmaceutical, Eli Lilly, Taiho Pharmaceutical, AstraZeneca, Nippon Boehringer Ingelheim, Pfizer, Bristol-Meyers Squibb, Clovis Oncology, Eisai, Takeda Bio, Dainippon Sumitomo Pharma, AbbVie, MSD, Merck Serono, Kyowa Hakko Kirin, Daiichi Sankyo, Servier, Kissei, Ignyta, and Astellas; and received personal fees from Ono Pharmaceutical, Novartis Pharma, Chugai Pharmaceutical, Eli Lilly, Taiho Pharmaceutical, AstraZeneca, Nippon Boehringer Ingelheim, Pfizer, MSD, Kissei, Clovis Oncology, and Bristol-Meyers Squibb. All the other authors have no conflicts of interest to declare.

Abbreviations

ALK	Anaplastic lymphoma kinase
EML4	Echinoderm microtubule-associated protein-like 4
NSCLC	Non-small-cell lung cancer
EGFR	Epidermal growth factor receptor
TKIs	Tyrosine kinase inhibitors
ICI	Immune checkpoint inhibitor
ABCP	Atezorizumab + bevacizumab + carboplatin + paclitaxel
PFS	Progression-free survival
OS	Overall survival
HR	Hazard ratio
CI	Confidence interval
PD-L1	Programmed death-ligand 1
KRAS	v-Ki-ras2 Kirsten rat sarcoma viral oncogene homolog
HER2	Human epidermal growth factor receptor type 2
BRAF	B-Raf proto-oncogene, serine/threonine kinase
ORR	Objective response rate
TIL	Tumor-infiltrating lymphocytes
CRP	C-reactive protein
LDH	Lactate dehydrogenase
NLR	Neutrophil-to-lymphocyte ratio
PCR	Reverse transcriptase
IHC	Immunohistochemistry
FISH	Fluorescence in situ hybridization

References

1. Solomon, B.; Varella-Garcia, M.; Camidge, D.R. ALK gene rearrangements: A new therapeutic target in a molecularly defined subset of non-Small cell lung cancer. *J. Thorac. Oncol.* **2009**, *4*, 1450–1454. [CrossRef]
2. Kwak, E.L.; Bang, Y.J.; Camidge, D.R.; Shaw, A.T.; Solomon, B.; Maki, R.G.; Ou, S.H.; Dezube, B.J.; Jänne, P.A.; Costa, D.B.; et al. Anaplastic lymphoma kinase inhibition in non-small-Cell lung cancer. *New Engl. J. Med.* **2010**, *363*, 1693–1703. [CrossRef]
3. Choi, Y.L.; Takeuchi, K.; Soda, M.; Inamura, K.; Togashi, Y.; Hatano, S.; Enomoto, M.; Hamada, T.; Haruta, H.; Watanabe, H.; et al. Identification of novel isoforms of the EML4-ALK transforming gene in non-Small cell lung cancer. *Cancer Res.* **2008**, *68*, 4971–4976. [CrossRef]
4. Soda, M.; Takada, S.; Takeuchi, K.; Choi, Y.L.; Enomoto, M.; Ueno, T.; Haruta, H.; Hamada, T.; Yamashita, Y.; Ishikawa, Y.; et al. A mouse model for EML4-ALK-Positive lung cancer. *Proc. Natl. Acad. Sci. USA* **2008**, *105*, 19893–19897. [CrossRef]
5. Pao, W.; Girard, N. New driver mutations in non-Small-Cell lung cancer. *Lancet Oncol.* **2011**, *12*, 175–180. [CrossRef]

6. Takeuchi, K.; Soda, M.; Togashi, Y.; Suzuki, R.; Sakata, S.; Hatano, S.; Asaka, R.; Hamanaka, W.; Ninomiya, H.; Uehara, H.; et al. RET, ROS1 and ALK fusions in lung cancer. *Nat. Med.* **2012**, *18*, 378–381. [CrossRef]
7. Nagashima, O.; Ohashi, R.; Yoshioka, Y.; Inagaki, A.; Tajima, M.; Koinuma, Y.; Iwakami, S.; Iwase, A.; Sasaki, S.; Tominaga, S.; et al. High prevalence of gene abnormalities in young patients with lung cancer. *J. Thorac. Dis.* **2013**, *5*, 27–30.
8. Chen, Z.; Teng, X.; Zhang, J.; Huang, K.; Shen, Q.; Cao, H.; Luo, H.; Yuan, Y.; Teng, X. Molecular features of lung adenocarcinoma in young patients. *BMC Cancer* **2019**, *19*, 777. [CrossRef]
9. Shaw, A.T.; Kim, D.W.; Nakagawa, K.; Seto, T.; Crinó, L.; Ahn, M.J.; De Pas, T.; Besse, B.; Solomon, B.J.; Blackhall, F.; et al. Crizotinib versus chemotherapy in advanced ALK-Positive lung cancer. *New Engl. J. Med.* **2013**, *368*, 2385–2394. [CrossRef]
10. Solomon, B.J.; Mok, T. First-Line crizotinib in ALK-positive lung cancer. *New Engl. J. Med.* **2015**, *372*, 782.
11. Katayama, R.; Khan, T.M.; Benes, C.; Lifshits, E.; Ebi, H.; Rivera, V.M.; Shakespeare, W.C.; Iafrate, A.J.; Engelman, J.A.; Shaw, A.T. Therapeutic strategies to overcome crizotinib resistance in non-Small cell lung cancers harboring the fusion oncogene EML4-ALK. *Proc. Natl. Acad. Sci. USA* **2011**, *108*, 7535–7540. [CrossRef]
12. Katayama, R.; Shaw, A.T.; Khan, T.M.; Mino-Kenudson, M.; Solomon, B.J.; Halmos, B.; Jessop, N.A.; Wain, J.C.; Yeo, A.T.; Benes, C.; et al. Mechanisms of acquired crizotinib resistance in ALK-Rearranged lung Cancers. *Sci. Transl. Med.* **2012**, *4*, 120ra17. [CrossRef]
13. Sasaki, T.; Koivunen, J.; Ogino, A.; Yanagita, M.; Nikiforow, S.; Zheng, W.; Lathan, C.; Marcoux, J.P.; Du, J.; Okuda, K.; et al. A novel ALK secondary mutation and EGFR signaling cause resistance to ALK kinase inhibitors. *Cancer Res.* **2011**, *71*, 6051–6060. [CrossRef]
14. Ou, S.H.; Ahn, J.S.; De Petris, L.; Govindan, R.; Yang, J.C.; Hughes, B.; Lena, H.; Moro-Sibilot, D.; Bearz, A.; Ramirez, S.V.; et al. Alectinib in Crizotinib-Refractory ALK-Rearranged Non-Small-Cell Lung Cancer: A Phase II Global Study. *J. Clin. Oncol.* **2016**, *34*, 661–668. [CrossRef]
15. Hida, T.; Nokihara, H.; Kondo, M.; Kim, Y.H.; Azuma, K.; Seto, T.; Takiguchi, Y.; Nishio, M.; Yoshioka, H.; Imamura, F.; et al. Alectinib versus crizotinib in patients with ALK-positive non-small-cell lung cancer (J-ALEX): An open-Label, randomised phase 3 trial. *Lancet* **2017**, *390*, 29–39. [CrossRef]
16. Peters, S.; Camidge, D.R.; Shaw, A.T.; Gadgeel, S.; Ahn, J.S.; Kim, D.W.; Ou, S.I.; Pérol, M.; Dziadziuszko, R.; Rosell, R.; et al. Alectinib versus Crizotinib in Untreated ALK-Positive Non-Small-Cell Lung Cancer. *New Engl. J. Med.* **2017**, *377*, 829–838. [CrossRef]
17. Romanidou, O.; Landi, L.; Cappuzzo, F.; Califano, R. Overcoming resistance to first/second generation epidermal growth factor receptor tyrosine kinase inhibitors and ALK inhibitors in oncogene-Addicted advanced non-Small cell lung cancer. *Ther. Adv. Med. Oncol.* **2016**, *8*, 176–187. [CrossRef]
18. Russo, A.; Russano, M.; Franchina, T.; Migliorino, M.R.; Aprile, G.; Mansueto, G.; Berruti, A.; Falcone, A.; Aieta, M.; Gelibter, A.; et al. Neutrophil-to-Lymphocyte Ratio (NLR), Platelet-to-Lymphocyte Ratio (PLR), and Outcomes with Nivolumab in Pretreated Non-Small Cell Lung Cancer (NSCLC): A Large Retrospective Multicenter Study. *Adv. Ther.* **2020**, *37*, 1145–1155. [CrossRef]
19. Dardaei, L.; Wang, H.Q.; Singh, M.; Fordjour, P.; Shaw, K.X.; Yoda, S.; Kerr, G.; Yu, K.; Liang, J.; Cao, Y.; et al. SHP2 inhibition restores sensitivity in ALK-rearranged non-Small-Cell lung cancer resistant to ALK inhibitors. *Nat. Med.* **2018**, *24*, 512–517. [CrossRef]
20. Socinski, M.A.; Jotte, R.M.; Cappuzzo, F.; Orlandi, F.; Stroyakovskiy, D.; Nogami, N.; Rodríguez-Abreu, D.; Moro-Sibilot, D.; Thomas, C.A.; Barlesi, F.; et al. Atezolizumab for First-Line Treatment of Metastatic Nonsquamous NSCLC. *New Engl. J. Med.* **2018**, *378*, 2288–2301. [CrossRef]
21. Gainor, J.F.; Shaw, A.T.; Sequist, L.V.; Fu, X.; Azzoli, C.G.; Piotrowska, Z.; Huynh, T.G.; Zhao, L.; Fulton, L.; Schultz, K.R.; et al. EGFR Mutations and ALK Rearrangements Are Associated with Low Response Rates to PD-1 Pathway Blockade in Non-Small Cell Lung Cancer: A Retrospective Analysis. *Clin. Cancer Res.* **2016**, *22*, 4585–4593. [CrossRef] [PubMed]
22. Bylicki, O.; Guisier, F.; Monnet, I.; Doubre, H.; Gervais, R.; Janicot, H.; Perol, M.; Fournel, P.; Lamy, R.; Auliac, J.B.; et al. Efficacy and safety of programmed cell-Death-Protein-1 and its ligand inhibitors in pretreated patients with epidermal growth-Factor receptor-mutated or anaplastic lymphoma kinase-Translocated lung adenocarcinoma. *Medicine (Baltimore)* **2020**, *99*, e18726. [CrossRef]

23. Schabath, M.B.; Dalvi, T.B.; Dai, H.A.; Crim, A.L.; Midha, A.; Shire, N.; Gimbrone, N.T.; Walker, J.; Greenawalt, D.M.; Lawrence, D.; et al. A Molecular Epidemiological Analysis of Programmed Cell Death Ligand-1 (PD-L1) Protein Expression, Mutations and Survival in Non-Small Cell Lung Cancer. *Cancer Manag. Res.* **2019**, *11*, 9469–9481. [CrossRef] [PubMed]
24. Rangachari, D.; VanderLaan, P.A.; Shea, M.; Le, X.; Huberman, M.S.; Kobayashi, S.S.; Costa, D.B. Correlation between classic driver oncogene mutations in EGFR, ALK, or ROS1 and 22C3-PD-L1 >/=50% expression in lung adenocarcinoma. *J. Thorac. Oncol.* **2017**, *12*, 878–883. [CrossRef] [PubMed]
25. Reck, M.; Rodriguez-Abreu, D.; Robinson, A.G.; Hui, R.; Csoszi, T.; Fulop, A.; Gottfried, M.; Peled, N.; Tafreshi, A.; Cuffe, S.; et al. Pembrolizumab versus chemotherapy for PD-L1-Positive non-Small-Cell lung Cancer. *New Engl. J. Med.* **2016**, *375*, 1823–1833. [CrossRef] [PubMed]
26. Herbst, R.S.; Baas, P.; Kim, D.W.; Felip, E.; Perez-Gracia, J.L.; Han, J.Y.; Molina, J.; Kim, J.H.; Arvis, C.D.; Ahn, M.J.; et al. Pembrolizumab versus docetaxel for previously treated, PD-L1-positive, advanced non-small-cell lung cancer (KEYNOTE-010): A randomised controlled trial. *Lancet* **2016**, *387*, 1540–1550. [CrossRef]
27. Borghaei, H.; Paz-Ares, L.; Horn, L.; Spigel, D.R.; Steins, M.; Ready, N.E.; Chow, L.Q.; Vokes, E.E.; Felip, E.; Holgado, E.; et al. Nivolumab versus docetaxel in advanced nonsquamous non-Small-Cell lung cancer. *New Engl. J. Med.* **2015**, *373*, 1627–1639. [CrossRef]
28. Dudnik, E.; Moskovitz, M.; Daher, S.; Shamai, S.; Hanovich, E.; Grubstein, A.; Shochat, T.; Wollner, M.; Bar, J.; Merimsky, O.; et al. Effectiveness and safety of nivolumab in advanced non-small cell lung cancer: The real-Life data. *Lung Cancer* **2018**, *126*, 217–223. [CrossRef]
29. Karatrasoglou, E.A.; Chatziandreou, I.; Sakellariou, S.; Stamopoulos, K.; Kavantzas, N.; Lazaris, A.C.; Korkolopoulou, P.; Saetta, A.A. Association between PD-L1 expression and driver gene mutations in non-small cell lung cancer patients: Correlation with clinical data. *Virchows Arch* **2020**. [CrossRef]
30. Mazieres, J.; Drilon, A.; Lusque, A.; Mhanna, L.; Cortot, A.B.; Mezquita, L.; Thai, A.A.; Mascaux, C.; Couraud, S.; Veillon, R.; et al. Immune checkpoint inhibitors for patients with advanced lung cancer and oncogenic driver alterations: Results from the IMMUNOTARGET registry. *Ann. Oncol.* **2019**, *30*, 1321–1328. [CrossRef]
31. Oya, Y.; Yoshida, T.; Kuroda, H.; Mikubo, M.; Kondo, C.; Shimizu, J.; Horio, Y.; Sakao, Y.; Hida, T.; Yatabe, Y. Predictive clinical parameters for the response of nivolumab in pretreated advanced non-Small-Cell lung cancer. *Oncotarget* **2017**, *8*, 103117–103128. [CrossRef]
32. Iivanainen, S.; Ahvonen, J.; Knuuttila, A.; Tiainen, S.; Koivunen, J.P. Elevated CRP levels indicate poor progression-Free and overall survival on cancer patients treated with PD-1 inhibitors. *ESMO Open* **2019**, *4*, e000531. [CrossRef]
33. Tanizaki, J.; Haratani, K.; Hayashi, H.; Chiba, Y.; Nakamura, Y.; Yonesaka, K.; Kudo, K.; Kaneda, H.; Hasegawa, Y.; Tanaka, K.; et al. Peripheral Blood Biomarkers Associated with Clinical Outcome in Non-Small Cell Lung Cancer Patients Treated with Nivolumab. *J. Thorac. Oncol.* **2018**, *13*, 97–105. [CrossRef]
34. Yoneshima, Y.; Ijichi, K.; Anai, S.; Ota, K.; Otsubo, K.; Iwama, E.; Tanaka, K.; Oda, Y.; Nakanishi, Y.; Okamoto, I. PD-L1 expression in lung adenocarcinoma harboring EGFR mutations or ALK rearrangements. *Lung Cancer* **2018**, *118*, 36–40. [CrossRef]
35. Koh, J.; Jang, J.Y.; Keam, B.; Kim, S.; Kim, M.Y.; Go, H.; Kim, T.M.; Kim, D.W.; Kim, C.W.; Jeon, Y.K.; et al. EML4-ALK enhances programmed cell death-Ligand 1 expression in pulmonary adenocarcinoma via hypoxia-inducible factor (HIF)-1α and STAT3. *Oncoimmunology* **2016**, *5*, e1108514. [CrossRef]
36. Im, S.J.; Hashimoto, M.; Gerner, M.Y.; Lee, J.; Kissick, H.T.; Burger, M.C.; Shan, Q.; Hale, J.S.; Lee, J.; Nasti, T.H.; et al. Defining CD8+ T cells that provide the proliferative burst after PD-1 therapy. *Nature* **2016**, *537*, 417–421. [CrossRef]
37. Chang, C.H.; Qiu, J.; O'Sullivan, D.; Buck, M.D.; Noguchi, T.; Curtis, J.D.; Chen, Q.; Gindin, M.; Gubin, M.M.; van der Windt, G.J.; et al. Metabolic Competition in the Tumor Microenvironment Is a Driver of Cancer Progression. *Cell* **2015**, *162*, 1229–1241. [CrossRef]
38. Teng, M.W.L.; Ngiow, S.F.; Ribas, A.; Smyth, M.J. Classifying Cancers Based on T-cell Infiltration and PD-L1. *Cancer Res.* **2015**, *75*, 2139–2145. [CrossRef]
39. Chen, N.; Fang, W.; Zhan, J.; Hong, S.; Tang, Y.; Kang, S.; Zhang, Y.; He, X.; Zhou, T.; Qin, T.; et al. Upregulation of PD-L1 by EGFR Activation Mediates the Immune Escape in EGFR-Driven NSCLC: Implication for Optional Immune Targeted Therapy for NSCLC Patients with EGFR Mutation. *J. Thorac. Oncol.* **2015**, *10*, 910–923. [CrossRef]
40. Champiat, S.; Dercle, L.; Ammari, S.; Massard, C.; Hollebecque, A.; Postel-Vinay, S.; Chaput, N.; Eggermont, A.; Marabelle, A.; Soria, J.C.; et al. Hyperprogressive disease is a new pattern of progression in cancer patients treated by anti-PD-1/PD-L1. *Clin. Cancer Res.* **2017**, *23*, 1920–1928. [CrossRef]

41. Kato, S.; Goodman, A.; Walavalkar, V.; Barkauskas, D.A.; Sharabi, A.; Kurzrock, R. Hyperprogressors after immunotherapy: Analysis of genomic alterations associated with accelerated growth rate. *Clin. Cancer Res.* **2017**, *23*, 4242–4250. [CrossRef]
42. Ferrara, R.; Mezquita, L.; Texier, M.; Lahmar, J.; Audigier-Valette, C.; Tessonnier, L.; Mazieres, J.; Zalcman, G.; Brosseau, S.; Le Moulec, S.; et al. Hyperprogressive Disease in Patients with Advanced Non-Small Cell Lung Cancer Treated With PD-1/PD-L1 Inhibitors or With Single-Agent Chemotherapy. *JAMA Oncol.* **2018**, *4*, 1543–1552. [CrossRef]
43. Lin, J.J.; Chin, E.; Yeap, B.Y.; Ferris, L.A.; Kamesan, V.; Lennes, I.T.; Sequist, L.V.; Heist, R.S.; Mino-Kenudson, M.; Gainor, J.F.; et al. Increased Hepatotoxicity Associated with Sequential Immune Checkpoint Inhibitor and Crizotinib Therapy in Patients with Non-Small Cell Lung Cancer. *J. Thorac. Oncol.* **2019**, *14*, 135–140. [CrossRef]
44. McCusker, M.G.; Russo, A.; Scilla, K.A.; Mehra, R.; Rolfo, C. How I treat ALK-Positive non-Small cell lung cancer. *ESMO Open* **2019**, *4*, e000524. [CrossRef]
45. Lee, J.O.; Kim, T.M.; Lee, S.H.; Kim, D.W.; Kim, S.; Jeon, Y.K.; Chung, D.H.; Kim, W.H.; Kim, Y.T.; Yang, S.C.; et al. Anaplastic Lymphoma Kinase Translocation: A Predictive Biomarker of Pemetrexed in Patients with Non-Small Cell Lung Cancer. *J. Thorac. Oncol.* **2011**, *6*, 1474–1480. [CrossRef]
46. Lisberg, A.; Cummings, A.; Goldman, J.W.; Bornazyan, K.; Reese, N.; Wang, T.; Coluzzi, P.; Ledezma, B.; Mendenhall, M.; Hunt, J.; et al. A Phase II Study of Pembrolizumab in EGFR-Mutant, PD-L1+, Tyrosine Kinase Inhibitor Naïve Patients with Advanced NSCLC. *J. Thorac. Oncol.* **2018**, *13*, 1138–1145. [CrossRef]
47. Kotake, M.; Murakami, H.; Kenmotsu, H.; Naito, T.; Takahashi, T. High incidence of interstitial lung disease following practical use of osimertinib in patients who had undergone immediate prior nivolumab therapy. *Ann. Oncol.* **2017**, *28*, 669–670. [CrossRef]
48. Brahmer, J.; Reckamp, K.L.; Baas, P.; Crinò, L.; Eberhardt, W.E.; Poddubskaya, E.; Antonia, S.; Pluzanski, A.; Vokes, E.E.; Holgado, E.; et al. Nivolumab versus Docetaxel in Advanced Squamous-Cell Non-Small-Cell Lung Cancer. *New Engl. J. Med.* **2015**, *373*, 123–135. [CrossRef]
49. Yoshida, T.; Oya, Y.; Tanaka, K.; Shimizu, J.; Horio, Y.; Kuroda, H.; Sakao, Y.; Hida, T.; Yatabe, Y. Differential Crizotinib Response Duration Among ALK Fusion Variants in ALK-Positive Non-Small-Cell Lung Cancer. *J. Clin. Oncol.* **2016**, *34*, 3383–3389. [CrossRef]

International Journal of
Molecular Sciences

MDPI

Communication

Beta3-Tubulin Is Critical for Microtubule Dynamics, Cell Cycle Regulation, and Spontaneous Release of Microvesicles in Human Malignant Melanoma Cells (A375)

Mohammed O. Altonsy [1,2], Anutosh Ganguly [1,3,4], Matthias Amrein [5], Philip Surmanowicz [1], Shu Shun Li [3], Gilles J. Lauzon [1] and P. Régine Mydlarski [1,*]

1 Division of Dermatology, Department of Medicine, University of Calgary, Calgary, AB T2T 5C7, Canada; mohammedomarahmed.mo@ucalgary.ca (M.O.A.); ganutosh@umich.edu (A.G.); philip.surmanowicz@gmail.com (P.S.); glauzon@ualberta.ca (G.J.L.)
2 Department of Zoology, Faculty of Science, Sohag University, Sohag 82524, Egypt
3 Department of Microbiology, Immunology and Infectious Diseases, University of Calgary, Calgary, AB T2N 4N1, Canada; shusli@ucalgary.ca
4 Department of Surgery, University of Michigan, Ann Arbor, MI 48105, USA
5 Department of Cell Biology and Anatomy, University of Calgary, Calgary, AB T2N 4N1, Canada; mamrein@ucalgary.ca
* Correspondence: regine.mydlarski@ucalgary.ca; Tel.: +1-403-955-8345; Fax: +1-403-955-8200

Received: 6 February 2020; Accepted: 25 February 2020; Published: 28 February 2020

Abstract: Microtubules (MTs), microfilaments, and intermediate filaments, the main constituents of the cytoskeleton, undergo continuous structural changes (metamorphosis), which are central to cellular growth, division, and release of microvesicles (MVs). Altered MTs dynamics, uncontrolled proliferation, and increased production of MVs are hallmarks of carcinogenesis. Class III beta-tubulin (β3-tubulin), one of seven β-tubulin isotypes, is a primary component of MT, which correlates with enhanced neoplastic cell survival, metastasis and resistance to chemotherapy. We studied the effects of β3-tubulin gene silencing on MTs dynamics, cell cycle, and MVs release in human malignant melanoma cells (A375). The knockdown of β3-tubulin induced G2/M cell cycle arrest, impaired MTs dynamics, and reduced spontaneous MVs release. Additional studies are therefore required to elucidate the pathophysiologic and therapeutic role of β3-tubulin in melanoma.

Keywords: melanoma; β3-tubulin; microtubules; microvesicles

1. Introduction

Class III beta-tubulin (β3-tubulin), encoded by the TUBB3 gene, is one of seven beta-tubulin isotypes in the human genome. It is constitutively expressed in neurons of the peripheral and central nervous system, regulating neuronal differentiation and development [1]. Microtubules (MTs), which comprise the cytoskeleton in almost all eukaryotic cells [2], are assembled through the dimerization of α- and β-tubulin proteins [3]. The dynamics of MTs, which involve phases of elongation, shortening, and pause [4], are relevant to intracellular trafficking, the formation of the mitotic spindle, cytokinesis, cell membrane blebbing, cell migration, and phagocytosis [2,5–7].

Microvesicles (MVs) are secreted by intact cells as microscopic membrane-enclosed sacs ranging in diameter from 100 to 1000 nm. They are formed by plasma membrane protrusions (buds), which then close and become separated from the progenitor cell, eventually to be released into the extracellular environment [8,9]. MVs are distinguished from the more homogeneous exosomes (40–100 nm), which are formed intracellularly within microvesicular bodies, and are thus of endosomal origin [10,11].

Membrane-bound proteins and luminal content determine the MV's function, which consequently relates to the cell of origin [9,12]. Melanoma cell-derived MVs are postulated to play a biological role in the process of carcinogenesis. For example, malignant transformation of melanocytes is associated with the increased production and release of MVs with procoagulant activity, leading to a hypercoagulable state, a major cause of death in cancer patients [13,14]. Murine melanoma cell-derived MVs promote metastasis and immunosuppression through regulatory T-cell expansion and apoptosis of tumor-specific $CD8^+$ cytotoxic T-cells, effects mediated by the MV-associated Fas ligand or TRAIL [15].

Furthermore, the remote delivery of peripheral mast cells originated-MVs to lymph nodes via the lymphatics would allow for more systemic modulation of inflammation and immunosuppression [10]. A similar mechanism may be involved in the invasion of melanoma cells into the dermis. Overall, MVs are postulated to promote cancer initiation, survival, and spread [12,13,16]. Herein, we hypothesize that β3-tubulin interferes with microtubule dynamics, cell cycle regulation, and the spontaneous release of MVs in A375 human malignant melanoma cells.

2. Results

2.1. A375 Cells Express β3-Tubulin mRNA and Protein

β3-tubulin is highly expressed in malignant as compared with normal melanocytes, and siRNA knockdown of β3-tubulin hinders the migration of melanoma cells [17]. Using RT-PCR and Western blot analysis, we demonstrated that the A375 human malignant melanoma cells express β3-tubulin mRNA and protein (Figure 1A). siRNA knockdown of β3-tubulin significantly reduced the mRNA and protein expression by 79.3% ± 7.7% ($p < 0.001$) and 83% ± 4.9% ($p < 0.01$), respectively. The effect of treatment with either β3-tubulin siRNA or control siRNA on the immunostaining of melanoma cells A375 with both β3-tubulin and a specific molecular marker for melanoma cells (melanoma-associated antigen, or MAA) confirmed the specificity of β3-tubulin knockdown by β3-tubulin siRNA (Figure 1B and Figure S1), with no observed effect on cell viability assessed by an MTT assay (Figure 1C).

2.2. siRNA Knockdown of β3-Tubulin Reduces the Spontaneous Release of MVs by A375 Cells

The spontaneous release of MVs by A375 cells was assessed using Alexa fluor 488-labeled wheat germ agglutinin (WGA) to stain cell membranes, followed by fluorescence microscopy. A375 cells release ring-like MVs with clear intravesicular spaces with sizes ranging from 200 to 1000 nm (Figure 1F). MVs were also detected in the culture medium (Figure 1G). The effect of β3-tubulin knockdown on MVs counts was assessed by flow cytometric quantification of MAA-labeled MVs in the culture media of naïve, control-, and β3-tubulin siRNA-treated cells. Knockdown of β3-tubulin induced a significant 48.63% ($p < 0.001$) reduction of released MVs, relative to naïve cells (Figure 1D,E), while control siRNA did not affect MVs release from cells. These data demonstrate that β3-tubulin modulates, at least in part, spontaneous MVs release from A375 melanoma cells.

2.3. siRNA Knockdown of β3-Tubulin Suppresses MTs Dynamics and Induces G2/M Cell Cycle Arrest

Altered cellular MTs dynamics, including elongation, shortening, and a pause, are salient to carcinogenesis, broad-spectrum chemotherapy resistance and cell survival [18]. The effect of β3-tubulin knockdown on MTs dynamics in A375 cells transfected with EGFP-microtubule-associated protein-4 cDNA (EGFP-MAP4) was assessed. A sequence of frames, five-second apart, of the life-history of MTs +ends transitioning between phases of elongation (growth), shortening, and pause is shown (Figure 2A). MTs +end displacements (growth/shortening) of naïve and control siRNA-treated cells were greater than those of β3-tubulin knockdown cells (Figure 2A, Figure S2, and Table S1). The rate of total growth and total shortening were reduced by 39.7% ± 5% ($p < 0.001$) and 52% ± 4.7% ($p < 0.001$), respectively, in β3-tubulin knockdown cells as compared to naïve cells, whereas control siRNA treatment had no effect on these parameters. Further, pause frequency was significantly higher in β3-tubulin knockdown cells by 19.04 % ± 4.2% ($p < 0.05$), whereas overall MTs dynamicity was significantly reduced by 46.53% ± 4.3% ($p < 0.001$).

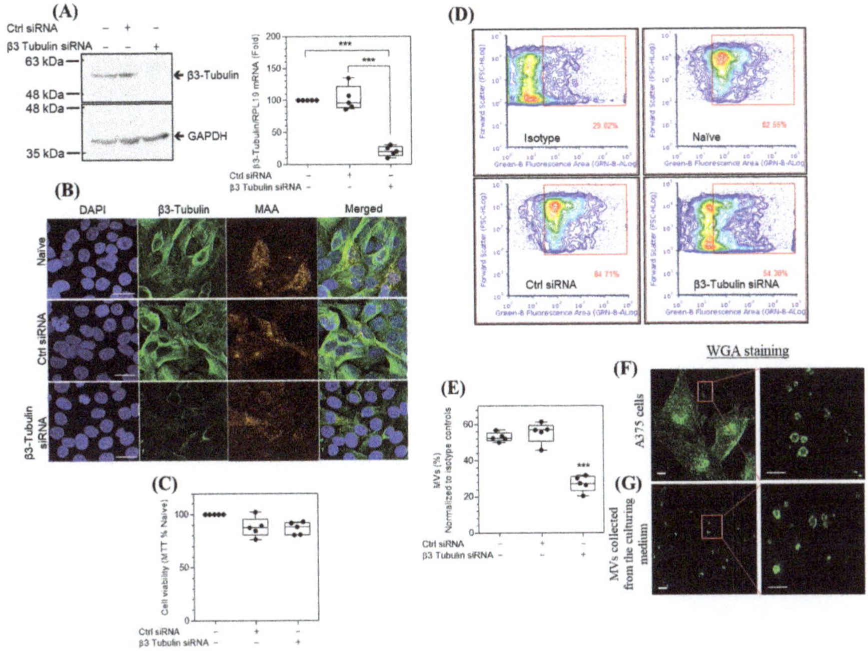

Figure 1. β3-tubulin knockdown reduces the numbers of microvesicles (MVs) spontaneously released by A375 melanoma cells under the standard culture conditions. (**A**) Western blots of β3-tubulin and GAPDH loading control (panel A, left), and RT-PCR quantification of β3-tubulin mRNA expression normalized to the C_T values of RPL19 (panel A, right). (**B**) Naïve, control (ctrl)–, and β3-tubulin siRNA-treated cells were immunostained with antibodies specific for β3-tubulin (green) and MAA (orange), DAPI (blue) was used as nuclear DNA counterstain, original magnification ×630 (bar, 20 µm). (**C**) MTT assay shows the effect of siRNA treatment on the cellular viability of A375 cells. Flow cytometric contour plots (**D**) and quantification (**E**) of MVs collected from the culture media of naïve, ctrl-, and β3-tubulin-siRNA treated cells and immunostained with an MAA antibody or isotype control, plots illustrate the percentage of positively stained MVs (D, red square gated events). Super-resolution microscopy with maximum intensity projection of a confocal stack micrographs of A375 cells (**F**), or MVs collected from culture media (**G**), stained with Alexa fluor 488-conjugated-WGA, red rectangles, in F and G, show randomly selected areas for higher magnification (right enlarged micrographs), illustrating ring-like microvesicles of varying sizes (200–1000 nm) with optically clear lumen. Original magnification ×1000, scale bar 5 µm (left graphs) or 1 µm (magnified right graphs). Statistical significance was determined between different groups using an ANOVA with Tukey's correction for multiple comparisons. $n = 5$, *** $p < 0.001$ versus naïve cells.

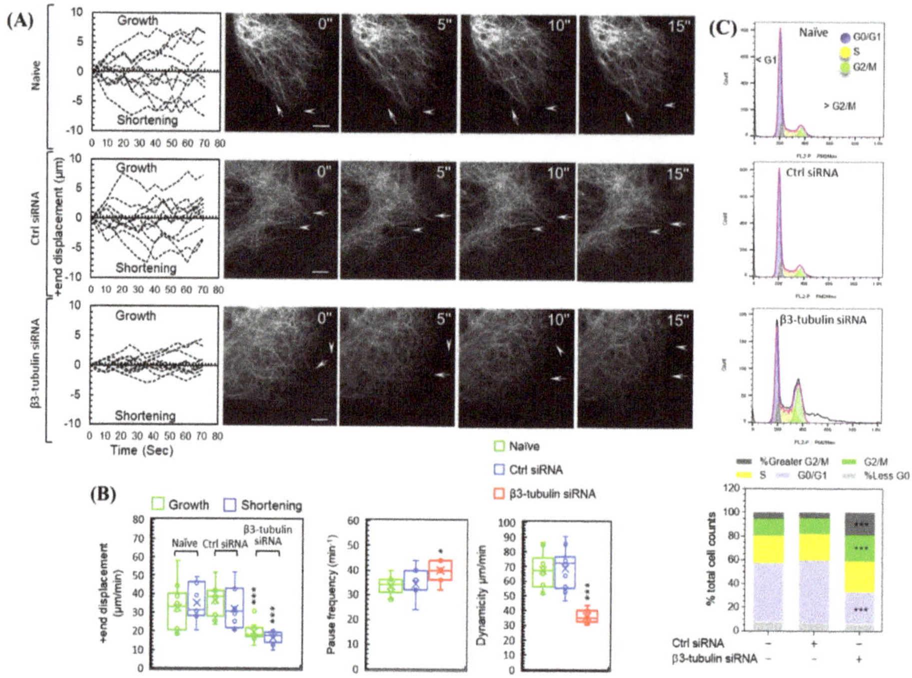

Figure 2. β3-tubulin knockdown suppresses microtubules (MTs) dynamics and disrupts cell cycle in A375 cells. (Panel (**A**), right) Micrographs of EGFP-MAP-4-labeled MTs in naïve, ctrl-, and β3-tubulin siRNA-treated A375 cells. Images were acquired every 5 s and show MTs +end growth (arrows) and shortening (arrowheads). Original magnification ×1000 (bar, 5 μm). (panel (**A**) left) Traces of ten representative MTs +end displacements (growth and shortening) measured over a period of 75 s. Quantification of dynamic growth and shortening (panel (**B**), left), and pause frequency (panel (**B**), middle) and dynamicity (panel (**B**), right). (**C**) Flow cytometric histograms (top panels), and cell cycle growth phase distribution analysis (bottom panel) demonstrate the effect of β3-tubulin knockdown on cell cycle in A375 cells. Statistical significance between naïve, ctrl-, and β3-tubulin-siRNA groups was determined using an ANOVA with Tukey's correction for multiple comparisons. $n = 10$ (MTs analysis) or = 6 (cell cycle analysis), *** $p < 0.001$, and * $p < 0.05$ versus naïve cells.

Cell cycle analysis revealed that growth phase redistribution in β3-tubulin knockdown cells as compared to naïve cells, where G2/M and polyploidy (> G2/M) cells were significantly increased by 53.12% ($p < 0.001$) and 245.74% ($p < 0.001$), respectively. Furthermore, G0/G1 cells were significantly reduced by 45.49% ($p < 0.001$). Similar effects were not seen in control siRNA-treated cells. There was no significant change in the less than G0 (< G0) and S populations (Figure 2C and Figure S3).

3. Discussion

By understanding the mechanisms of carcinogenesis at the cellular and molecular level, current research focuses on targeting strategies for melanoma prevention and treatment. With essential regulatory roles in a myriad of inter- and intra-cellular activities, MTs dynamicity is an important target for many anti-cancer therapeutics [19–22]. β-tubulin is a structural protein that maintains the microtubule cytoskeleton. Distinct from the other six beta-tubulin isotypes in the human genome, β3-tubulin has a unique molecular structure that facilitates its binding to factors involved in the oxidative stress and nutrient deprivation response, thereby bolstering the survival of stressed cells [23]. For example, induced expression of β3-tubulin promotes resistance to the widely used chemotherapeutic agent, paclitaxel, in Hela and MCF-7 cell lines [24]. Furthermore, increased levels of β3-tubulin expression have been linked to malignant transformation and migration of melanocytes [17]. In tandem with these studies, our Q-PCR, Western blot, and immunofluorescence data confirmed that β3-tubulin was highly expressed in the human malignant melanoma cell line (A375).

To understand the relevance of β3-tubulin on melanoma progression, we used siRNA to knockdown β3-tubulin in the A375 cell line. We studied several factors known to contribute to metastasis and drug resistance, namely MVs release, MTs dynamics, and cell cycle regulation [19–22,25]. Our data demonstrated significant suppression of MTs dynamic parameters (e.g., growth, shortening, and pause frequencies), implying that β3-tubulin is critical for MTs dynamicity. As normal MTs dynamics are essential for mitotic spindle formation and function, disrupting MTs dynamics would lead to the activation of the cell cycle checkpoints and force the malignant cells to initiate growth arrest and apoptotic cell death [26]. Despite the importance of β-tubulin isotypes in resistance to antimitotic drugs [27,28], there are no studies to date examining the role of β3-tubulin in cell cycle regulation in malignant melanoma. Thus, we investigated the effect of β3-tubulin siRNA on the cell cycle in A375 melanoma cells, and the data demonstrated robust induction of G2/M arrest and a significant increase in polyploidy population in β3-tubulin siRNA-treated cells. Such data provide direct evidence that β3-tubulin alters melanoma cell cycle regulation. G2/M arrest, which may be the result of defective mitotic spindle formation, was shown to enhance the cytotoxic effect of chemotherapy in melanoma cell lines [29,30]. These results are consistent with previous work on Hela cells [31], where the reduction of β3-tubulin expression by siRNA resulted in partial inhibition of cell growth. Interestingly, in H460 (non-small cell lung cancer), β3 tubulin siRNA increased vincristine-and paclitaxel-induced suppression of microtubule dynamicity and cell death at low and high drug concentrations, respectively, while there was no effect on the G2/M population [32]; such data might propose varied and cell-specific roles of β3 tubulin in different cell types.

Despite the importance of MVs as potential therapeutic targets in cancer (i.e., angiogenesis, matrix degradation and invasion, metastasis, and immunosuppression) and thrombosis [9,12,33], the mechanisms of their biogenesis and release remain poorly characterized. Several lines of evidence demonstrate that MVs are important for melanoma progression [13,14], coagulation [12], and inflammatory skin conditions (i.e., psoriasis) [34]. The correlation between MTs dynamics and MVs release was previously studied in mastocytoma P815 cell [35], where the authors demonstrated that microvesicle formation and shedding require MTs disruption. The role of MTs in controlling other cellular activities, such as movement and intracellular vesicle transportation was also reported [36,37]. However, the mechanisms of MVs biogenesis and release in melanoma cells remain controversial. Herein, we demonstrated that β3-tubulin knockdown-induced repression of MTs dynamicity was associated with a reduction in the numbers of the MVs released by A375 cells, implying that β3-tubulin may play a role in MVs production via regulating the MTs dynamics in melanoma cells. Further studies are required to elucidate the mechanistic role of the α- and β- tubulins isotypes in the treatment and prevention of melanoma progression.

4. Materials and Methods

4.1. Cell Culture EGFP-MAP-4 Transfection and siRNA Silencing

A375 human malignant melanoma cells (ATCC® CRL-3224) were propagated in Dulbecco's Modified Eagle's Medium (DMEM—catalog no. 11995-065, Life Technologies, Grand Island, NY, USA) containing 4.5 g/L D-glucose, L-glutamate, 110 mg/L sodium pyruvate, and supplemented with 10% fetal bovine serum (FBS—catalog no. 098105, MULTICEL) and 1% penicillin-streptomycin (catalog no. 15140-122, Life Technologies). EGFP-MAP-4 transfection (provided by Dr. Joanna Olmsted, University of Rochester, USA) was performed as previously described [24]: briefly, cells were seeded in twelve-well plates at a density of 150,000 cells per well and transfected with 1 μg plasmid DNA using lipofectamine 2000 (catalog no. P/N52887, Invitrogen, Carlsbad, CA, USA) following the manufacturer's instructions.

Transfected cells were selected by growing in a growth medium containing G418, 2 mg/mL (catalog no. A1720; Sigma-Aldrich, Oakville, ON, Canada). For siRNA silencing, A375 cells were transfected with 25 nM of β3-tubulin siRNA (catalog no. sc-105009, Santa Cruz Biotechnology, Dallas, TX, USA) or FlexiTube Lamin A/C non-targeting siRNA (catalog no. SI03650332, Qiagen, Valencia, CA, USA), using transfection reagent lipofectamine RNAiMAX (catalog no. 13778-075, Thermo Fisher Scientific, Carlsbad, CA, USA) as per the manufacturer's protocol. Western analysis, RNA isolation, immunofluorescent microscopy, and MVs purification were performed 48 h after transfection, unless otherwise stated.

4.2. RNA Isolation, cDNA Synthesis, and qPCR

Total RNA was extracted using NucleoSpin RNA purification kits (catalog no. 740955- 250, D-Mark Biosciences, Düren, Germany) and 500 ng was used for cDNA synthesis (qScript cDNA Synthesis kit, catalog no. CA101414-098, Quanta Biosciences, Gaithersburg, MD, USA). Real-time PCR (RT-PCR) was performed with a StepOne Plus PCR machine (Applied Biosystems, Foster City, CA, USA) using the fast SYBR Green master mix (catalog no. 4385618, Life Technologies). The amplification conditions were 95 °C for 20 s followed by forty cycles at 95 °C for 3 s and 60 °C for 30 s. Primer sequences were as follows: β3-tubulin, forward 5′-CGA AGC CAG CAG TGT CTA AA-3′, reverse 5′-GGA GGA CGA GGC CAT AAA TAC-3′; ribosomal protein L19 (RPL19), forward 5′-ATC GAT CGC CAC ATG TAT CA-3′, reverse 5′-GCG TGC TTC CTT GGT CTT AG-3′.

4.3. Electrophoresis, Western Analysis, and Fluorescent Microscopy

Fifty micrograms of A375 melanoma cell lysate protein in radioimmunoprecipitation assay buffer (RIPA) containing 1% of Halt™ protease inhibitor cocktail (catalog no.1861279, ThermoScientific) was separated by SDS-PAGE and transferred onto nitrocellulose membranes (catalog no. rpn203d, GE Health). Membranes were immunoprobed with rabbit monoclonal anti-human β3-tubulin (catalog no. d71g9-xp, Cell Signaling), or mouse anti-human glyceraldehyde-3-phosphate dehydrogenase (GAPDH; catalog no. 4699-9555, Biogenesis). The secondary antibodies were peroxidase-conjugated goat anti-rabbit IgG (catalog no. 111-035-003, Jackson ImmunoResearch) and peroxidase-affini-pure goat anti-mouse IgG (catalog no. 115-035-003, Jackson ImmunoResearch). Immune complex visualization was carried out using ECL TM prime WB reagents (catalog no. rpn2232sk, GE Health). For fluorescence microscopy, A375 melanoma cells were fixed in pre-warmed culturing medium containing 3.7% formaldehyde (catalog no. f-8775, Sigma) for 15 min at 37 °C. Cellular and vesicular membranes were labeled with Alexa 488-conjugated wheat germ agglutinin (WGA, catalog no. W11261, Invitrogen) 5 μg/mL diluted in PBS for 10 min at room temperature followed by two washes in PBS. For cytoplasmic and inner membrane-bound protein immuno-detection, cells were permeabilized for 5 min at room temperature with 0.2% tween X-100 (catalog no. t9284, Sigma) in PBS and blocked for 1 h at room temperature using a blocking buffer (PBS containing 5% albumin bovine serum, catalog no. a-4503, Sigma). The cells were probed with antibodies against β3-tubulin, melanoma associated antigen (MAA, mouse monoclonal anti human, catalog no. ab34165, Abcam), isotype controls (rabbit IgG (monoclonal, catalog no. ab172730, Abcam), or mouse IgG 2b (catalog no. 557351, BD Pharmingen)) diluted in blocking buffer and incubated overnight at 4 °C. The secondary antibodies used were rabbit Alexa fluor 488 goat anti-rabbit (catalog no. A11034, Invitrogen) and mouse Alexa fluor 555 goat anti-mouse (catalog no. A21424, Life technologies). 4′,6-diamidino-2-phenylindole (DAPI, catalog no. d21490, Molecular probes) was used as a nuclear counterstain marker. Immunoprobed cells were mounted using prolong gold anti-fad reagent (catalog no. p36930, Invitrogen) and visualized by confocal microscopy and structured illumination super-resolution microscopy (Zeiss Elyria).

4.4. MVs Purification and Flowcytometric Quantification

MVs purification was carried out following the referenced protocol by Lima et al. [13]. Briefly, A375 melanoma cell culture supernatants were centrifuged at 800× *g* for 10 min to exclude floating dead cells and debris. Another centrifugation for 15 min at 14,000× *g* yielded a pellet, which was resuspended and recollected by repeat centrifugation (this and subsequent centrifugations were at 14,000× *g* for 15 min at 4 °C). The washed pellet containing MVs was suspended in 100 µL ice-cold PBS. MVs were fixed by adding 900 µL of 100% methanol at −20 °C dropwise with vigorous vortexing. This suspension was incubated at −20 °C for 5 min followed by the addition of 1 mL of ice-cold PBS. Fixed MVs were pelleted by centrifugation, resuspended in 250 µL blocking buffer and incubated at room temperature for 1 h. MVs were collected by centrifugation and immunostained overnight at 4 °C in 250 µL of the blocking buffer containing a mouse monoclonal MAA antibody or mouse IgG 2b isotype control at a concentration of 1 µg/mL. MVs were then washed once in PBS, incubated one hour at room temperature in a blocking buffer containing Alexa fluor 488 goat anti-mouse (catalog no. a11001, Life Technologies) and washed once in PBS. MVs quantification was assessed using a flow cytometry (Guava® easyCyte; MilliporeSigma). Data analysis was carried out using FCS express 6 plus research edition software.

4.5. *Analysis of Microtubules Dynamics*

A375 cells were transfected with EGFP-MAP-4 (provided by Dr. Joanna Olmsted, University of Rochester, USA) as previously described [24]. MTs dynamics analysis was performed as previously published [38]: briefly, MTs +ends were tracked using ImageJ software (Manual tracking plugin, https://imagej.nih.gov/ij/plugins/track/track.html). Only 0.5 µm changes in growth/shortening traces were plotted for each MT. Any growth or shortening event below the 0.5 µm threshold was considered a pause. The growth/shortening rate was calculated by dividing the sum of all growth (positive) or the sum of all shortening (negative) by the total time spent growing or shortening respectively. Pause frequencies were calculated by dividing the total number of pauses by the total measuring time. MTs dynamicity was extracted by dividing the sum of total length (growth and shortening) by the total measuring duration. Data was analyzed using GraphPad Prism 6, and MS-Excel software. Data Significances were calculated using a student *t*-test and one-way analysis of variants (ANOVA) with Tukey's correction for multiple comparison.

4.6. *Propidium Iodide (PI)-Staining and Cell Cycle Analysis*

A375 cells were fixed for 30 min in 70% ethanol at 4 °C and washed 2× in PBS. RNase (100 µg/mL) was added, and the cells were incubated for 20 min at 37 °C, followed with 2× washes in PBS. Cells were then incubated in 3 µM PI (catalog no. p4170; Sigma) in the staining solution (100 mM tris, PH 7.4, 150 mM NaCl, 1 mM $CaCl_2$, 0.5 mM $MgCl_2$, and 0.1% Nonidet P-40) for 15 min at room temperature. Cell cycle analysis was conducted using a flow cytometer (Guava® easyCyte; MilliporeSigma). Data analysis was carried out using FlowJo v10.6.1 software.

5. Conclusions

β3-tubulin knockdown suppresses MTs dynamics, decreases MVs release, and induces G2/M cell cycle arrest in human malignant melanoma cells (A375). β3-tubulin, an important microtubular protein, may therefore be used to study the role of MTs and MVs in the pathogenesis and treatment of melanoma.

Supplementary Materials: Supplementary materials can be found at http://www.mdpi.com/1422-0067/21/5/1656/s1.

Author Contributions: Research design: M.O.A., A.G., P.R.M., Conducted experiments and data acquisition: M.O.A., A.G., P.S., P.R.M., Contributed new reagent or analytical tool: S.S.L., G.J.L., M.A., Performed data analysis: M.O.A., A.G., G.J.L., S.S.L., M.A., P.R.M., Wrote or contributed to the drafting, writing and critically reviewing of the manuscript: M.O.A., A.G., G.J.L., M.A., S.S.L., P.S., P.R.M., Given final approval of the version to be published: M.O.A., P.R.M. All authors have read and agreed to the published version of the manuscript.

Funding: This research received no external funding.

Acknowledgments: The authors gratefully acknowledge the support provided by Susanne V. Gibson and the Canadian Dermatology Foundation.

Conflicts of Interest: The authors declare no conflict of interest.

Abbreviations

MTs	Microtubules
MVs	Microvesicles
WGA	Wheat germ agglutinin
MAA	Melanoma associated antigen

References

1. Katsetos, C.D.; Herman, M.M.; Mork, S.J. Class III beta-tubulin in human development and cancer. *Cell Motil. Cytoskelet.* **2003**, *55*, 77–96. [CrossRef] [PubMed]
2. Kueh, H.Y.; Mitchison, T.J. Structural plasticity in actin and tubulin polymer dynamics. *Science* **2009**, *325*, 960–963. [CrossRef]
3. Desai, A.; Mitchison, T.J. Microtubule polymerization dynamics. *Annu. Rev. Cell Dev. Biol.* **1997**, *13*, 83–117. [CrossRef] [PubMed]
4. Karp, G. *Cell and Molecular Biology: Concepts and Experiments*; John Wiley & Sons: Hoboken, NJ, USA, 2005; p. 355.
5. Sugiyama, T.; Pramanik, M.K.; Yumura, S. Microtubule-Mediated Inositol Lipid Signaling Plays Critical Roles in Regulation of Blebbing. *PLoS ONE* **2015**, *10*, e0137032. [CrossRef] [PubMed]
6. Ganguly, A.; Yang, H.; Sharma, R.; Patel, K.D.; Cabral, F. The role of microtubules and their dynamics in cell migration. *J. Biol. Chem.* **2012**, *287*, 43359–43369. [CrossRef] [PubMed]
7. Knabbe, J.; Nassal, J.P.; Verhage, M.; Kuner, T. Secretory vesicle trafficking in awake and anaesthetized mice: Differential speeds in axons versus synapses. *J. Physiol.* **2018**, *596*, 3759–3773. [CrossRef]
8. Kobayashi, T.; Okamoto, H.; Yamada, J.-I.; Setaka, M.; Kwan, T. Vesiculation of platelet plasma membranes. Dilauroylglycerophosphocholine-induced shedding of a platelet plasma membrane fraction enriched in acetylcholinesterase activity. *Biochim. Biophys. Acta* **1984**, *778*, 210–218.
9. Théry, C.; Ostrowski, M.; Segura, E. Membrane vesicles as conveyors of immune responses. *Nat. Rev. Immunol.* **2009**, *9*, 581–593. [CrossRef]
10. Kunder, C.; John, A.L.S.; Li, G.; Leong, K.W.; Berwin, B.; Staats, H.F.; Abraham, S.N. Mast cell-derived particles deliver peripheral signals to remote lymph nodes. *J. Exp. Med.* **2009**, *206*, 2455–2467. [CrossRef]
11. Latifkar, A.; Hur, Y.H.; Sanchez, J.C.; Cerione, R.A.; Antonyak, M.A. New insights into extracellular vesicle biogenesis and function. *J. Cell Sci.* **2019**, *132*, jcs222406. [CrossRef]
12. Leroyer, A.; Anfosso, F.; Lacroix, R.; Sabatier, F.; Simoncini, S.; Njock, M.-S.; Jourde, N.; Brunet, P.; Camoin-Jau, L.; Sampol, J.; et al. Endothelial-derived microparticles: Biological conveyors at the crossroad of inflammation, thrombosis and angiogenesis. *Thromb. Haemost.* **2010**, *104*, 456–463. [CrossRef] [PubMed]
13. Lima, L.G.; Chammas, R.; Monteiro, R.Q.; Moreira, M.E.C.; Barcinski, M.A. Tumor-derived microvesicles modulate the establishment of metastatic melanoma in a phosphatidylserine-dependent manner. *Cancer Lett.* **2009**, *283*, 168–175. [CrossRef] [PubMed]
14. Laresche, C.; Pelletier, F.; Garnache-Ottou, F.; Lihoreau, T.; Biichlé, S.; Mourey, G.; Saas, P.; Humbert, P.; Seilles, E.; Aubin, F. Increased levels of circulating microparticles are associated with increased procoagulant activity in patients with cutaneous malignant melanoma. *J. Investig. Dermatol.* **2014**, *134*, 176–182. [CrossRef] [PubMed]
15. Wieckowski, E.U.; Visus, C.; Szajnik, M.; Szczepanski, M.J.; Storkus, W.; Whiteside, T.L. Tumor-derived microvesicles promote regulatory T cell expansion and induce apoptosis in tumor-reactive activated CD8+ T lymphocytes. *J. Immunol.* **2009**, *183*, 3720–3730. [CrossRef] [PubMed]
16. Muhsin-Sharafaldine, M.-R.; Saunderson, S.; Dunn, A.C.; Faed, J.M.; Kleffmann, T.; McLellan, A. Procoagulant and immunogenic properties of melanoma exosomes, microvesicles and apoptotic vesicles. *Oncotarget* **2016**, *7*, 56279–56294. [CrossRef]

17. Orfanidis, K.; Wäster, P.; Lundmark, K.; Rosdahl, I.; Öllinger, K. Evaluation of tubulin beta-3 as a novel senescence-associated gene in melanocytic malignant transformation. *Pigment Cell Melanoma Res.* **2017**, *30*, 243–254. [CrossRef]
18. Rodrigues-Ferreira, S.; Molina, A.; Nahmias, C. Microtubule-associated tumor suppressors as prognostic biomarkers in breast cancer. *Breast Cancer Res. Treat.* **2019**, *179*, 267–273. [CrossRef]
19. Cirillo, L.; Gotta, M.; Meraldi, P. The Elephant in the Room: The Role of Microtubules in Cancer. *Adv. Exp. Med. Biol.* **2017**, *1002*, 93–124.
20. Dumontet, C.; Duran, G.E.; Steger, K.A.; Murphy, G.L.; Sussman, H.H.; Sikic, B.I. Differential expression of tubulin isotypes during the cell cycle. *Cell Motil. Cytoskelet.* **1996**, *35*, 49–58. [CrossRef]
21. Jordan, M.A.; Wilson, L. Microtubules as a target for anticancer drugs. *Nat. Rev. Cancer* **2004**, *4*, 253–265. [CrossRef]
22. Dumontet, C.; Jordan, M.A. Microtubule-binding agents: A dynamic field of cancer therapeutics. *Nat. Rev. Drug Discov.* **2010**, *9*, 790–803. [CrossRef] [PubMed]
23. Cicchillitti, L.; Penci, R.; Di Michele, M.; Filippetti, F.; Rotilio, D.; Donati, M.B.; Scambia, G.; Ferlini, C. Proteomic characterization of cytoskeletal and mitochondrial class III beta-tubulin. *Mol. Cancer* **2008**, *7*, 2070–2079. [CrossRef] [PubMed]
24. Ganguly, A.; Yang, H.; Cabral, F. Class III beta-tubulin counteracts the ability of paclitaxel to inhibit cell migration. *Oncotarget* **2011**, *2*, 368–377. [CrossRef] [PubMed]
25. Parker, A.L.; Kavallaris, M.; McCarroll, J.A. Microtubules and their role in cellular stress in cancer. *Front. Oncol.* **2014**, *4*, 153. [CrossRef] [PubMed]
26. Jordan, M.A.; Kamath, K. How do microtubule-targeted drugs work? An overview. *Curr. Cancer Drug Targets* **2007**, *7*, 730–742. [CrossRef] [PubMed]
27. Bhattacharya, R.; Cabral, F. A ubiquitous beta-tubulin disrupts microtubule assembly and inhibits cell proliferation. *Mol. Biol. Cell* **2004**, *15*, 3123–3131. [CrossRef] [PubMed]
28. Burkhart, C.A.; Kavallaris, M.; Band Horwitz, S. The role of beta-tubulin isotypes in resistance to antimitotic drugs. *Biochim. Biophys. Acta* **2001**, *1471*, O1–O9.
29. Priebe, M.K.; Dewert, N.; Amschler, K.; Erpenbeck, L.; Heinzerling, L.; Schön, M.P.; Seitz, C.S.; Lorenz, V.N. c-Rel is a cell cycle modulator in human melanoma cells. *Exp. Dermatol.* **2019**, *28*, 121–128. [CrossRef]
30. Liu, J.; Jiang, G.; Mao, P.; Zhang, J.; Zhang, L.; Liu, L.; Wang, J.; Owusu, L.; Ren, B.; Tang, Y.; et al. Down-regulation of GADD45A enhances chemosensitivity in melanoma. *Sci. Rep.* **2018**, *8*, 4111. [CrossRef]
31. Shibazaki, M.; Maesawa, C.; Akasaka, K.; Kasai, S.; Yasuhira, S.; Kanno, K.; Nakayama, I.; Sugiyama, T.; Wakabayasi, G.; Masuda, T.; et al. Transcriptional and post-transcriptional regulation of betaIII-tubulin protein expression in relation with cell cycle-dependent regulation of tumor cells. *Int. J. Oncol.* **2012**, *40*, 695–702.
32. Gan, P.P.; McCarroll, J.A.; Po'Uha, S.T.; Kamath, K.; Jordan, M.A.; Kavallaris, M. Microtubule dynamics, mitotic arrest, and apoptosis: Drug-induced differential effects of betaIII-tubulin. *Mol. Cancer* **2010**, *9*, 1339–1348. [CrossRef] [PubMed]
33. Castellana, D.; Toti, F.; Freyssinet, J.M. Membrane microvesicles: Macromessengers in cancer disease and progression. *Thromb. Res.* **2010**, *125* (Suppl. 2), S84–S88. [CrossRef]
34. Pelletier, F.; Garnache-Ottou, F.; Angelot, F.; Biichlé, S.; Vidal, C.; Humbert, P.; Saas, P.; Seilles, E.; Aubin, F. Increased levels of circulating endothelial-derived microparticles and small-size platelet-derived microparticles in psoriasis. *J. Investig. Dermatol.* **2011**, *131*, 1573–1576. [CrossRef] [PubMed]
35. Liepins, A. Possible role of microtubules in tumor cell surface membrane shedding, permeability, and lympholysis. *Cell. Immunol.* **1983**, *76*, 120–128. [CrossRef]
36. Downing, K.H.; Nogales, E. New insights into microtubule structure and function from the atomic model of tubulin. *Eur. Biophys. J.* **1998**, *27*, 431–436. [CrossRef]
37. Nogales, E. Structural insights into microtubule function. *Annu. Rev. Biochem.* **2000**, *69*, 277–302. [CrossRef]
38. Gierke, S.; Kumar, P.; Wittmann, T. Analysis of microtubule polymerization dynamics in live cells. *Methods Cell Biol.* **2010**, *97*, 15–33.

MDPI
St. Alban-Anlage 66
4052 Basel
Switzerland
Tel. +41 61 683 77 34
Fax +41 61 302 89 18
www.mdpi.com

International Journal of Molecular Sciences Editorial Office
E-mail: ijms@mdpi.com
www.mdpi.com/journal/ijms

www.ingramcontent.com/pod-product-compliance
Lightning Source LLC
Chambersburg PA
CBHW061459210926
51850CB00008B/1765
9783036579276